Hubble, Humason and the Big Bang

The Race to Uncover the Expanding Universe

Ron Voller

Hubble, Humason and the Big Bang

The Race to Uncover the Expanding Universe

Foreword by John S. Mulchaey

 Springer

Published in association with

Chichester, UK

Ron Voller
Brooklyn, NY, USA

SPRINGER-PRAXIS BOOKS IN POPULAR ASTRONOMY

Popular Astronomy
ISSN 2626-8760   ISSN 2626-8779   (eBook)
Springer Praxis Books
ISBN 978-3-030-82180-7   ISBN 978-3-030-82181-4   (eBook)
https://doi.org/10.1007/978-3-030-82181-4

Project Editor: David M. Harland

This Springer imprint is published by the registered company Springer Nature Switzerland AG
The registered company address is: Gewerbestrasse 11, 6330 Cham, Switzerland

For Ann

Preface

A full recounting of the partnership between Edwin Hubble and Milton Humason has been wanting, one that immerses us in the history that created them, that surrounded them, and that helped define the nature of their characters as men of science and the world. The fact that the discovery of the Big Bang experimentally has been linked so indelibly with Hubble, while not an accident of history, is certainly misleading for a number of reasons, first and foremost as it regards Humason's contribution to that legacy. Linking them inexorably to the discovery for all time is a central focus of this book and, it seems, a just and worthy cause.

Their individual biographies tend to romanticize and amplify their lives at the expense of the other to one degree or another. The necessary focus on the life of the individual gives a narrow view of his association with, and relationship to his partner, robbing the reader of insights into what makes successful partnerships work. These stories, though vibrant and revealing in their own right, lack a real depiction of the dependence each had on the other, and the strain of outside and inside pressures on the outcome of their work and professional relationship.

How these two arrived at Mount Wilson and their partnership, how they managed to keep it together and drive on in their work in the face of bold and persistent doubt, contrary theories, and calamitous world events is as intriguing a story as their groundbreaking work on universal evolution. It is a story inextricably interwoven with Albert Einstein, Willem de Sitter and other theoretical physicists of the era who strained to correctly understand and define the meaning of their eye-popping results.

Most readers, even those with a fairly deep knowledge of the lives and careers of Edwin Hubble and Albert Einstein, are not aware of the confusion and uncertainty the revelations of the Hubble-Humason partnership wrought on themselves and others through the 1930s and beyond. It is hoped that these stories will be as fun and interesting to read as they were to discover and to write.

The book is intended for readers of science history from any age group, and for generalists and enthusiasts as well as academic and professional readers. The discussion of their combined story may offer insights into the human condition, professional and social ethics, the power of purpose, the passion and glory of discovery, and the pitfalls that await those whose concerns and ambitions are not well managed.

We kick off with a brief overview of the subject and an abbreviated history of roughly two centuries of development in astronomy and physics prior to the arrival on the scene of Hubble and Humason. It is meant to immerse the reader in the subject and the science that surrounds it and may be read in full or dipped into for historical context while reading the remainder of the book.

Chapters 2, 3 and 4 present the early lives of both Hubble and Humason, detailing the important events, family history and personal traits that helped to shape their personalities leading up to their arrival at Mount Wilson. Their biographies should be consulted for a full accounting of their individual stories. In the case of Hubble, Gale Christianson's book, *The Mariner of the Nebulae*, is a fantastic read, and *The Muleskinner and the Stars*, by this writer, is the first and only "complete" biography of Humason to date.

Next, Chapters 5, 6 and 7 review their early careers as each man made his mark at the observatory and on astronomy. Their insight, vision and sheer will compelled them to the heights of their individual disciplines while their personal moral and ethical character traits made indelible impressions on others in the field that would follow them throughout their careers and beyond. These public perceptions, positive and negative, great or small, persist to this day.

Chapters 8, 9 and 10 show how their partnership came together in fits and starts to reveal the first comprehensive, practical and compelling evidence of what would later be named the Big Bang theory of universal evolution. The groundbreaking revelation prompts world renowned physicists to seek out the Mount Wilson duo for review and discussion of their results, in the process bestowing upon them international scientific fame. Here again, their personal natures, tendencies, egos and ambitions are revealed, while the limitations of their abilities are met with startling outcomes. Arguments are revealed for counterclaims to the discoveries at the center of the story and discussed in view of the historical developments as well as personal attributes that affected them.

Finally, in Chapters 11 through 14 we detail the team's struggle to maintain momentum as the world devolves into war, delaying the resumption of their assault on the expansion theory with the 200-inch telescope at Palomar

Mountain. Newcomers to the observatories like Walter Baade and Rudolf Minkowski as well as those outside the observatory like Fred Hoyle present new and revealing evidence that clouds and shapes the perception of their earlier results. This section necessarily continues the Hubble-Humason connection into the years and decades after their departure from the scene in an effort to bring the central focus of their mission to its present conclusion.

An epilogue provides conclusions and final thoughts on the subject of this interesting and dynamic partnership. And to wrap up there are appendices with suggestions for further reading, including professional publications by Hubble and Humason.

John Donne once famously wrote, "No man is an island entire of itself; every man is a piece of the continent, a part of the main." We all have our horizons. Only by working together do we see beyond them.

Foreword

A Timely Tribute

Most people imagine astronomers as lone individuals silently looking through their telescopes night after night. Indeed, many of history's most famous scientists labored in solitude. By contrast, modern astronomy is a highly collaborative enterprise. Many of us work with teams of scientists from around the world. Indeed, today it is not uncommon for a paper in The Astrophysical Journal to feature tens or even hundreds of authors. The 2017 paper on the first electromagnetic detection of a gravitational wave event famously had several *thousand* authors.

Between these extremes is the special collaboration that occurs when two scientists team up to address a problem. In these partnerships, each individual provides his or her specific strengths to address the challenges that they face, often complementing areas where the other researcher may have less expertise. The dual-person team also allows theories and interpretations to be vetted prior to publication. The resulting conclusions represent the concurrence of both person's research methods and solutions. A two-author paper carries the implicit assumption that both contributed substantially to the research. Generally this is true. I have had the pleasure of being part of a scientific duo many times in my career, and I look back on these papers particularly fondly. They represent some of my best work as a professional astronomer.

Among renowned scientific duos, Edwin Hubble and Milton Humason rank at the top. Together, this unusual pairing of the famous scientist and the former janitor provided the conclusive observational evidence that the universe is expanding. The implications of this discovery could not be greater. An expanding universe must have had a beginning. This led inevitably to the Big Bang theory. Much of our modern understanding of the universe derives from the

Hubble-Humason collaboration. While Hubble continues to be recognized for the telescope that bears his name and is seen by many as the most important astronomer of the 20th century, much of his best work would not have been possible without Humason, whose patience allowed him to collect data at Mount Wilson that perhaps no other person at the time was capable of obtaining.

While much has been written about each man individually, Ron Voller's fascinating book explores their unlikely collaboration as never before. He makes the convincing case that left to their own devices, neither man would likely have uncovered the evidence for the Big Bang. While others would eventually have stumbled on this evidence, the unique combination of their skills and the best telescope of their era accelerated our understanding of the universe by many years. The Hubble-Humason partnership epitomizes the power of the two-person scientific team to unlock some of nature's biggest mysteries.

This book is timely for a number of reasons. Today, nearly a century after Hubble and Humason's revolutionary work, astronomy is in a golden age. Every day, new discoveries expand and reshape our understanding of the cosmos. These findings are a tribute to the creativity and ingenuity of scientific collaboration, and a crucially important validation of science itself.

As director of The Carnegie Observatories, I have the pleasure of working in the same historic building in Pasadena where Hubble and Humason carried out their important work. Almost every day, I enjoy seeing in our library a large photograph from February 1931 commemorating a talk given by Albert Einstein that was attended by many other famous astronomers. Several of these luminaries were selected to pose with Einstein in the portrait. Hubble towers above them, and directly by his side stands Humason. The message couldn't be clearer.

John S. Mulchaey
Science Deputy, Carnegie Institution for Science
Crawford H. Greenewalt Chair and Director, Carnegie Observatories
January 2021

Acknowledgements

I would like to thank Dan Lewis at the Huntington Library for his steadfast support throughout the many years of my research on this subject. Dan opened the door wide before I could even knock, and when I was stuck in the mud, he pulled me out and set me on firm footing. Also, to the team at the Huntington's Munger Research Center for their cheerful assistance during my infrequent but time-intensive and time-sensitive drive-by study sessions. They always came to my aid and offered additional tips on possible research connections. They really get it! And Eun-Joo Ahn, a graduate student who offered her support and whose general enthusiasm for the subject (especially Humason) was a real shot in the arm when I needed it.

Thanks to Mark Hurn and the University of Cambridge Institute of Astronomy; the Field Museum Photo Archives; the Huntington Library Photo Archives; Daniel Meyer and the Hanna Holborn Gray Special Collections Research Center at the University of Chicago; Stacey Christen and the Lowell Observatory Archives; Loma Karklins at the Caltech Archives; Edward Copenhagen at the Harvard University Archives; Kit Whitten at the Carnegie Institution for Science; Luisa Haddad at the University of California Santa Cruz Special Collections & Archives; Anke Vollersen and the Hamburg Observatory Archives; and the Humason family for supplying the images that contributed greatly to the telling of this story.

On the mountain, I have to thank, posthumously, Don Nicholson who was the first person to really set me on the path to an understanding of the incredible history of Mount Wilson and the Observatory that bears its name. Don's gift for storytelling and willingness to share his vast knowledge and personal accounts helped me immeasurably.

More recently, to Gale Gant and Larry Webster for their guidance and for sharing their own knowledge of the mountain, the buildings, the instruments, and anecdotes of those who worked there during my visit with Ann Humason Bernt and her family in 2017. To Cindy Hunt at the Carnegie Observatories for helping to arrange that meeting and for the excellent display that she organized for Ann and her family for the 100th anniversary of the 100-inch telescope. It was a memorable visit!

To John Mulchaey at the Carnegie Observatories for his contribution to and enthusiastic support for the book (and for the excellent tour of the observatory offices), I will always be grateful.

Thanks also to Deborah Shapley for sharing her thoughts and anecdotes about her grandfather, Harlow Shapley, and for contributions to the details of this story.

To Clyde Blackman and his wife, Jane, for inviting me into their home and sharing details of his mother, Alice Gross Blackman's life. Tragically, Clyde died before this work was completed but I was fortunate to have met him, as I am sure many others who knew him would agree. His family's connection to the Humason story had always been a mystery to me, and I appreciate the Blackman's kind hospitality and sharing a bit about Clyde's mother (And I enjoyed Clyde's skill at the grill!).

To my sister-in-law, Sarah, for her assistance in my research, and Brian Voller and Emily Weitz for their brilliant advice and suggestions, which helped me make vital adjustments to the storyline.

To my editors, former and present, Maury Solomon and Hannah Kaufman, Maury for her belief in my work and Hannah for spearheading the arrival of this book. I am especially grateful to David M. Harland for his thorough, expert, and tenacious editing of the manuscript. The book is better for his efforts. It's been a brilliant ride and I can't thank the production team at Springer enough.

To my mother for her belief in me and to my family for their often sarcastic, always on-point and loving support of my work. They know how to help keep the rubber on the road.

Finally, to my extended family, the Bernts, who continue to invite me into their fold and to whom I am immensely grateful for the support and for making their family history, which has been so vital to the completion of this work, available to me. And, to Ann Humason Bernt, without whom this book would not be possible.

Many heartfelt thanks to all!

Contents

Acknowledgments

Abbreviations

AAAS	American Association for the Advancement of Science
AAOHC	Annals of the Astronomical Observatory of Harvard College
AAS	American Astronomical Society
AO	Allegheny Observatory
ApJ	Astrophysical Journal
APS	American Physical Society
ASA	Acoustical Society of America
ASP	Astronomical Society of the Pacific
BAIN	Bulletin of the Astronomical Institutes of the Netherlands
BBC	British Broadcasting Corporation
BCE	Before Common Era
Caltech	California Institute of Technology
CCD	Charge-coupled devices
CE	Common Era
CIW	Carnegie Institution of Washington
CIWY	Carnegie Institution of Washington Yearbook
CMB	Cosmic Microwave Background
CN	Catalogue of Nebulae and Clusters of Stars
CNSC	Commission on Nebulae and Star Clusters
CO	Carnegie Observatories
CTIAO	Cerro Tololo Inter-American Observatory
DIJA	Dow Jones Industrial Average
FAS	French Academy of Sciences
FDIC	Federal Deposit Insurance Corporation
GC	General Catalogue of Nebulae and Clusters
GSA	Geological Society of America
HAG	Hubble Atlas of Galaxies

HCO	Harvard College Observatory
HD	Henry Draper Catalogue
HMS(C)	Humason-Mayall-Sandage Catalogue
IAU	International Astronomical Union
IC	Index Catalogue to the NGC
IUCSR	International Union for Cooperation in Solar Research
JHA	Journal for the History of Astronomy
KNAW	Royal Netherlands Academy of Arts and Sciences
KPNO	Kitt Peak National Observatory
kps	kilometers per second
KSA	Kapteyn Selected Areas
LMC	Large Magellanic Cloud
LO	Lowell Observatory
LON	League of Nations
M	Messier object
MIT	Massachusetts Institute of Technology
MNRAS	Monthly Notices of the Royal Astronomical Society
MWC	Mount Wilson Contributions
MWO	Mount Wilson Observatory
MWOA	Mount Wilson Observatory Association
MWPO	Mount Wilson and Palomar Observatories
MWSO	Mount Wilson Solar Observatory
NAS	National Academy of Sciences
NASA	National Aeronautics and Space Association
NGC	New General Catalogue
NGPSS	National Geographic and Palomar Sky Survey
NRC	National Research Council
OAO	Orbiting Astronomical Observatory
OSA	Optical Society of America
parsec	Parallax second of arc
PASP	Publications of the Astronomical Society of the Pacific
PNAS	Proceedings of the National Academy of Sciences
PO	Palomar Observatory
Quasar	Quasi-stellar objects
RAS	Royal Astronomical Society
RASC	Royal Astronomical Society of Canada
RBS	Royal Bohemian Society
RCT	Ritchey-Chrétien Telescope
RPS	Revolutions per second
RSA	Revised Shapley-Ames Catalogue
RSAS	Royal Swedish Academy of Sciences

SAC	Shapley-Ames Catalogue of Bright Galaxies
SEC	Securities Exchange Commission
SMC	Small Magellanic Cloud
TVA	Tennessee Valley Authority
UCB	University of California at Berkeley
UCLA	University of California Los Angeles
U of C	University of Chicago
USNO	United States Naval Observatory
USSR	United Socialist Soviet Republic
WPA	Works Progress Administration
WWI	World War I
WWII	World War II
YO	Yerkes Observatory

1

Two Centuries Of Astronomical Discovery

A general introduction to the subject of the book will be followed by a brief, focused history of the advances made in telescope design and optics, scientific methods of research, and discovery in the 200 years before the turn of the 20th century. It is meant to offer some context for beginners to the main body of the book and a refresher for those with a broader understanding of this history. From Huygens and Newton in the 17th and 18th centuries to Einstein, Planck and Hale in the 19th and 20th centuries this summary will set the stage for the historic partnership between Edwin Hubble and Milton Humason that altered our perception of the universe in graphic and epic detail.

The word "horizon" descends from Ancient Greek, its root, horos, meaning "boundary or limitation." The Greek phrase horizon (kyklos) or "bounding (circle)" was used to describe what the Greeks observed as the point at which the Earth and sky were connected.[1] Looking out at the horizon on a clear night, it is easy to imagine the Earth as a flat disk surrounded by a dome like a snow globe, the stars like tiny flakes of snow floating above.

For the majority of human history, the sky existed as a two-dimensional backdrop to our three-dimensional world, majestic, serene, sometimes terrifying, other times peaceful, but always mysterious. As darkness fell and the golden veil of daylight faded, the sky revealed a wondrous confusion of twinkling points and dusty clouds of light, stretching the limits of our imagination.

Our sense of sight can be as deceptive as our sense of humor can be elusive. We tend to take for granted that what we see exists in a whole or true state. For many centuries we assumed the Earth was the center of the universe. The

© Springer Nature Switzerland AG 2021
R. Voller, *Hubble, Humason and the Big Bang*, Springer Praxis Books,
https://doi.org/10.1007/978-3-030-82181-4_1

Sun revolved around us, appearing in the eastern sky each morning and disappearing in the same spectacular fashion in the west at dusk.

Over time, observers, philosophers, mathematicians, and artists began to wonder how the great glowing orb made its way back to the other side of the sky to appear again each morning. The Moon waxed and waned and disappeared altogether in the night sky only to return again. Conspicuous visitors, the planets, appeared in the otherwise uniform patterns of light in the nighttime sky, their motions compelling us to try to understand the true nature of the firmament.

Eventually a different, less obvious, notion of the universe began to unfold. In the third century B.C.E., Aristarchus of Samos constructed a new Sun dominated solar system where the Earth and other planets revolved around our parent star.[2] But this new idea was resisted in favor of the geocentric model put forth by Plato, Aristotle, and others, then developed by Claudius Ptolemy in the shape of epicyclical motions of the Sun, Moon, and planets.

Aristarchus's idea was finally picked up again by the Polish Renaissance polymath Nicolaus Copernicus and corroborated by Galileo Galilei a century later. Galileo paid dearly for his support of Copernican theory. He was found guilty of being "vehemently suspect of heresy" in 1633 for crimes against the Holy Scripture and eventually sentenced to house arrest for the remainder of his life. Galileo died in 1642 at age 77, in Tuscany.[3] He was reburied in the Basilica of Santa Croce in Florence 100 years after his death, at which time three of his fingers and a tooth were removed from his body. The middle finger of his right hand is displayed today in the Galileo Museum in Florence [4] flipping the eternal bird to the Italian church for 25 years of persecution, censure, and incarceration.

Despite his hard luck, Galileo's assertion of the intrinsic reality of heliocentrism was echoed by his contemporary, the German astronomer Johannes Kepler, who used the theory in his development of planetary motions. These two men in turn inspired the development of theoretical and mechanical physics by Christiaan Huygens and Isaac Newton during the early stages of the scientific and industrial revolution.

At its core, science is the language of skeptics developed by skeptics to make believers of skeptics. Innovation develops out of critical thought and the reinterpretation of ideas and knowledge, leading the way to new technology, experimentation and discovery that pushes us beyond our horizons into the realm of the previously unknown. It may be said that this cycle has remained constant throughout our history.

In today's world, as satellite, telescope, computer, and photographic technologies allow us to view the universe in ever-increasing detail, we have come to take these basic truths for granted – the motion of the Earth, the planets, and their satellites around the Sun. Our understanding of the world around us has grown dramatically, but the eye's ability to see has evolved little, if at all, during the past 3,000 years. Standing on the Earth today reveals no more visible proof that the world is round, that it revolves around the Sun, or that the Milky Way is one of hundreds of billions of galaxies in a universe nearly 14 billion years in the making than it did in the time of Aristarchus.

In fact, the limits of sight are breathtakingly near when we consider how much our sense of sight actually tells us about how a bird flies, how and what we breathe or the biochemical reactions this involves, or the atomic and sub-atomic machinations taking place constantly that make our very existence possible. At some point not very far from our noses the limits of pure vision are woefully shallow.

Developing a better understanding of the processes by which the most fundamental characteristics and functions of our world take place required an entirely different kind of vision, one that imagined a world beyond the horizons available to the senses, linked with the subconscious sense of intuition, fueled by curiosity, and built over centuries of invention and experimentation.

Many of these most fundamental discoveries were made not by an individual but by two or more people working together in collaboration toward a common end. Marie and Pierre Curie, William and Margaret Huggins, the Wright brothers, James Watson and Francis Crick, Antoine and Marie-Anne Lavoisier, Robert Bunsen and Gustav Kirchhoff, William, Caroline and John Herschel; these and many others together made scientific breakthroughs built on a foundation of scientific discovery that significantly increased our knowledge of the processes that make our world go around. Their collective vision combined with that of countless contributions from scientists, philosophers, and thinkers through the ages, have expanded the limits of our collective sight far beyond the human sensory horizon, from the infinitesimal inner workings of molecular science to the vast outer reaches of space.

In many cases, these men and women were scientists brought together either out of love for one another or for their common scientific pursuit or both. Once in a while, however, a scientific duo of a less refined, less predictable and less collegial orientation bucks the trend and transcends the limits of human experience to achieve a fundamental discovery despite and, in fact, exceeding their personal limitations.

Such were Edwin Hubble and Milton Humason, a maverick duo of epic proportions who nevertheless, through their partnership of 25 years at the Mount Wilson Observatory (later the Mount Wilson and Palomar

Observatories and presently the Carnegie Observatories), thoroughly established the Big Bang Theory of universal evolution and exploded Einstein's concept of spacetime, putting the field of cosmology on a firm foundation. Their work sent ripples of confusion and doubt through the science world, causing the greatest theoretical minds of the day to question the true meaning of their results, while reimagining the depths of the known universe and pushing the limits of the four-dimensional realm out hundreds of millions of light years.

From our connection to the Earth, to the Earth's connection to the heavens and to the connection of the known universe with the origin of space and time, our curiosity, our need to explore the limits of our world is perhaps the single most defining expression of human existence. The revelation of our limitations, physical and mental, and our desire to press forward to meet and, whenever possible, exceed them perhaps most uniquely defines the human spirit.

The confluence of these two ideas is expressed compellingly through the extraordinary tale of Edwin Hubble and Milton Humason, in their individual life stories as well as their decades-long professional partnership. Building upon the history of astronomical discovery and scientific theory, they were able to extend our common horizon from the edge of the Earth to the outer reaches of the known universe, compelling us forward in pursuit of the last horizon, the Big Bang and the origin of space and time. What follows is a short history of the relevant developments in the science, research techniques and instrumentation that set the stage for their historic run at history.

A Biographical Timeline

250 Years of Development in Physics and Astronomy

Christiaan Huygens And Isaac Newton: The First Unification Of Physics

The legacy of thought and experiment set forth by Galileo, who has been called the father of modern science,[5] was carried forward into the 18th century by Christiaan Huygens, the first theoretical physicist,[6] and Isaac Newton, the co-inventor of calculus, the modern reflecting telescope, and classical mechanics. They were essentially contemporaries, although Huygens's entry into the world of physics preceded Newton's by a dozen or more years and actually inspired his slightly younger counterpart.

Huygens was born on April 14th, 1629, in The Hague to a Dutch diplomat and advisor to the House of Orange, Constantijn Huygens, and his wife Suzanna van Baerle, a wealthy heiress, intellectual and poet who died in childbirth in 1637.

Constantijn Huygens was also a poet and musician who counted among his friends Galileo, Marin Mersenne and René Descartes and he made sure his children received the finest liberal education. In addition to math, logic and rhetoric, Christiaan studied French and Latin, fencing and dancing, music, history, and geography.[7] At age 15, his father hired Jan Jansz de Jonge Stampioen, the former tutor to William II, Prince of Orange, to tutor the boy in mathematics.[8] His early mastery of the subject led his father to glowingly compare young Christiaan to Archimedes.[9]

One of Christiaan Huygens's greatest influencers was his father's old friend Galileo, whose study of everything from projectile motion and velocity to gravity and objects in free fall, to the properties of pendulums and hydrostatics and astronomical observations captivated the young problem solver. Although Galileo was under house arrest by the time Huygens came of an age to profit from it, his Italian predecessor's work was nevertheless an inspiration.

Huygens became interested in telescopes in the early 1650s and began grinding and polishing his own objective telescope lenses with his older brother Constantijn. In 1655, while studying the peculiar disc-shaped ring of Saturn, Huygens discovered the planet's moon Titan, the sixth moon to be discovered in the solar system after our own and the four so-called Galilean moons of Jupiter: Europa, Callisto, Io and Ganymede.[10]

A year later, while studying Galileo's work on pendulums and isochronism, Huygens invented a pendulum clock, which he patented in 1657[11] (Fig. 1.1). A visually blind Galileo had designed a clock escapement while under house arrest, which he and his son Vincenzio had started to develop into a working clock 20 years earlier, but never completed.[12]

Huygens' patent followed an alternate design. The clock used a lateral "crown" gear at the top of the mechanism called a verge escapement. The clock featured a long, weighted pendulum moving back and forth in harmonic oscillation that produced a restorative force to the pendulum, keeping its swing rate in balance.[13] Pendulum clocks became the standard around the world until the 1930s, keeping time in homes, offices and train stations, enabling the increasing pace of life to begin throughout the industrial revolution. Huygens might be blamed posthumously for unwittingly setting the stage for the advent of "rush hour," but then by that same logic he is also

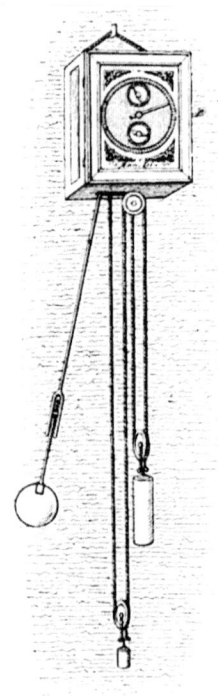

Fig. 1.1 Diagram of the first pendulum clock by Christiaan Huygens in 1657. The clock's drive was propelled using Huygens's ingenious "endless rope" system of weights and pulleys. (PD-US Expired)

behind the creation of "happy hour." His book, *Horologium Oscillatorium*, was published in 1673 and became one of the most noteworthy volumes on 17th century mechanics.[14]

In 1659, Huygens capped an incredible decade of theory, discovery, and invention by producing the standard mathematical formula for centrifugal force, the force that governs a body's movement along a curved path. Published in a later edition of *Horologium*, it was later cited by Newton as influencing his own work in celestial mechanics.[15]

Around the same time, across the English Channel, 17-year-old Newton was returning home to Woolsthorpe after spending five years at The King's School in Grantham where he had been taught mathematics, Latin, Greek, and various other subjects.[16]

Newton's childhood was more troubled than his Dutch counterpart. His father, after whom he was named, had died three months before his

premature birth on Christmas Day in 1642, the same year Galileo died.[1]
Young Isaac was weak and sickly and barely survived infancy. He was told later
he was so small at birth he could have fit in a quart pot.[17]

After three years his mother, Hannah Ayscough, remarried a Reverend
Barnabas Smith and left young Isaac in the care of his grandmother, Margery
Ayscough. Following Smith's premature death, she returned to her parents'
when Isaac was 11. A year or so later he was sent to the boarding school in
Grantham, leaving his mother to care for the three children from her marriage
to the Reverend Smith. His mother's decisions, though perhaps virtuous given
her own circumstances, left a stinging resentment that Isaac would carry
throughout his life.[18]

In recent years it has been suggested by those in the medical community
that Newton suffered from Asperger's Syndrome. If so, this might explain
both his social aloofness and highly inquisitive and curious nature.[19] From an
early age his interests were in devouring the books left to him by his stepfather
and copying much of the contents of what he called Smith's "waste book"
containing accounts of his theories and experimentations.[20]

Shortly after his return home, Isaac's mother attempted to get him into
farming, but he refused to work, fell into fits of rage and was insufferable.
According to the servants, Isaac was "fit for nothing but the 'versity.'"

His former schoolmaster Henry Stokes agreed, pleading with his mother to
send Isaac back to school, which she did in June 1661 with the financial assistance of her brother, the Reverend William Ayscough, who had also studied
at the school.[21]

He was enrolled at Trinity College of Cambridge University in 1661. The
five years that followed were to become Isaac's eternal spring. By the end of
1666 he would revolutionize mathematics, unify Galilean and Keplerian
mechanics, make important contributions to the understanding of light and
color, and establish his position among the leading names in all of science for
all time.

This period started off innocently enough. To pay for his schooling, Newton
became a servant to a man named Humphrey Babington, a local rector and
fellow of the college who was also the uncle of a young woman who might
have been young Isaac's first and only love.[22] He was awarded a scholarship in
1664 and while taking his Bachelor of Arts degree from 1664 to 1665, Newton
began to scribble down his notes and questions on the work of Galileo,

---

[1] The December 25th, 1642, date of birth for Newton is by the Julian calendar which England retained
after Catholic countries adopted the Gregorian calendar in 1582, according to which he was born on
January 4th, 1643. England didn't adopt the new calendar until 1752.

Descartes and Kepler as well as his thoughts on the writings in his stepfather's notebook.

The college had to close in 1665 due to the outbreak of the bubonic plague that would eventually wipe out nearly a quarter of London's population, and the young graduate went home to ride out the storm of disease and death.[23]

While the plague ravaged London, Newton applied his formidable powers of perception and invention to the various questions that had been foremost in his mind. In a series of feats of extraordinary insight and ingenuity he developed integral and differential calculus, devised a theory of the nature of color and light and conceived a mathematical system to explain gravity and Keplerian motion, drawing upon centuries of thought and discovery (inspired, perhaps, by one of his mother's apple trees) that he would later develop into his three universal laws of motion and the law of gravitation. On his return to college in 1667 the 25-year-old was beginning to prepare the work that would unify the earthly and celestial physics.

His first investigation concerned light and color. In an earlier experiment Newton had shown that "pressure was transmitted uniformly through the whole mass of a fluid in equilibrium," contradicting Descartes' description of the propagation of light. He agreed with the English physicist Walter Charleston that matter was made up of atoms and further postulated that light was made up globules or corpuscles which moved together in waves.[24]

While observing light through a prism he noted that the dispersion of light through a refracting lens would cause the colors in the light to separate, one of the first determinations of chromatic aberration as it is called today. On that premise, he began developing his own precisely shaped mirrors from speculum metal for use in a telescope of his own design in an attempt to eliminate the effects of refraction.

He finished the new telescope the following year as he was awarded his master's degree and became a fellow at Trinity College. The instrument, instead of using a light refracting objective lens to focus light from an object down to the eye of the observer, reflected light off a primary mirror at the bottom of the tube to another smaller mirror near the top of the tube and then to an eyepiece mounted on the side of the tube.

The result was a more defined image of the object viewed and supreme light reducing power in a telescope that was a mere 8 inches long. Although it appeared to be smaller, the passage of light from one mirror to the next actually increased the focal length to that of a refracting telescope two to three times its length, with greater image clarity.

The wooden tube was captured at either end by an iron armature connected to a round oaken ball, itself set between two arched iron keepers for the rotation of the instrument. It was 1668 and the 26-year-old Isaac Newton had created the first Newtonian reflector (Fig. 1.2).

The next year, Newton inherited the post of Lucasian Professor of Mathematics, which was yielded by Isaac Barrow on his behalf.[25]

By this time, Christian Huygens had made his way to Paris (arriving in 1666) where he earned a position in the French Academy of Sciences, newly founded by King Louis XIV. Gottfried Leibniz was a German diplomat, mathematician and inventor whose binomial number system permeates modern computing technology. On finding the deficiency of his knowledge of analytic geometry, he took tuition from Huygens in 1673. Huygens, in turn, initially struggled to see the value of calculus, which Leibniz had invented independently of Newton, but later began to appreciate his former mentees "contributions" to "the beauty of geometry."[26]

Fig. 1.2 Replica of Isaac Newton's first reflecting telescope from 1668. (Image courtesy of the Observatories of the Carnegie Institution for Science Collection at the Huntington Library, San Marino, California)

Newton and Leibniz would fall into a prolonged battle over the invention of the new mathematics after Leibniz published his treatise on the subject in 1684, three years before Newton. Newton's notebooks would later establish the English genius probably developed his calculus before Leibniz, but the latter's nomenclature was generally considered superior and was preferred to Newton's.

In 1675, Christiaan and his brother Constantijn jointly conceived of a tubeless telescope designed for use with new objective lenses of their own making. By now, Christiaan had been shaping and polishing lenses for over twenty years and he designed his own polishing machine.

The "aerial telescope," as Huygens referred to it, consisted of a single objective lens mounted in a small, round iron tube that was set aloft on a tall wooden mast resting on a ball joint so that it would swivel. The lens tube was connected to a handheld eyepiece – of Huygens' design – by use of a thin cord that helped to keep the two in line. A lantern held close to his ear projected light through the eyepiece to guide him to the objective atop the mast and on toward the object he sought to observe. An engraving of the tubeless telescope appeared in Huygens 1684 book, *Astrocopia Compendiaria*.[27]

Although his telescopes were rejected in favor of more user-friendly types, Huygens's wave theory of light, which he first proposed in 1678 and published in a *Treatise on Light* in 1690, would become commonly accepted theory.

Newton had published his own *Hypothesis of Light* in 1675 in which he postulated the existence of the infamous universal ether, rarer in the pores of bodies than in free space, that caused light corpuscles to move and change directions by its pressure.[28] Given his standing in the scientific community at the time, Newton's *Hypothesis* was more broadly accepted than Huygens's account. But the pendulum would swing back and forth between theories over time as different experiments provided contradictory evidence.

Augustin Fresnel, building on work by Thomas Young, would rescue Huygens's wave theory in 1818, reporting on results from his experiments with the diffraction patterns made by the shadows of 25 different objects when illuminated by a light source. The fringes of light and darkness created by this process, Fresnel determined, could not be created by pressure on Newton's predicted particles of light, but must, instead, result from waves of light being interrupted in their path by the obstructive object. If the peak of one wave was aligned perfectly with the trough of the other, destructive interference would cause them to cancel out and the light would disappear. If the same waves of light were aligned at their peaks, constructive interference occurred where the light peaks were twice as tall.[29]

For his remarkable insights into the theory of light, Fresnel received the 1819 Grand Prix from the French Academy of Sciences (FAS).[30] This work later influenced James Clerk Maxwell in his research on electromagnetism. In this way, Fresnel was to Maxwell what Maxwell was to Albert Einstein, who would use Maxwell's revelations on electromagnetism to extend Newtonian mechanics almost a hundred years later.

Nevertheless, Isaac Newton's realization that white (visible) light was actually made up of an array of hues across a spectrum from red to violet, overturned centuries of Aristotelian doctrine.

In 1704, Newton published the results of his experiments with light in a book called *Opticks*, in which he asserted his long-held view that light was made up of very small particles or corpuscles whose color variants were harmonically balanced. As has been seen above, this explanation was abandoned in favor of Huygens's propositions from his *Treatise on Light* (with help from Young and Fresnel). The color of light is the expression of its wavelength. The revelations about the spectral color of visible light and its movement in waves would become building blocks in the development of astrophysics and ultimately enable the work of Edwin Hubble and Milton Humason.

Even more striking in its fundamental provocations to science was Newton's *Principia*, known by its full name as *Philosophiae Naturalis Principia Mathematica* (*Mathematical Principles of Natural Philosophy*), in which he brought forth his long-concealed universal laws of motion and the law of gravitation.

De motu corporum (*On the motion of bodies*), the first in what would become a three-volume masterpiece, was delivered by Newton himself to the Royal Society in 1686. It had immediate appeal, and the Society's governing body decided a month later to publish the rest of it.

Edmond Halley, a member of the Society and particularly enthusiastic proponent of the revelations in the first book because it held out the prospect of a new understanding of the universe, began to press the Cambridge professor for the rest of it.

Newton sent Halley the second volume, which was an extension of the first, a year later, and promised Halley the third shortly thereafter. But a dispute between Newton and his longtime foe Robert Hooke, who devised demonstrations of experiments for the Society, caused Newton to renege momentarily. Halley was unwavering in his support of Newton and after some fawning and reassurance convinced him to provide the final volume, *De mundi systemate* (*On the system of the world*), which he did in April of 1687.

All the pieces were in place but there was one more tiny obstacle to overcome. The year before, the Society had lent its full support and financial backing to that great centerpiece of scientific thought, *The History of Fishes*. Inexplicably, the book had flopped and now the Society had no money for the publication of Newton's groundbreaking work.

Halley, who had recently become the clerk to the Society, was not a wealthy man but he had the means to pay to have the *Principia* published. His pay for his role as clerk was a meager 50 pounds, and even that had been swept away by the poor performance of *Fishes*. With no funds available, the society decided to pay Halley in copies of the failed book, a tidy sum not worth its weight, I'm sure he would've agreed.[31]

Newton's first law of motion observed simply that a body at rest will remain at rest and a body in motion will remain in motion unless acted upon by an external force. The second law stated that the acceleration of a body was dependent on the force acting upon it and the mass of the body as given by the equation $F = ma$. The third law of motion stated that for every action there is an equal and opposite reaction.

In other words, an apple didn't just decide to fall from a tree, it was compelled there by the gravitational attraction – and a stiff wind, perhaps – between its mass and that of the Earth, breaking its connection to the tree at the stem and sending it plunging toward the ground with a force measurable by multiplying its mass by the increase in its velocity. The force that detached the apple from the tree branch was counteracted by slightly moving the branch in a direction opposite to that in which the apple dropped, proportional to the force that caused the apple to fall in the first place. This phenomenon, Newton asserted, applied not only on Earth but throughout the universe.

In creating his law of gravitation, Newton drew on the work of both Galileo and Kepler, who speculated on the movement of planets around the Sun without a mechanical system in which this could be tested. Newton's theory utilized Huygens's formula for centrifugal force in bringing Galilean mechanics on the movement of objects on Earth into the same realm as Kepler's bodies in space, the Copernican model. The universal law of gravitation said that objects were attracted to each other with a force that was directly proportional to the product of their masses and inversely proportional to the square of their distance, in an equation written as:

$$F_g = G\left(m_1 \; x \; m_2\right) / r_2$$

where F_g is the gravitational force, G is the gravitational constant, m equals the masses of the two objects and r is the distance between them.

This was the first unification of modern physics and it made Isaac Newton a star. He had devised a solution that explained why what goes up must come down, why the Moon travels around the Earth, why the Earth and planets travel around the Sun, and the trajectory of cannonballs fired from the king's cannons at his enemy's ramparts.

Newton's fundamental laws, called collectively classical mechanics to distinguish them from subsequent relativistic and quantum mechanics, still work very well in explaining the behavior of most bodies larger than a molecule and smaller than a planet as long as they're at or near room temperature and are not traveling anywhere near the speed of light.

Newton and Huygens lived very different lives in some respects, but with coincidental similarities. Where Newton was irascible and defensive, Huygens was inclined to cordial debate in keeping with his pedigree. Neither man ever married. Newton evidently suffered from some mental illness or breakdown later in life that caused him to exhibit odd behavior, according to those who knew him. On returning to The Hague in 1681 Huygens seemed to suffer from depression. Nevertheless, the outpouring of scientific thought and experiment that the two brought to the field of physics would occupy science for the next 200 years.

The two men met in England in 1689 on the Dutch physicist's third visit to the country. At the time, Newton was trying to get himself hired on as a professor at King's College, Cambridge, and enlisted his fellow from the east (whose brother Constantijn was secretary to King William III) to assist in his appointment. An audience was arranged, and Newton was granted the king's blessing. But when Newton refused to take holy orders, the college protested so vehemently that the king withdrew the appointment.

Huygens died in The Hague on July 8th, 1695, at age 66. His final years were marked by poor health, loneliness, and melancholy. He was buried in an unmarked grave near his father who had passed just eight years earlier.

Newton was knighted by Queen Anne in 1705, becoming the first scientist to receive the honor. He passed away on March 20th, 1726, at his home in London. He was laid to rest at Westminster Abbey in a royal sendoff. The title of "Master and Worker of the Mint" which Newton had held for almost three decades was bestowed on his nephew, John Conduitt, by King George II.[32]

Revolutionaries: Lavoisier, Messier And The Herschels

Antoine Lavoisier, Charles Messier and the Herschels – William, Caroline, and John – propelled a revolution in astronomy and chemistry in the 18th and

19th centuries while living through political revolutions that ensnared their home countries in debilitating wars. Well, almost all of them lived…

Lavoisier was born into French nobility on August 26th, 1743, the son of a Parliamentary lawyer. He was educated in Paris, earning a degree in law. However, his real interests were in the field of scientific exploration, and he would never practice law, preferring instead to involve himself in the questions that puzzled scientists of the day. In 1771 he married the 13-year-old daughter of the Farmer-General, Marie Anne Pierrette Paulz. She became his lifelong companion, confidant, and colleague in his scientific endeavors.[33]

Much of his early life was spent in pursuit of social reform, where he made contributions to problems with urban street lighting and attempted to make reforms to improve conditions for prison inmates. He made philanthropic contributions to science and led an improvement in the performance and manufacture of gunpowder, improving the French economy while aiding the American Revolutionary cause in the 1770s.

A member of the French bourgeois class during the past two decades of Bourbon rule, Lavoisier leveraged his already considerable wealth and privilege to reap millions of livres by investing in the *Ferme Générale* (General Farm) that lent the French monarchy money which taxpayers would repay with interest.[34] As a member of the administrative committee in the 1780s he led the commissioning of a six-foot-high wall around the Paris perimeter with 65 ornate gateways to allow the flow of traffic into and out of the city. These gates were manned by tax collectors who drew tariffs for all goods entering the city, reaping still more tax money for Lavoisier and his fellow investors. Although he didn't live lavishly on his income and spent his time working to better his community and working on solutions to scientific questions of the day, Lavoisier was hated by his countrymen for his part in this ruthless taxation scheme.[35]

Antoine and Marie Anne Lavoisier were lucky in love and partners in the field of science they endeavored to impact. Starting in 1772, with his loving young wife at his side, Lavoisier began an investigation of chemical elements that would lead him to publish the first text on the subject of chemistry, *Traité élémentaire de chimie* (*Elementary Treatise on Chemistry*) in 1789.

It detailed how water could be produced by burning hydrogen (the word he coined for Henry Cavendish's "inflammable air") and oxygen (another word he coined) in a jar over mercury. A key outcome of this experiment was Lavoisier's finding that the weight of the water produced was equal to that of the two exhausted gases. When elements were altered from one state to another, their composition would change but no mass would be lost. The law of the conservation of mass had been discovered by Lavoisier.[36]

In addition to his groundbreaking work in combustion, the book also details Lavoisier's attack on the long-held belief that a combustible substance, phlogiston, existed in chemical elements and that this was released by combustion. This 2,000-year-old idea had now been proven false by Lavoisier.

The 33 elements in Lavoisier's "Table of Simple Substances" included 10 suspicious characters. "Simple substances belonging to all the kingdoms of nature, which may be considered as the elements of bodies." Those elements named in the treatise are light, caloric oxygen, azote (nitrogen), hydrogen, sulfur, phosphorus, charcoal muriatic radical (chloride), fluoric radical (fluoride), boracic radical, antimony, arsenic, bismuth, cobalt, copper, gold, iron, lead, manganese, mercury, molybdena (molybdenite), nickel, platina (platinum), silver, tin, tungstein (tungsten), zinc, lime, magnesia (magnesium), barytes (baryte), argil (clay), and silex.[37]

The advent of the qualitative and quantitative discovery of some of the basic elements would play an important role in the development of astronomy and astrophysics in the years to come, as the colors in Newton's spectrum gained further definition. As we will see, these revelations would lead to the establishment of the field of spectroscopy that would prove to be an important tool in discovering the expanding universe.

Antoine Lavoisier was seen as a leading figure of science and highly respected among the French aristocracy for his contributions to the country and to their pocketbooks. He led a rich life, literally and figuratively, and he died in the most dramatic of fashions. During the year-long "Reign of Terror" at the height of the French Revolution in 1794 he was accused of tax fraud and ended his life eight inches shorter, having been guillotined in front of an angry mob at the age of 50. But his work led the first chemical revolution, and he is regarded by many as the father of modern chemistry.

One of the first astronomers to attempt to catalogue the distant objects in the night sky was Charles Messier, a French astronomer and compatriot of Lavoisier who was born June 26th, 1730. Like his well-healed and ultimately ill-fated chemistry counterpart, Messier was born into a wealthy family, the 10th of a dozen children to a court usher in the tiny French Principality of Salm. Unfortunately, six of his siblings died during childhood and his father passed away when Charles was 11 years old.

At age 21, Messier took a position as an assistant to Joseph Nicolas Delisle, Astronomer of the Navy at the Royal College of France in Paris. The observatory was equipped with a four-inch refracting telescope housed at the Hotel de Cluny. Through his association with Delisle, Messier learned the finer points of observing and documentation, and became a trusted and respected assistant. After Delisle retired in 1765, Messier was appointed Astronomer of the Navy where he remained throughout his life.

When a comet returned in 1758 after an interval of 76 years to fulfill the prediction by Edmond Halley based on Newton's law of gravitation, hunting comets became something of a sport, with Messier one of the most enthusiastic participants. Frustrated at being misled by fuzzy patches in the sky that proved to be nebulae and star clusters he started to compile a list of objects which resembled comets but weren't. He was assisted in this task by Pierre Méchain, the Computer of the Depot of the Navy, an avid observer of comets who would later become director of the Paris Observatory.

One hundred three of the Messier objects were published in his *Catalogue of Nebulae and Star Clusters* in 1781, but the modern count includes seven objects identified after the catalogue was published (in total designated M1 to M110).

It was the third and final publication of the catalogue that brought Messier broad appeal around Europe. He was made a fellow of the Royal Society, a foreign member of the Royal Swedish Academy of Sciences and a member of the FAS in Paris.

Messier lost his pension and salary during the Revolution, but at least he got to keep his head. Still, penniless and in his sixties, he was so poor he had to borrow lamp oil from his friend, the French astronomer Jerome Lalande.[38] After a year, his financial prospects improved, and he was able to continue observing on his own accord.

He was awarded the Cross of the Legion of Honor by the French Emperor Napoleon Bonaparte in 1806. In honor of the award, Messier chose a comet he discovered in 1769 to commemorate the year of Napoleon's birth. In a memoir on the subject, Messier entertained the possibility that the comet's appearance amounted to a sign or a harbinger of Napoleon's future greatness.

The appearance of an astrological tone in the memoir shattered Messier's reputation and his health began to deteriorate. He had a stroke in 1815 and died two years later at age 86.[39]

Friederich Wilhelm Herschel, a contemporary of Lavoisier and Messier, was born in Hanover, Germany on November 15th, 1738, to a Jewish oboist named Isaac Herschel and a German Lutheran named Anna Ilse Moritzen. Hanover was united under King George II of England at the time and young Wilhelm, as he was known, was playing the oboe in the Hanover Military Band with his father and older brother Jakob.[40] In 1757 a combined force of the Guard and British military were routed by the overwhelming military might of the French army during the Seven Years War. Wilhelm was accused of desertion, so his father sent him to England to seek refuge.[41] The English language came easily, and he soon anglicized his name as Frederick William Herschel.

After gaining a foothold in his adopted country William quickly became a composer of note and was appointed director of the orchestra in Bath in 1780. During his long life as a musician and composer Herschel was prolific, composing 24 symphonies, 14 concertos, six sonatas for violin, cello and harpsichord and dozens of smaller works for organ, violin, oboe, and other instruments.

In 1772 William rescued his younger sister Caroline from a life of servitude caring for her aging mother in Hanover. William and Caroline had always been close, and he brought her to Bath to live with him. She blossomed under his tutelage and support, first as a singer and later as a celestial observer.[42]

It was around this time that William Herschel began to involve himself in the study of the stars that would lead him to become one of the founding fathers and leading innovators of modern astronomy. His readings of philosophy and astronomy fueled a growing interest in charting objects in the night sky. Since this occupation required the use of instruments of the highest quality and he was not endowed of a large fortune to spend on his hobby, his interests were extended to the field of optics and telescope design.

Herschel spent many long hours learning to make and polish speculum metal mirrors for use in the big reflecting telescopes of that time. As he became proficient in the craft, he began making his own telescopes as well, casting and polishing his own mirrors and selling both mirrors and telescopes for extra income to support his staff and observational research. As his abilities became more widely known, his hobby developed into a tidy business.

Over four hundred mirrors were cast, ground, and polished by Herschel and his assistants in this capacity with diameters between six and 48 inches. In addition, the team sold at least 60 complete telescopes, most of them with focal lengths in the range seven to 10 feet.[43]

Owing to the poor reflecting power of the speculum mirrors of the day, Herschel soon dispensed with the Newtonian secondary mirror and instead angled his primary toward an eyepiece positioned at the front of the tube. This Herschelian focus became his preferred viewing position, but improved mirror-making methods in the latter part of the 19th century would render it obsolete.[44]

Herschel also pioneered the creation of large reflecting telescopes, completing his first instrument, a telescope of 20-feet focal length, in 1774. After several failed attempts, he finally finished a mirror of sufficient quality for the instrument and began observing with the telescope two years later, just as the war of rebellion was breaking out between England and her colonies across the Atlantic.

The telescope consisted of a 12-inch primary mirror resting on a pivot over a small base, supported at the front of the tube by a pulley system attached to a tall wooden post. The observer used a tall and rather ungainly looking ladder to climb up to the eyepiece to view the stars. It was an awkward-looking contraption but with the superior focal power Herschel was able to see much more detail in the Messier objects and extend his research well beyond them.

On March 13, 1781, the year that Messier published the final version of his now famous catalogue, Herschel discovered the planet Uranus. This discovery caught the attention of King George III, who appointed him Court Astronomer. He was elected as a fellow of the Royal Society and granted a yearly salary by the king to devote himself to the construction of new telescopes, several of which were purchased by the king as gifts to other dignitaries. As a result of his enormous success, Herschel was becoming a man of means and he soon moved his family and his facilities to Slough, in the county of Buckinghamshire 60 miles west of London.

Herschel created two more large reflectors, a more refined version of his 20-footer with an aperture of 18.7-inches and a 40-foot instrument (Fig. 1.3) with its 49.5-inch speculum mirror constructed with donations from the British crown. It was completed in 1789, 100 years before Edwin Hubble's birth. Nevertheless, the great reflector was an inspiration to the young Missourian, who would write a paper detailing the instrument's design and capabilities while in college in the early 1900s.

The giant telescope's design followed that of its half-sized twin. The tube and primary mirror of the behemoth rested on its base inside a small shed on a concrete and brick slab several feet above the ground. The scaffolding that held the front of the tube and observing platform extended skyward some 60 feet in an A-frame shape. A set of large caster wheels rotated the monstrosity on a circular track. In the first month of using it, in 1789, Herschel discovered the Saturnian moons Mimas and Enceladus. However, although the king was pleased with his creation, Herschel found it too cumbersome (downright dangerous, in fact) for real exploration, and so for the most part he used his 20-footers for deep sky searches.[45] Despite its weaknesses for use in systematic research of the stars, the great reflector inspired generations of telescope builders and would remain the largest telescope in the world for half a century.[46]

Using their improved observing capabilities to the fullest extent, Herschel and his sister Caroline, who was his constant, diligent and very competent assistant, compiled notes and drawings of thousands of new objects while redefining many of the 110 Messier objects. They published the first set of objects in the *Catalogue of One Thousand new Nebulae and Clusters of Stars* (CN) in 1786.[47]

Fig. 1.3 The Herschel 40-foot telescope of 1789, with its complex tube elevation and rotation systems, was a marvel in its time, although Herschel preferred his smaller aperture telescopes. (Image courtesy of the University of Cambridge, Institute of Astronomy)

It was followed three years later by second edition with another 1,000 objects. In the opening to this volume, Herschel stated:

These curious objects, not only on account of their number, but also in consideration of their great consequence, as being no less than whole sidereal systems, we may hope, will in future engage the attention of astronomers.

At this point in his career Herschel thought that a sufficiently powerful telescope would be able to resolve stars in all nebulae and they were what we would nowadays call galaxies.[48] However, he would later step back from this bold assertion.

The final update, *Catalogue of 500 new Nebulae, nebulous Stars, planetary Nebulae, and Clusters of Stars*, was assembled by Herschel together with Caroline and his son, John, and published in 1802.[49] In the space of 20 years William and Caroline Herschel had extended Messier's catalogue 25-fold. But the celebrated telescope maker and observer was not finished yet.

On February 11th, 1800, Herschel made a discovery while working in his lab that would pave the way for the field of spectroscopy. He was testing various filters for use in studying sunspots and noticed there was a lot of heat produced when using a red filter. He decided to try an experiment wherein he directed the Sun's light through a prism and measured the temperature of the various colors spanning its spectrum. As he transferred the thermometer from the blue to the red end of the spectrum the temperature rose slightly. He repeated this experiment several times, allowing the thermometer to return to room temperature before each attempt. On one pass he saw that the temperature continued to rise as the thermometer moved past the red end into the area where it appeared the sunlight was no longer refracted. Herschel was shocked when he attained the same result on retrial and continued to verify this peculiarity. He concluded that there must be light energy outside the visible end of the spectrum of sunlight. Herschel had discovered infrared radiation.[50]

William Herschel continued observing with his beloved sister by his side until his death in 1822. Along the way he found the first moons of Uranus, Titania and Oberon,[51] determined the tilt of Mars on its axis,[52] made the first determination that the Sun was in motion and offered the first glimpse at the structure of the Milky Way Galaxy.[53]

When the Royal Astronomical Society (RAS) was created in 1820, Herschel became its first president. He was elected to the American Academy of Arts and Sciences (AAAS) and knighted in the Royal Guelphic Order. He helped to create the Astronomical Society of London and was a member of the Royal Swedish Academy of Sciences.

In view of his accomplishments, Frederick William Herschel had proved himself more than worthy of his titles. The questions he posed through his observations and conclusions would help to establish new fields of research that would improve our understanding of the broader universe. He was to 18th century astronomy what George Hale and Edwin Hubble were to 20th century astronomy combined.

Caroline Herschel's childhood was marred by illness. Smallpox and typhus had stunted her growth and left her scarred. She was deemed ineligible for marriage and condemned to the service of her family. Having been an outcast in her own home she was forever grateful to William for offering her the opportunity of a better life. Still required to do housework and make the family home in Bath, she was repaid by her loving brother by tutoring in English, music, mathematics, and astronomy.[54]

But it was the pursuit of the stars that most captivated Caroline. In addition to her detailed coordination of her brother's famous nebular studies she made significant discoveries of her own, chiefly of comets. She was adamant that she should earn her own living and became one of the first women to receive a salary as a scientist. She was involved in the creation of the nomenclature for all of the catalogues that she and William produced and spent many years after his death doing the same for the catalogues she created with her nephew.

In 1828 Caroline Lucretia Herschel was awarded the gold medal from the RAS for her contributions to astronomy.[55] The front of the coin bore the face of Isaac Newton and on the back was the image of Herschel's 40-foot telescope.[56]

William Herschel's greatest gift to the cause of scientific evolution, the development of modern culture and the advancement of astronomy was his son. The only child of William and Mary (Baldwin-Pitt) Herschel, John Frederick Herschel was born in Slough on March 7th, 1792, and grew up on the grounds of the family observatory.

Having been born in the time when the family's fortunes were centered largely on the business of astronomy and the creation of astronomical instruments, John Herschel followed in his father's footsteps, excelling in the fields of chemistry, optics and telescope building while helping his father and aunt in locating, naming and documenting stars and nebulae for their catalogues. He would later help to establish (with William and others) the RAS and would be awarded its gold medal twice, in 1826 and again 1836 for his contributions to astronomy.

Having mastered the 20-foot reflector, which his father had built some 40 years earlier, Herschel now conceived of a plan to investigate the stars and nebulae of the Southern Hemisphere. He set sail with his family in 1833 for the English colony of South Africa. He returned four years later having made a sweep of the southern skies. In 1847 he published *Results of Astronomical Observations made at the Cape of Good Hope*. In addition to the new survey of stars, double stars, clusters, and nebulae of the Southern Hemisphere he had compiled on the expedition, He had also taken it upon himself to name the seven moons of Saturn as Mimas, Enceladus, Tethys, Dione, Rhea, Titan, and Iapetus.[57] He was awarded a second Copley Medal for his work in South Africa that year.

While he was busy working on his sweep of the southern heavens, Herschel apparently still found time to create an extensive illustrated guide to the flora of the Cape with his wife and to read on a broad range of subjects.

In 1864 John Herschel published the *General Catalogue of Nebulae and Clusters* (GC) by combining the observations made by himself and by his father, with the help of his aunt, Caroline. This newest edition of the family catalogue included more than 5,000 objects, doubling the previous count. The CN and GC were later amalgamated into the *New General Catalogue*

(NGC) by John Dreyer in 1888, which has almost 8,000 objects. The NGC remains one of the most widely used catalogues for the location and appearance of deep space objects to this day.

Like his father before him, John Herschel would make his own mark on the development of instrumentation, in his case in the new field of photography. In 1819 he discovered that sodium thiosulfate (originally called hyposulfite) could be employed as a solvent for silver halides. Working with this new technique, he invented a process of adhering photographic images to glass using this "hypo" as a fixer. The earliest known photograph on glass was taken by Herschel on September 9th, 1839, of his father's 40-foot telescope (Fig. 1.4). In a paper to the RAS in March 1839, Herschel coined the term "photography" to describe the method of capturing images.[58] Several years later, in 1842, he invented the cyanotype process, a contact print process using light-sensitive paper that became known as the blueprint.[59]

Quite unlike his father, John Herschel was a prolific procreator, an industry one strains to understand how he had time for given his myriad pursuits, save

Fig. 1.4 A computer generated positive of John Herschel's negative image from 1839 - the first picture developed on glass – showing his father William Herschel's 40-foot telescope at Sough. (PD-US Expired)

for the fact that his involvement in the matter was relatively time insensitive. In 1829 he married his cousin Margaret Brodie Stewart, 18 years his junior.[60] They must have really liked each other because they proceeded to have 12 children.[61]

In *A Preliminary Discourse on the Study of Natural Philosophy* published in 1831, John Herschel set forth the idea of an inductive approach to scientific exploration that increases the likelihood of a valid outcome by using the process of experimentation.[62] It was a tremendously influential piece that reverberated throughout the science world. In his autobiography, Charles Darwin remembered how Herschel's work "stirred…a burning zeal to add even the humblest contribution to the noble structure of Natural Science."[63] In the same volume, Darwin remembered being "delighted to dine" with Herschel "at his charming house at the C. of Good Hope and afterwards at his London house." Darwin added that Herschel "never talked much, but every word which he uttered was worth listening to" and that he "often had a distressed expression." A mutual friend and admirer of John's "said that he always came into a room as if he knew that his hands were dirty, and that he knew that his wife knew that they were dirty."[64]

Herschel lived the remainder of his life in Collingwood near Hawkhurst in the county of Kent until dying in 1871, aged 79. Over his career he received the Hanoverian Order of Guelphic Knighthood from King William IV while working on the Cape of Good Hope in the 1830s and was made a baronet at the coronation of Queen Victoria in 1838. He also led the RAS as its president three times from 1827 to 1829, 1839 to 1841, and 1847 to 1849.

John Frederick William Herschel, "the most eminent scientist in Britain," was buried at Westminster Abbey near the tomb of Isaac Newton after a national funeral. When Charles Darwin died in 1882, he was laid to rest a few feet away.[65]

The Dawn Of Astrophysics: Fraunhofer And His Curious Lines

As he lay beneath the pile of wooden roof beams and shingles that had made up the walls and roof of the workshop that he was working in moments before, 14-year-old Joseph von Fraunhofer must have thought life couldn't get much worse.[66]

Fraunhofer was born March 6th, 1787, in Straubing, Bavaria, the 11th child born into a family whose expertise in glassmaking went back generations on both his mother's and father's side. Tragedy struck when his mother died in 1797 and his father died a year later.[67] The children were separated and Joseph was sent to Munich to begin an apprenticeship with the glassmaker Philipp Anton Weichelsberger, a harsh disciplinarian who insisted on perfection from his young pupil.

On this fateful evening, July 21ˢᵗ, 1801, young Fraunhofer had been working at the shop, toiling away, and trying to avoid the wrath of his mentor and master, when the roof suddenly collapsed. As if Weichelsberger's overweening persona wasn't enough, now he (or his house at least) was literally on top of the young apprentice. Joseph could hear his master calling out for help, apparently in need of a doctor to attend to his wife who was seriously injured in the collapse of the house, which was attached to the shop.[68]

The boy could see the shimmering lamplight flickering through the clapboards which, resting on the heavy roof beams, had somehow held back the roof from crushing him. The commotion at the foot of the rubble told him someone, he hoped, had heard his faint cries for help and was coming to rescue him.

The clamor of men descending on the heap rose in the minutes that followed as the townspeople gathered and began the difficult task of removing debris to extract the boy within. The scraping of boards joined the call of the men sifting for loose planks to remove from the pile. When prompted the boy answered as the group, compelled by its leader, began to home in on his position. Finally, after what seemed like an eternity, the last few boards were removed and the Joseph's face at last met the night air.

A cheer went up amid the crowd of men as the boy stood before them in the lamplight, relatively unharmed by the avalanche of wood and brick and mortar. As unlikely as it seemed to him that he was still alive, what happened next confused the young apprentice even more.

The Prince Elector Maximilian Joseph IV of Bavaria arrived in the village soon after the event to tour the site of the accident and meet the miracle boy who had escaped it with his life. So moved was the future king that he ordered one of his advisors, Joseph von Utzschneider to see to the boy's wellbeing. In addition to this charity the king also provided Fraunhofer with a small stipend and requested Utzschneider to instruct him in the art of glassmaking.

For the next several years Fraunhofer split his time, continuing his apprenticeship with Weichelsberger while also studying under Utzschneider. However, he became increasingly disenchanted living under at Weichelsberger shop, where he was prevented from attending a local school and denied the use of a lamp to light his study of optics at night.[69]

By age 19, Fraunhofer could not stand the abuse being directed toward him any longer and sought the advice of Utzschneider about what he could do to change his situation. By this time Utzschneider had taken up his own practice and convinced Fraunhofer to use his inherited salary from the prince to buy out his apprenticeship and join him to assist in the manufacture of achromatic lenses. The plan worked and in 1806 Fraunhofer was accepted by the Benediktbeuern Institute, a secular Benedictine monastery that focused on the craft of glassmaking.[70]

Utzschneider, seeing the young man's potential, put Fraunhofer under the guidance of a Swiss-born instructor named Pierre Guinand. A watchmaker by trade, Guinand excelled at creating quality uniform glass discs for large optical telescopes.[71] Under his guidance, Fraunhofer's skill blossomed and in just three years he had taken control of the mechanical department at the institute.

A few years later, after Utzschneider's partner left the firm, Fraunhofer became a partner and functioned as director of the institute. In the years that followed, Fraunhofer's superior glassmaking and polishing skills transformed the Bavarian glassmaking industry, enabling it to overtake England as the leader in optics manufacturing at the time. He designed new machines to assist in the grinding and polishing of glass and used them to produce the best refracting telescopes in the world. In 1846 the German astronomer Johann Gottfried Galle in Berlin discovered the planet Neptune using a Fraunhofer telescope with a lens 9 inches in diameter.[72]

After years of toil, Michael Faraday finally concluded that British glassmakers were not up to the standards of Fraunhofer and his fellow artisans in Bavaria. So threatened were the British by the supremacy of Bavarian glass that they sent none other than John Herschel to visit Fraunhofer at the Benediktbeuern in the hopes of discovering a few of his secrets. As was the case with all who sought Fraunhofer in this manner, Herschel left Bavaria empty handed.[73]

In 1814 Fraunhofer began to tinker with diffraction gratings, building upon the earlier work of Young and Fresnel, and developed an especially precise grating with thousands of slits to be used in place of a prism, along with a spectrometer of his own design (Fig. 1.5).

A spectrometer is a device used to separate and measure the spectral components of physical phenomena. The phenomenon most readily available to Fraunhofer at that time was fire, so he experimented using a lamp in his lab. The grating significantly improved the clarity of the spectrum and he found in it a single bold vertical line in the orange range. He may have been burning sodium in whale oil or perhaps sulfur in alcohol, whatever he had at his disposal to create as bright a light as possible.[74]

Intrigued by this result using artificial light he decided to turn his grating spectrometer toward the Sun expecting to find the same bright line, but to his amazement what he found was "an infinite number of vertical lines."[75]

Fig. 1.5 Fraunhofer (center) demonstrates his spectrometer for Utzschneider (seated at left) and von Reichenbach (right) from an 1897 painting by Richard Wimmer.

After confirming that the lines were due to the light being dispersed by the Sun and not by a fault of the equipment or other anomaly, Fraunhofer was left with only one conclusion: the Sun's light was far more complex than fire light.

Fraunhofer continued to study these dark fixed lines, discovering 574 in all (the Sun's spectral lines are today known to count in the thousands or even millions) and devising a wavelength scale (measured in nanometers or billionths of a meter) which permitted their location on the solar spectrum to be verified by others. The boldest of these lines he labeled A through K (Fig. 1.6). Thanks to his pioneering spirit and brilliant technical abilities the field of spectroscopy was taking its first fledgling steps. Several of Fraunhofer's bold lines would become indispensable to Milton Humason in his quest to determine radial velocities for galaxies.

Glassmaking was a dangerous business with its practitioners inhaling the toxic fumes of lead oxide and furnace heat. As a result, Fraunhofer contracted tuberculosis and died at his home in Munich on June 7th, 1826.[76]

Nevertheless, in his short but remarkable lifetime he made significant improvements to optical lenses, telescopes, heliometers, micrometers, achromatic microscopes, diffraction gratings and of course spectrometers. He

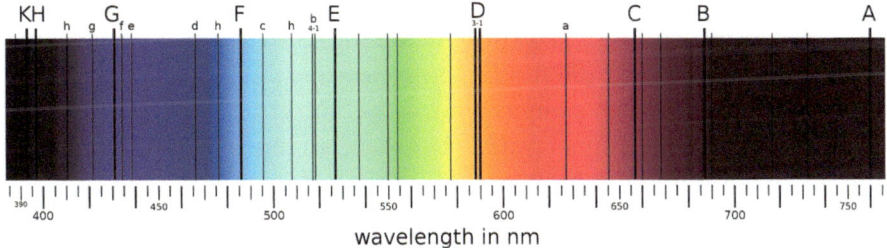

Fig. 1.6 Diagram showing the solar spectrum with Fraunhofer lines. The capitalized letters A-K indicate the bold lines in the spectral array. The H and K lines of calcium at left were important tools used by Humason in his research on nebular velocities.

helped to lay the groundwork for virtually the entire field of astrophysics.[77] For his contributions he was knighted by King Maximillian I and made a member of the Bavarian royal court with the title of Joseph Ritter von Fraunhofer.

The Dawn Of Spectroscopy: From Bunsen To Balmer

Robert Bunsen finally had the partner he had been looking for to aid in his research. The theoretical physicist Gustav Kirchhoff had joined the University of Heidelberg, Germany in 1854. At that time Bunsen was at work on a new, hotter and cleaner-burning laboratory burner (yes, that burner) with Peter Desaga, a university mechanic who would produce fifty or so duplicates of the newly refined burner the following year. A dutiful, good-humored and attentive teacher who never married but was beloved by his students, Bunsen led by example, often working alone in his laboratory while making vital contributions to science.[78]

Kirchhoff meanwhile had formulated his now universal circuit laws governing currents in electrical circuits while a student at the University of Königsberg in 1845. He married Clara Richelot, the daughter of his mathematics professor, Friedrich Richelot in 1857, the same year he calculated that an electrical signal traveling through a non-resistant wire must move at the speed of light.[79] Unlike Bunsen, who was said to be a warm, funny and insightful professor, Kirchhoff's lectures were dull and monotonous. But he was a gifted theorist who would contribute much to the future development of the two pillars of modern physics in the coming years.

Bunsen's new burner burned air with natural gas which produced a hot nearly colorless flame. He was using it to determine the color of heated minerals. Upon seeing what he was up to, Kirchoff suggested that they pass the light from the vapor created from heating these elements through a prism.

Working together they built a spectroscope, the first of its kind (Fig. 1.7). A trapezoidal box, its inner walls blackened, sat upon three legs. In it a 60-degree prism filled with carbon disulfide rested on a brass plate affixed to a vertical axis that could be rotated to focus light through the prism to create a fine image of the spectrum. A mirror attached to the vertical axis reflected the image to a small telescope placed a short distance away for viewing. A pair of telescopes were set through the non-parallel sides of the box so that their objectives were pointed at the prism. The eyepiece of one of these was replaced by an adjustable brass slit to focus the light through the tube to the prism inside the box. A burner was placed just outside the slit with its flame at the axis of the slit and tube. Finally, a platinum wire held a small piece of the mineral to be burned for the investigation.[80]

Using their spectroscope, Kirchhoff and Bunsen systematically tested various minerals and found that each of the elements they studied had its own spectral signature. That is, the lines produced in its spectrum appeared in the same place every time it was tested. A metal such as incandescent tungsten showed a continuous spectrum like a rainbow. Low-density vapors, on the other hand, showed only certain wavelengths along the spectrum specific to

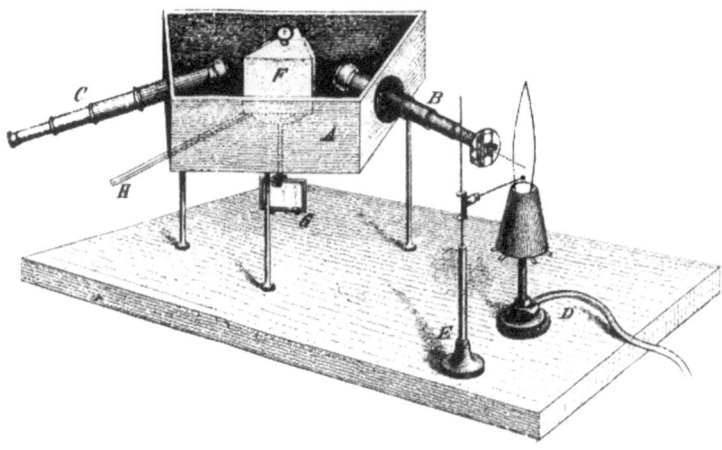

Fig. 1.7 The Bunsen-Kirchhoff spectroscope. The first instrument designed to study the spectral properties of the elements as it appeared in 1860. (PD-US Expired)

the element producing the vapor. One of these signatures first identified by Kirchhoff and Bunsen was that of sodium (Na), which produced a pair of bold lines in the yellow end of the spectrum. These were designated with a capital D on the Fraunhofer scale, and they are commonly referred to today as the sodium doublet – one of the most recognizable features on the visible spectrum. The signature of these elements was caused by photons being emitted at this specific wavelength by sodium in the hot gas or vapor, producing what is known as an emission spectrum. That is, the dark field where the continuous spectrum would otherwise be is illuminated only by the signature lines of the vapors.

Acting on suspicion, Kirchhoff next allowed sunlight to pass through the sodium vapor prior to passing it through the prism. This produced the solar spectrum with its constituent Fraunhofer lines and the sodium lines of the vapor's signature clearly visible within it. The relatively cool vapor (to the heat of the sunlight) had absorbed photons of sodium from the solar surface at the same wavelength, creating a bold dark double line in the same yellow part of the spectrum that they occurred in the emission spectrum of sodium vapor. This is called an absorption spectrum, and it proved for the first time that sodium exists in the Sun.[81]

The conclusion of this study led Kirchhoff to devise his three laws of spectroscopy. (1) a solid, liquid, or dense gas excited to emit light will radiate at all wavelengths, producing a continuous spectrum. (2) a low-density gas excited in the same fashion will produce an emission spectrum with lines characteristic of the elements. (3) if light from a source that issues a continuous spectrum is passed through a relatively cool low-density gas the result will be an absorption spectrum of the elements present in the gas.[82]

A lightbulb burning a tungsten filament will produce a continuous, unbroken spectrum of tungsten. A fluorescent bulb filled with mercury vapor illuminating a phosphorescent coating will produce an emission spectrum of mercury. The same fluorescent bulb lit also from behind by the hotter incandescent bulb burning tungsten will produce an absorption spectrum showing the continuous spectrum of tungsten with the signature lines of mercury vapor exposed within it (Fig. 1.8).

Once a scale had been devised to measure the positions of the lines in the spectrum, the field of spectroscopy was born, giving scientists a means of examining light from distant objects in a way that would reveal their chemical composition and physical characteristics.

In the years that followed, Bunsen and Kirchhoff correctly identified the elements whose emission or absorption were found in the other Fraunhofer lines (Fig. 1.9). Chief among these, as far as our story is concerned, were the

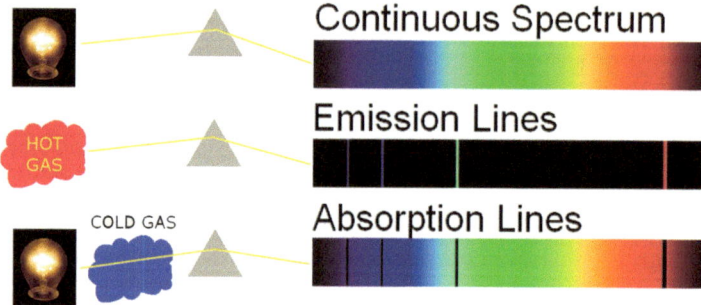

Fig. 1.8 Diagram showing the three ways the spectrum of visible light is exposed. Continuous spectra are created when light from hot metal, liquid, or dense gas is shined through a prism. The "emitted" light from s hot low-density gas exposes only the spectral signature of the illuminated gas in an emission spectrum. Absorption spectra are created when light from a dense, hot source is shown through a cool low-density gas.

H and K lines of singly ionized calcium (identified chemically as Ca II or Ca^{II}), which were frequent targets of Humason during his deep space campaign. He favored these because of their relatively wide features and used them in his comparison spectra to gauge redshifts or velocities.

In 1868 Norman Lockyer, the British astronomer and founder of the journal Nature, saw a yellow line in the solar spectrum at a wavelength which did not correspond to any known substance, so he claimed the discovery of something new to science and named it "helium" after the Greek word for the Sun, Helios.[83]

The American amateur astronomer and pioneer of astrophotography Henry Draper took the first picture of a star's spectrum in 1872. Thus, the field of stellar spectroscopy opened with the brightest star in the northern sky, Vega in the constellation of Lyra. Draper was a respected physician and graduate of New York University School of Medicine. In 1840 his father, John Draper, had taken the first clear photograph of the Moon.[84]

Unfortunately, Henry Draper died at the age of 45, very likely of pneumonia. His widow, the wealthy socialite Mary Anna (Palmer) Draper who had become his assistant after their marriage in 1867, created a fund for a Henry Draper Medal to be awarded by the National Academy of Sciences (NAS) to persons making outstanding contributions to astrophysics. In addition, she

Fig. 1.9 Kirchhoff and Bunsen diagram showing the comparison of elements with the Fraunhofer spectrum (top). Note the bold (white) D sodium doublet (Na) (third from top). Conspicuously absent from this chart are the H and K lines of calcium (Ca) (second from bottom). These would be better identified later as emulsion techniques improved photographic images toward the ultraviolet and infrared ends of the spectrum.

funded the Henry Draper Catalogue, which today contains spectral data for more than a quarter million stars.[85]

In 1885 a 60-year-old Swiss physicist named Johann Jakob Balmer devised an empirical formula that predicted the transition of electrons from higher to lower energy levels in the emission spectra of hydrogen with a high degree of precision. This was important because hydrogen is by far the most abundant element in the universe and is found in the spectra of numerous types of objects. Using his formula, Balmer predicted the increased energy levels of the four spectral lines of hydrogen that appear in visible light – hydrogen alpha (red), hydrogen beta (cyan), hydrogen gamma (blue) and hydrogen delta (violet). Although these four lines are part of the broader Fraunhofer line series they are usually designated by an h and the appropriate Greek letter: h-α, h-β, h-γ an h-δ. Along with the H and K lines of calcium, these hydrogen lines were often used by Humason and other astronomical spectroscopists.[86]

Doppler: Pairing The Perception Of Light And Sound

Standing before a crowded room of scientists at the meeting of the Royal Bohemian Society (RBS) in Prague on Wednesday, May 25th, 1842, a 38-year-old Austrian physicist named Christian Andreas Doppler started a dissertation that would forever link his name with a central premise of wavelength dynamics.

Doppler was born into a family of stonemasons in Salzburg, Austria, on November 29th, 1803. Studying at Vienna University of Technology he gained a bachelor's degree in 1825 and an advanced degree in mathematics, mechanics, and astronomy in 1829. He worked as an assistant to Adam von Burg at the university for some years toiling in obscurity. In 1836 he married a girl from his hometown named Mathilde Sturm and the couple had five children together. Shortly after the birth of his first child, Mathilde, in 1837, Doppler was appointed Supplementary Professor of Higher Mathematics and Practical Geometry at the Prague Technical Institute and it was there that he began to develop his theory of the nature and perception of both sound and light waves. Well regarded by students, he received a full professorship in 1841.[87]

Doppler had accepted membership to the RBS in 1840 and it was there, not far from the Estates Theater where a fellow Salzburgian Wolfgang Amadeus Mozart had premiered his famous opera *Don Giovanni* 55 years earlier, that he spoke on this date in 1842. The crux of his argument was that the perception of light and sound was not merely based on "what spaces of time and with what levels of intensity the creation of the waves [took] place," but also "at which time intervals and with what strength these waves…[were] perceived by the eye or ear of any observer."[88]

Furthermore, he went on to say, "the color and intensity of a perception of light or the pitch and volume of any sound…would become shorter if the observer [wa]s hurrying towards the oncoming wave, and longer if he [wa]s fleeing from it…"[89]

The color and intensity or pitch and volume emitted by a particular source varied with respect to the velocity of approach or recession of the observer as well. Today, we take this basic fact of the similarity between light and sound waves for granted. Doppler himself stated, "…nothing seem[ed] to be more intelligible than that to an observer" but prior to that time no one had realized its existence.[90]

The concept was easily demonstrated for sound waves (famously by arranging horn players on a train to issue a single tone while it drove past an observer) but Doppler simply didn't have the math skills to prove his theory for light, so he merely predicted its viability in his lecture. It was proven independently by the French physicist Hippolyte-Louis Fizeau whose math corrected and solidified the principles for both sound and light. As Karl Marx and Friedrich Engels were drafting the *Communist Manifesto* in 1848, Fizeau was predicting the blueshift or redshift in light waves from objects in space.[91] In recognition of his contribution, the theory is nowadays widely referred to as the Doppler-Fizeau Effect.

William Huggins: The Father Of Modern Spectroscopy

William Huggins swung open the door to his observatory on the grounds of his home at Tulse Hill in London. He opened the steel door to the stove near the entrance and loaded it with firewood and tinder. Holding the flint close to the tinder he switched the grate until a spark produced a tiny flame. In a matter of moments, a fire was growing in the belly of the stove. Huggins closed the door and warmed his hands over the range top.

It was a cool, damp English morning in 1859 and the 35-year-old astronomer was eager to make a start on his new project involving investigating the spectra of stars. Like many astronomers and physicists around the world, he had been intrigued by the work of Robert Bunsen and Gustav Kirchhoff on the spectra of elements in gaseous forms. He wondered if the spectrum analysis the two men had created could be applied to the stars.[92]

Huggins was born February 7th, 1824, in the village of Stoke Newington in London. His father, William Thomas Huggins was a successful silk and linen merchant and his mother Lucy, a devoted housewife, kept the residence above the family business on Gracechurch Street and helped to manage the business during the busy seasons.[93]

Homeschooled mostly by a variety of the best tutors available for hire with a middle-class merchant's income, William had an inventive young mind, and his parents encouraged his various interests. He spent his formative years attending lectures at the Adelaide Gallery and was inspired by demonstrations of new technologies in chemistry and physics.[94]

By the early 1850s Huggins was running his father's business and had joined the Royal Microscopial Society. There he met Warren De La Rue, a published amateur microscopist who had recently become a fellow of the RAS. Thrilled by the thought of studying the stars Huggins sought election to the RAS, which he achieved in 1854.[95] Sometime that year or the next, he sold the family silk business and was building his own observatory on the grounds of his new home at Upper Tulse Hill, eager to explore the heavens.[96]

The small three-story observatory was attached to the main house by a passageway on the second floor. A Dollond telescope with a 5-inch lens was mounted on a platform on the third floor, 16-feet above ground and high enough to allow unobstructed views of the night sky. There was ample space for other astronomical equipment and a study for the contemplation of findings and direction of research. It was an eminently suitable facility for an anxious astronomy enthusiast to make his start in the field. His rapid evolution from curious businessman to starry-eyed observer to a fully-fledged amateur astronomer in the decade leading up to 1863 was due to his family's relative wealth, his joining the RAS and to his mentor, William Dawes, befriended at an RAS meeting in the late 1850s. He sold his 5-inch Dollond in 1858 in favor of an 8-inch (Fig. 1.10) with optics supplied by Alvan Clark, one of the leading telescope makers of the day.

Fig. 1.10   Picture of William Huggins from 1910.

When Bunsen and Kirchhoff published their findings on spectra of various elemental gases in 1859, this piqued Huggins's interest. By now a serious amateur dedicated to setting a course for research, he bought a spectroscope and the lab apparatus he would require to compare the spectral characteristics of the Sun with those of the stars.[97]

After several years of careful study and painstaking trial and error with his neighbor and collaborator Allen Miller, a chemistry professor from King's College, Huggins published his initial findings, which provided some of the earliest evidence that the stars were in fact distant suns like our own and composed of the same elements in various proportions. These investigations, made with Miller, were done nearly concurrently with those by other names in the field, in particular the American Lewis Rutherfurd and the Italians Giovanni Donati and Father Angelo Secchi.[98]

In 1864 Huggins turned his attention to nebulae. It had long been debated whether these were dusty clouds of dust or gas inherent to the Milky Way Galaxy or distant independent galaxies comparable to ours. Upon examining a planetary nebula in Draco, Huggins found an emission spectrum indicating the nebula was a luminous gas. Other objects were more likely to be composed of stars, although his telescope could not resolve them individually. He had discovered that the spectroscope could distinguish between types of nebulae.[99] It was another layer in the development of spectroscopy as a tool for understanding the nature of stellar objects.

With his next project Huggins would set the table for the discovery of the Big Bang some 60 years later. In June 1867 he bought a micrometer and upgraded his spectroscope for greater dispersion in the spectrum so that he could accurately measure wavelengths and investigate the radial velocities of stars in terms of the Doppler effect. In 1868 he presented results, showing for the first time that stars were either moving toward (a Doppler blueshift) or away (a Doppler redshift) from the Earth.[100]

Astronomers could now use spectroscopy to analyze the chemical composition of stars relative to the Sun and measure their movements relative to the Earth. They could also now determine whether a particular nebula was gaseous or should contain stars. The research by Huggins and others through the latter part of the 19th century set the stage for the pioneering work on nebular velocities by Slipher, Humason and others in the 20th century which would help to unravel some of the biggest mysteries being posed by theoretical physics.

In 1875 William Huggins married Margaret Lindsay Murray, the young heiress to a wealthy Irish family of bankers and politicians born in Dublin. They shared a passion for stellar research and in 1899 jointly published the *Atlas of Representative Stellar Spectra*.

After Johann Balmer discovered his now famous (scientifically anyway) formula for the spectral lines of hydrogen, William and Margaret Huggins became the first to identify them in a star, once again Vega, the brightest star in the northern sky. Margaret was later recognized for her contributions as a spectroscopist in collaboration with her husband. In addition to their scientific work, they shared an interest in music and Margaret would often accompany William on piano while he played his violin.[101] William Huggins died in 1910 of complications from a hernia operation. He was 86 years old, and Margaret died five years later at the age of 66.

Along with William and John Herschel, William Huggins is one of the leading figures in astronomy of the 200 years prior to 1900, when the explosion of theoretical and practical physics and astronomy started. For his groundbreaking investigation of the spectra of stars and nebulae Huggins has become known as the father of modern spectroscopy. His work, building as it did upon that of the Herschels, Fraunhofer, Bunsen and Kirchoff and others, as well as that of his contemporaries in the field, was pivotal to the development of modern cosmology.

Parallax: Measuring The Distance To The Stars

Friederich Bessel stood on the viewing platform peering through the eyepiece of the 6.2-inch Fraunhofer achromatic heliometer at the Königsberg Observatory in Prussia in 1838. As its name implies, the instrument was designed to measure the Sun using the principle of parallax.

Stellar or trigonometric parallax is the most fundamental tool used by astronomers to measure distances to objects in space. It uses geometry and the Earth's orbit to form either an isosceles or right triangle whose base is formed by the diameter (or radius) of the Earth's orbit at two points separated by six months.

To simulate this effect, peek at your extended thumb with an object (any object will do) in the background with first one eye open, then with the other eye open. Your thumb will appear to change positions relative to the background. Measuring the change in positions and knowing the distance between your eyes will reveal the distance to your thumb.[102] Presto!

For decades after Johannes Kepler popularized the idea that the Earth and planets are in orbit around the Sun, parallax was used to attempt to ascertain everything from the Sun's diameter to its distance from the Earth and the nature and distance of the planets and stars.

The problem for astronomers, especially early on, was that the Sun and stars are a lot farther away than the average thumb. This meant that the angles to these objects would be too small to measure accurately. They needed a really long baseline for their measurements and one way to achieve one was to measure the diameter of the Earth's orbit. Down through the ages, everyone from Aristarchus to Kepler had attempted it. Christiaan Huygens got a reasonably good distance measurement using the transit of Venus across the Sun in 1653 to triangulate its position and distance. His technique used a fair amount of guesswork. In particular, determining the size of Venus. Then in 1672 Giovanni Cassini used parallax to discover first the distance to Mars and consequently the distance to the Sun.

In the 19th century, the instrument of choice for those seeking to make the most exacting measurements was the heliometer. It used two semi-lenses (a lens cut in half) set opposed with their flat edges against each other and adjusted together or apart using a finely tuned micrometer. By adjustment, an object could be brought into focus as a single image or split into two images by sliding the two semi-lenses. The micrometer reading of an object at its zero point (as a single image) and its outer point (in the case of the Sun, when two images of the Sun are touching side to side) in combination with the focal length of the objective enabled the astronomer to calculate the angular diameter of the object.[103]

Using these instruments astronomers were able to attain a more accurate measurement of the Earth's distance from the Sun, which today stands at 93 million miles, give or take. This distance from the Earth to the Sun is referred to in astronomy as the astronomical unit (AU). Dividing it by the tangent (angular measure) of one second of arc (0.00000484813) comes out to about 19.2 trillion miles or 3.26 light years. A light year is the distance light travels in a year, roughly 5.9 trillion miles. This basic unit, which astronomers call a parallax second of arc, or parsec, is used to specify interstellar and intergalactic distances.[104]

Right! What is a second of arc?

To make sense of the night sky, astronomers use the Moon's diameter to divide the sky into 360 degrees. As you probably learned in grade school or thereabouts, the Sun and the Moon appear to be about the same size to us here on Earth. But looking at the Sun for long periods of time would make for a bunch of blind astronomers and we wouldn't make any progress at all. So, the Moon it is.

Here are the basics. Two moons equal one degree (720 moons to go around once). There are 60 minutes of arc in a degree and each minute of arc contains

60 seconds of arc. Hence one degree equals 3,600 seconds of arc (60 x 60 = 3,600). It's that simple.

There is a fun way to gauge rough distances between stars by employing your outstretched hand for scale. The distance across the tip of your outstretched index finger is roughly 60 minutes of arc (1 degree). Your clenched fist (knuckles forward) is roughly 600 minutes of arc (10 degrees). And if you make the hand signal that you use to tell someone to call you, then the distance between the tips of your thumb and little finger is equal to 1,200 minutes of arc (20 degrees). Place 18 such calls and you go around the sky once!

A second of arc therefore equals an extremely small section of the night sky, about five millionths of the entire circumference when scaled using the size of the Moon as it appears to us from Earth.

To measure the distance to a nearby star (and we are talking only up to about 100 parsecs out, here) astronomers use these known variables and a bit of trigonometry. This can be done using either an isosceles or right triangle to form one of two equations: $d = 2AU/p$ for an isosceles triangle with the diameter of the Earth's orbit or $d = 1AU/p$ for a right triangle using the radius of its orbit. In either case, d equals distance and p equal the parallax angle. In the case of the right triangle the parallax angle must be divided by two to get the correct figure in degrees.

For instance, let's say a star we measured at two different points of the year gives us a parallax angle of 1.45 seconds of arc. Plugging this number into the formula above we get *2/1.45 = 1.379 parsecs*. Dividing our parallax angle by two and plugging it into the second equation will yield the same number. If we want to convert parsecs into light years, we just multiply that number by 3.26 to get 4.5ly. To convert light years to miles, multiply by the number of miles light travels in a year (5.9 trillion miles). The star is therefore 26.5 trillion miles from the Earth (Fig. 1.11).

Is this form of measurement as precise as using a caliper to get the diameter of a length of steel rod? Not by a long shot. Is it more than sufficient for determining the rough distance and scale of objects in space? Absolutely!

About 1817, Bessel approached Joseph von Fraunhofer and company about building a new and much more accurate heliometer for use in measuring the distance to a nearby star. To get an accurate measurement of such a distant object would require a telescope with an objective lens of at least 6 inches diameter that was free of the off-axis aberrations such as astigmatism and coma that impaired the heliometers of the day.

The result of much back and forth over many years was the instrument that Bessel was standing in front of at that very moment. It had been started by

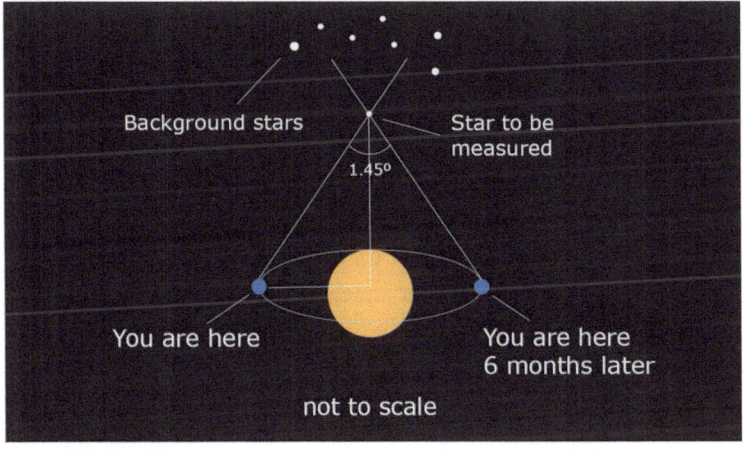

d = 2AV/p   d = 1AV/p
= 2/1.45   1.45/2 = 0.725
=1.379 parsecs   = 1/0.725
1.379 x 3.26ly   = 1.379 parsecs
=4.5 light years   - or -   1.379 x 3.26ly
4.5 x 5.9T (miles)   = 4.5 light years
= 26.5 trillion miles   4.5 x 5.9T (miles)
                            = 26.5 trillion miles

Fig. 1.11 Astronomers use the distance of the Earth's orbit as a baseline to determine the distance to a star by trigonometric parallax.

Fraunhofer himself, then finished by Georg Merz after Fraunhofer unexpectedly succumbed to tuberculosis in 1826.[105]

Now the 54-year-old observatory director, astronomer, mathematician and physicist was in a race to become the first person to determine the distance to a star. He knew other expert observers like Friedrich von Struve at Pulkovo Observatory in St. Petersburg, Russia, and Thomas Henderson at the City Observatory in Edinburgh were seeking this prize, although they had all independently chosen different targets.

Born the son of a civil servant in Westphalia in present day Germany in 1784, Bessel had excelled in mathematics as a student and was apprenticed to a shipping firm to assist in navigation. There he developed an interest in using astronomy as a means of determining longitude. He soon earned a reputation as a precise and detail-oriented experimenter. After two years at the Lilienthal Observatory where he measured the celestial positions of over 3,000 stars, he was appointed director of the newly created observatory at Königsberg in

1813, a post he held until his death from Ormond's disease in 1846. There he would determine positions for more than 50,000 stars and produce new atmospheric refraction tables (known as Bessel functions) to assist in the determination of positions of stars relative to the Earth including the Sun. In 1832 he calculated the Earth's ellipticity, providing the first determination that our planet is not perfectly spherical.[106]

But on this evening in 1838, Bessel's telescope (and his attention) was fixed on a double star in the constellation Cygnus called 61 Cygni, a star system displaying one of the largest proper motions (the degree of parallax or displacement from one position to another in the sky) then recorded, suggesting it was relatively close by. He had started his research four years prior by comparing it with a pair of stars of the 10th magnitude only to learn later that these stars were obfuscated by atmospheric conditions to a degree that rendered even the most careful analysis unreliable.[107] For a little over a year he had tracked 61 Cygni using his heliometer, measuring its angular variations to within a hundredth of a second of arc, night after night and up to sixteen times per night to allow for the greatest possible sample size.[108]

Within a year, Bessel was ready to release his results. He communicated them in a long letter dated October 23, 1838, to none other than John Herschel. In a discursive letter to the venerable Englishman, he unveiled his conclusions, based on the mean averages from 75 of his best measurements, indicating 61 Cygni to be 10.3 light years away.[109] This marked the first time anyone had determined a precise distance to a star.[2] At much the same time, Struve measured the parallax for Vega and Henderson for Alpha Centauri (which he measured while in South Africa). At last, astronomers were able to gain a sense of the depths of solar system.

In time, stars farther away than several hundred light years would render stellar parallax an increasingly unreliable tool until Walter Adams advanced the technique of spectroscopic parallax early in the 20th century at Mount Wilson.

The Ole Rømer Empire: Measuring The Cosmic Speed Limit

Of course, there would be no natural world, at least not the one we perceive, without visible light. For centuries science grappled with the nature of light. What was light? What was it made of? How did it travel? Was it faster or

---

[2] We now know that Bessel's distance estimate is 10% too small.

slower in different mediums? These and many more questions captured the imagination and fed the frustration of scientists the world over for millennia.

In 1676, a hundred years before the signing of the Declaration of Independence in the United States, a man named Olaus "Ole" Rømer, a Danish astronomer at the University of Copenhagen, was studying the cycle of eclipses of the planet Jupiter and its four Galilean moons. Jupiter eclipsed the moons from every day or so for the closest ones to every couple of weeks for the one farthest away, and Rømer was timing each of the events using one of Huygens's pendulum clocks when he made a startling realization. The eclipses were up to eight minutes earlier than expected during the half of the year when the Earth was closest to Jupiter in its orbit around the Sun. Conversely, during the second half of the year, as the Earth drifted farther away in its orbit, the eclipses occurred up to eight minutes later than expected. His calculations were correct, and the clock was the most accurate timekeeper in existence. The only variable that had changed was the Earth's distance to its Jovian cousin. That meant that the light from the eclipses must be taking longer to travel the extra distance as the Earth moved farther away. Light, he discovered, traveled at a finite speed.[110]

From that point on, the velocity of light would preoccupy science. Everyone including Huygens tried to get the most accurate measurement. It was only toward the middle of the 19th century that measurements approaching the modern standard were achieved, which is when the means of determining the velocity of light came down to Earth.

Hippolyte-Louis Fizeau, the man who improved upon Doppler's work, was the first to provide a more accurate measurement for the speed of light. He was already hailed by his countrymen for improvements that he made to the daguerreotype method of photography in 1840 and 1841.[111]

In 1849, Fizeau set up an experiment between the peaks of two mountains in France. The peaks were five miles apart and he placed a mirror on top of one of them. On the other he placed a small apparatus that consisted of a toothed wheel rotating at high speeds and a second mirror set above a light source. The idea was to shine a beam of light off the first mirror through a gap in the wheel and on to the peak five miles away, where it would reflect back to the wheel. Fizeau increased the speed of rotation of the wheel until the light moved through a gap in the wheel, reflected off the mirror five miles away, and returned through the next gap in the wheel. Since Fizeau knew how fast the wheel was rotating, the size of the gaps in the wheel, and the distance the light was traveling he could now easily calculate the speed the light.[112] The result was a velocity of 315,000 km/sec or 196,000 miles/sec, which was only about 10,000 miles/sec faster than the modern value.[113]

Around that time, a fellow Parisian and former friend of Fizeau's named Léon Foucault was creating a light speed measuring apparatus of his own. The longtime friends had fallen out in the pursuit of the best set up for the experiment. For his trial, Foucault equipped his wheel with a rotating mirror rather than teeth and caught the light at a distance that allowed him to easily measure its velocity without it traveling more than seventy feet. The light was reflected off the rotating mirror to a fixed flat mirror and back toward the rotating mirror, where it reflected off a different point on the mirror. Knowing the speed of rotation and the distance the light traveled provided the velocity. What is more, the size of his experiment allowed him to repeat his measurements in a lab, make tiny adjustments to it and even test whether the speed of light varied in different mediums.

By 1862 Foucault got a repeatable velocity measurement of 185,000 miles/sec, within 1,000 miles/sec of the modern value. Back in 1853 he had shown that light travels slower through water than through air. This was evidence in favor of Huygens's wave theory and against Newton's particle theory of light. Oh…and along the way he introduced his famous pendulum in 1851 as a physical demonstration of the effect of the Earth's rotation.[114]

Although Fizeau and Foucault came as close as anyone had come to an accurate measurement of the speed of light, the undisputed king of the field in the 19th century was Albert Michelson. He started working on the issue about 15 years after Foucault published his report, and he improved on his predecessor's apparatus by increasing its sensitivity. As a result, Michelson was able to displace the spot of returning light on his apparatus not by 1/40th of an inch as Foucault had achieved, but by 5 inches, allowing him to measure with far greater precision. In 1879 Michelson reported the speed of light as 186,355 miles/sec, a mere 73 miles/sec greater than the accepted value today.[115]

With two decades remaining in the 19th century, astronomers now had an array of tools with which to study the Sun and stars for answers to the universe's secrets. Advancements in telescopes, lenses and observing techniques enabled accurate distance measurements to nearby stars, providing a sense of the depth of space. Spectroscopy was giving insights into the chemical composition and physical characteristics of celestial objects. And discoveries about the nature of light had revealed that it traveled at a finite speed that depended on the medium through which it passed.

Still, major questions remained to be answered. The next 25 years would bring about a revolution in scientific thought and experimentation that would open the door to one of the greatest discoveries in all of human history.

Electromagnetism: Faraday, Maxwell And The Second Unification Of Physics

Visible light, the rainbow of color most of us can see with our eyes, makes up only a small, but integral portion of the electromagnetic spectrum. Indeed, if the entire electromagnetic spectrum were scaled to the length of a mile, then the visible component would occupy just a few feet.

As science unraveled the "dark" inner workings of the broader spectrum, the nature of our world became increasingly complex. The development of techniques to investigate and exploit the invisible, highly radiative parts of the spectrum began in the mid-19th century.

In 1845, while Doppler was puzzling the experimental and theoretical proof of his wave explanation of light, the 53-year-old English chemist and autodidact Michael Faraday was at work in his laboratory in London.

Born into a blacksmith's family in the London suburb of Newington Butts on September 22nd, 1791, Faraday's less formal education began as his working life did, as an apprentice bookbinder. He was illiterate but eager to learn and rise above the city streets he had known all his young life. It was there he became interested in understanding the mysterious force known through the ages as electricity, after reading about it in one of the books in the shop.[116] His eagerness and incredible aptitude had enabled him to become chemical assistant to Humphry Davy at the Royal Institution in 1813, and in 1825 he was appointed Director of the Laboratory.

After the discovery of electromagnetism by Hans Christian Oersted in 1821, Faraday discovered electro-magnetic rotations and developed the first electric motor by using wire in a bath of mercury to convert electrical energy to mechanical energy. Ten years later he discovered electromagnetic induction, the curious inductive electromotive force produced by moving a magnet in and out of a coiled wire, that led to the development of transformers, electric motors and generators.[117]

Through his experiments, Faraday demonstrated that all matter was subject to the push and pull of diamagnetic (inversely magnetized) and paramagnetic (weakly attracted by magnetic poles) forces and became convinced that the properties that governed electricity and magnetism must also apply to light. What confounded him was that he had as yet been unable to find a way to prove this theory experimentally. It was this goal that occupied his mind in the laboratory in 1845. That year, by the use of a device that sent a beam of light through a dense piece of leaden glass, Faraday was able to rotate the poles of the light beam by applying a magnetic force.[118] It was a remarkable experimental achievement, years in the making, and would have wide ranging ramifications for science and industry in the future. He reproduced this experiment using liquids, solids, and gases as conductors and varying strengths of magnets. It was clear to him that, contrary to scientific opinion of the time, light was subject to the same physical laws of magnetism as any other form of matter.

Michael Faraday was probably the preeminent experimental physicist of his generation, and a spirited speaker who was by accounts brilliantly succinct in writing of his discoveries. But he didn't have the background in trigonometry he needed to prove his experiments in theory.

Twenty years later, the Scottish theoretical physicist James Clerk Maxwell published a paper that extended Faraday's fundamental work in a new theory of electromagnetism: the second unification of physics.

Born in Edinburgh, June 13th, 1831, to a family of means, Maxwell received the finest education available in Scotland. A brilliant student, he finished his studies at Edinburgh Academy at age 16 and enrolled initially at the University of Edinburgh from 1847 to 1850 and then at Trinity College of Cambridge in England until 1856.

He was elected as a fellow of the Royal Society of Edinburgh in 1856 at age 24, and appointed Professor of Natural Philosophy at Marischal College in Aberdeen that year. The next year he won the Adams Prize for his essay *The Stability of Saturn's Rings*, in which he predicted that the rings of Saturn were neither whole nor fluid but instead consisted of small particles orbiting the planet. He published *Illustrations of the Dynamical Theory of Gases* with probability equations to explain the thermodynamic behavior of gases and was awarded a Rumford Medal in 1860 for his work on light and color which established the mathematics behind the Maxwell red, green, blue (RGB) color triangle that is employed in television, computer and smartphone monitors today. By 1861, before his 30th birthday, he was already a fellow of the Royal Society in London.[119]

While at King's College between the years 1860 and 1865 Maxwell finally met one of his great heroes, Michael Faraday. The two had been corresponding since Maxwell sent a copy of an essay about Faraday's lines of force, he had read at the Cambridge Philosophical Society in 1855, and they shared ideas and approaches. They met in London in 1860, shortly after Maxwell arrived with his family in tow.[120]

It is hard to imagine what it must have felt like for Maxwell to be hanging out with one of his childhood idols, even if he was getting a bit long in the tooth and suffering mental stress.

In 1861 Maxwell published *On Physical Lines of Force,* a more fully realized form of the paper given at Cambridge six years prior, in which he explained how both electric and magnetic effects travel at the speed of light and first theorized that light was a phenomenon born of the same medium as electricity and magnetism. Indeed, light and electromagnetism were different manifestations of the same physical phenomenon and must be governed by the same laws of physics.

In *A Dynamical Theory of the Electromagnetic Field* in 1865 Maxwell introduced the formal mathematics governing electromagnetism, electricity and optics that are employed today for use in a variety of industries ranging from communications to power generation.[121] Faraday died in 1867, having refused the offer of a formal burial at Westminster Abbey. Maxwell republished the paper in 1873 as *A Treatise on Electricity and Magnetism.* It gave four fully realized and highly complex differential equations, later to be known as Maxwell's equations, which drew upon the laws of electricity and magnetic radiation devised by Faraday and his contemporaries decades earlier, including the French physicists André-Marie Ampère and Charles-Augustin de Coulomb. Maxwell gave much credit to Faraday for the impact his experimental work had on the whole of physics.

One of the great scientific careers of all time came to an early close when Maxwell died of stomach cancer at his home in Cambridge on November 5th, 1879. He was just 48 years old. His body was removed to his Scottish homeland and laid to rest in his family's plot at Parton Kirk Cemetery in Castle Douglas, Galloway.

Planck And Einstein: The Dawn Of Modern Physics

As the first light fell upon the Earth to usher in the 20th century, three branches of physics comprised the whole of Galileo's legacy of science: Newtonian mechanics governing the realm of gravity and moving bodies; Boltzmann's thermodynamics, the realm of heat, gas and kinetic energy; and Maxwell's electromagnetism, that of light, electrical and magnetic energy.

The discovery of x-rays in 1895, radioactivity a year later and the electron the year after that were disturbing signals that questions remained to be answered about what was going on inside the physical world. Within five years of the turn of that century, two men would present epoch-making corrections to each of those governing branches of physics.

Max Planck and Albert Einstein were born during the second half of the 19th century, and both lived well into the 20th. Just after the turn of the century, at the height of Planck's theoretical powers and the beginning of Einstein's rise to the pinnacle of science, the two presented ideas that would upset many established beliefs and eventually subsume them in two coequal branches of physics called quantum and relativistic mechanics.

Max Karl Ernst Ludwig Planck was born on April 23rd, 1858, in Kiel, Germany, then known as the Duchy of Holstein, the northernmost state of the Holy Roman Empire, which was under Danish rule between the First and Second Schleswig Wars. It was transferred back to the German Kingdom of Prussia at the close of the second war in 1864 and by Max Planck's 13th birthday, Otto von Bismarck had unified the first German Empire, becoming its chancellor, and ruling until 1890.[122]

Young Max had been born into a family of intellectuals. His father was a professor of law at the University of Kiel until 1867, when he transferred his position to the University of Munich. There Max began his studies at the Maximilian gymnasium school where he first learned the principle of conservation of energy. A gifted student, he graduated early and matriculated to his father's university to begin his graduate study in physics. In 1877 he enrolled at the University of Berlin to study under Hermann von Helmholtz and Gustav Kirchhoff for a year. Although he didn't think much of Kirchhoff's teaching style, Planck undoubtedly studied intently his work on thermal radiation, which would later inform his greatest scientific achievement.[123]

By 1889, Planck was married with three children and had replaced the retiring Kirchhoff as professor in Berlin. His fourth child with his wife Marie Merck was born in 1893. The next few years were very happy for Planck and his family.

Planck began his historic approach to the problem of black-body radiation in 1894 by homing in on Kirchhoff's central question of the relationship between the intensity of the heat radiated by a black-body (so-called because it is a perfect absorber) and its color (its frequency) and temperature. Several well-known physicists including Wilhelm Wien of Germany and British mathematicians Lord Rayleigh and James Jeans had tried and failed to arrive at a solution to the problem which applied at all frequencies. They even gave their failure a spectacular name, the ultraviolet catastrophe! It was one of the biggest headaches plaguing physicists at the close of the 19th century and Planck was eager to see if he could figure it out.

On his first attempt in 1899 Planck managed to duplicate Wien's failed attempt to unify the solution. By making some alterations he eventually landed on a solution that worked only in specific frequencies. However, his solution assumed that light would be emitted in packets he called "quanta," the energy of which was given by the equation $E=hf$ where f equals the frequency of the light and *h* equals the Planck constant.[124] No, Planck didn't give it that name! It relates the amount of energy (electromagnetic action) given off by a photon to its frequency.

Planck wasn't overwhelmed with his solution at first, regarding it as something of an extremely educated fudge, but his breakthrough was a pioneering step that planted the seed for particle physics or quantum mechanics and would yield discoveries and predictions like wave-particle duality and probability and uncertainty theory through the 1920s and beyond. He had started to explain the physics of the very small, from molecules to atoms to even smaller particles, and how they interact in nature.

The published results of his work made Max Planck the world's most famous physicist overnight. But he wouldn't remain king of the physics hill for very long. Six hundred miles southwest of Berlin, Albert Einstein was acquiring his Swiss citizenship, having spent five years stateless after renouncing his German citizenship in 1896. He was a recent graduate of the polytechnic school in Zurich and was in love with a fellow student, a brilliant young Serbian physicist named Mileva Marić.

Einstein was a unique physicist, with a distinctive way of looking at problems, living in an unprecedented period of scientific development when even long-established theories were showing signs of stress, making the physics world desperate for answers. His guiding principle was to understand God's thoughts when He was creating the universe. In today's vernacular, he might have been referred to as a disruptor. Einstein was a classic, original, iconoclastic disruptor of epic proportions whose central desire to attack contradictions in scientific theories was combined with an innate ability to visualize different scenarios and outcomes to issues that baffled his contemporaries. His work almost unified the whole of physics. Ironically, the realm that would ultimately elude him in his quest to unify physics, quantum mechanics, was a field he helped to create.

Although he had shown great promise as a student, young Einstein could not find work in his chosen field, so in 1902 he went to work at the Federal Office of Intellectual Property in Bern as a patent examiner. Albert and Mileva married in 1903 and Hans Albert Einstein was born the following year. Eduard Einstein would be born in 1910 in Zurich.[125]

Having finally found some financial footing to provide for his young family, Einstein began to delve into several key questions of physics. In a lecture to the Royal Institution in April 1900 titled *Nineteenth Century Clouds over the Dynamical Theory of Heat and Light*, William Thomas (Lord Kelvin) of the University of Glasgow gave an indication of the two puzzles, namely Planck's conceptualization of black-body radiation and the outcome of the Michelson-Morley experiment.

In the latter, performed by the Polish American physicist Albert Michelson (of velocity of light fame) and the American physicist Edward Morley, an apparatus was designed and built in Cleveland, Ohio in 1887 to compare the speed of perpendicular beams of light for the detection of the relative motion through Newton's predicted luminous ether, postulated in his 1704 *Opticks*. The negative result of their experiment left Newton's theory without a solidifying proof of concept.

A book by the French physicist Henri Poincaré in 1904 also pointed to the two unsolved mysteries mentioned above, and added a third, namely the unexplained random movement of pollen particles suspended in fluid.[126]

Surrounded by a small group of brilliant friends that included his wife, who served as a sounding board for his seemingly radical ideas, Einstein started to develop a new concept of light, mass and energy that he published in four separate papers in 1905 for the German science journal Annalen der Physik. His solutions to these long unresolved problems would fundamentally alter our understanding of the world around us for all time.

The first of Einstein's papers, *On a Heuristic Viewpoint Concerning the Production and Transformation of Light*, published on June 9th, amplified Planck's work on black-body radiation by suggesting that light does not move in a seamless continuous wave but "as a finite number of energy quanta which are localized in space, which move without dividing, and which can only be produced and absorbed as complete units." Although light appeared as a wave to an observer, Einstein was suggesting (as had Planck) that it was actually made up of tiny bits of energy that moved together through space and could only be emitted or absorbed in their miniscule form. This statement was considered by many to be "the most 'revolutionary' sentence written by a physicist of the 20th century."[127]

At the time, conventional wisdom held that light moved as a wave through Newton's luminiferous ether, despite the lack of evidence from Michelson-Morley in 1887. The same year, an experiment by Heinrich Hertz had shown that when light was shined on a metal surface, particles that Hertz called photoelectrons were sometimes ejected. The brighter the light, the greater number of electrons released. Crucially, brightness didn't alter the kinetic energy of the electrons (their energy of motion), which never exceeded a given limit. Hertz referred to this phenomenon as the photoelectric effect.[128]

What Einstein proved was that this effect couldn't be achieved by a wave of light, whose energy is spread evenly along the wave, because it lacked the energy at the point where it interacted with an electron on the surface of the metal to immediately eject photoelectrons. Therefore, Einstein explained, wave theory broke down at the point of contact between the wave of light and the surface of the metal, where individual particles of the light produced the observed effect.

Today, many physicists believe Einstein's theory of light was the most revolutionary idea of the 20th century, but it was so prescient that it would take two decades to convince most of the physics community of its reality. In particular, Nobel laureates Max von Laue (1914), Max Planck (1918), Niels Bohr (1922) and Robert Millikan (1923) were skeptical well into the 1920s. Proof finally came in 1923 with an experiment by Arthur Compton at Washington University.[129]

Although Einstein was reluctant to accept quantum theory – he would famously assert in a letter to the theorist Max Born in 1926, "[God] does not play dice with the universe"[130] – he unwittingly gave oxygen to ignite the theory in the decades to come. In the paragraph preceding the one mentioned above, he acknowledged the reality of both light waves and quanta, paving the way for the development of a key constituent of particle or quantum physics known as wave-particle duality:

> The wave theory of light…has worked well in…purely optical phenomena and will probably never be replaced…however…the optical observations refer to time averages rather than instantaneous values. In spite of…confirmation of the theory as applied to diffraction, reflection, refraction, dispersion, etc.…the theory…may lead to contradictions…when…applied to phenomena of emission and transformation of light.[131]

Although light performed well as a wave in the realm of optics, it must also be regarded as being made up of energy quanta. The measurement of light as waves (spectral analysis) would be of paramount importance to the success of Edwin Hubble and Milton Humason in their practical discovery of the Big Bang through their redshift-distance program in the 1930s.

The same year as Einstein's letter was sent to Born, the ill-fated American chemist and dean of the school of chemistry at the University of California at Berkeley, Gilbert Newton Lewis, coined the term "photon" for Einstein's corpuscular light.[132] Is the fact that the middle name of the man who named Einstein's light particle is the same as the man who first suggested its existence over two hundred years prior purely a coincidence, or part of some mysterious force of nature? This mystery will likely never be resolved.

Due mainly to the difficulty in successfully testing the propositions made in Einstein's groundbreaking third and fourth papers of 1905 he was awarded the Nobel Prize in Physics in 1921 for his work on the particle theory of light, even as notables such as Bohr remained skeptical. Although he was confident in the soundness of his theory, Einstein would remain perplexed by the meaning and true nature of photons for the rest of his life.

In the July 18th issue of Annalen der Physik, Einstein published *On the Motion of Small Particles Suspended in a Stationary Liquid, as Required by the Molecular Kinetic Theory of Heat*, in which he presented a model for Brownian motion, the phenomenon discovered by the Scottish botanist Robert Brown in 1827 while pondering the motion of pollen grains in water viewed through a microscope. Working from ideas of thermodynamics in Ludwig Boltzmann's kinetic theory of gases, Einstein showed that the pollen in a water droplet was actually being shuffled by collisions with individual water molecules that were randomly whizzing around, propelled by thermal energy. He developed this theoretical solution into a more fully realized *Elementary Theory of Brownian Motion* in April of 1908. According to Max Born, this was convincing evidence for the "existence of atoms and molecules, of the kinetic theory of heat, and of the fundamental part of probability in the natural laws." Its experimental confirmation by the French physicist Jean Perrin in Paris in 1909 finally ended the long-running dispute on whether matter exists in seamless fluidity or is the sum of a multitude of individual constituent parts.[133]

Einstein would have been considered one of the greatest physicists of all time based on the merits of these two seminal papers alone, but his third and fourth papers in September and November 1905 cemented his name in the history of science in emphatic fashion and set in motion events that would alter the world order with both beneficial and catastrophic effects.

The first of these, *On the Electrodynamics of Moving Bodies*, introduced a radically new theory that linked space and time in a fourth dimension to form what he called spacetime, and balanced these with Maxwell's equations which made light, electricity and magnetism distinct forms of the same physical phenomenon. Einstein based his new theory on two founding principles. (1) The laws of physics appear the same to all observers. (2) The speed of light is constant. He then went on to show how there were no fixed reference points in the universe and that everything was moving relative to everything else. It was later named the Special Theory of Relativity because the rules only applied to frames of reference which were in constant and unchanging motion.

This paper, for a 26-year-old who had had trouble finding work in his chosen field, was a brash and daring bit of physics fireworks that flew in the face of 200 years of Newtonian mechanics. In his words, translated from paragraph two of the paper, "The introduction of a 'luminous ether' will prove to be superfluous inasmuch as the view here to be developed will not require an 'absolute stationary space' provided with special properties..."[134] In other words, if his theory was correct it was the end of a central tenet of Newtonian theory. Einstein's assertion that the speed of light is constant and therefore not relative to the movement of the observer was simply impossible in Newtonian physics.

This was strange and unwelcome news to the physics community at large, and nobody knew how to respond. Absolute time and absolute space, fundamental tenets of Newtonian physics, were gone. The ether was no longer necessary because light, unlike sound waves, for instance, did not require a medium to travel through.

According to Einstein, an object that accelerated closer and closer to the speed of light would gain mass. At the speed of light its mass would become infinite, which is impossible since infinite mass would require infinite energy to move. The speed of light was a limit of velocity unattainable by matter.

At the same time, a person riding on that object would experience time at a slower pace than someone on Earth – an effect referred to as time dilation in the equations developed by the Dutch physicist Hendrik Lorentz. Furthermore, a person viewing the person riding on the object near the speed of light would perceive the object to be much shorter in length than it would be if it was sitting in front of them.[135]

The effects of Einstein's new theory were so revolutionary that it remains controversial to some scientists even today. Nevertheless, relativity was first confirmed experimentally by a German physicist named Alfred Bucherer in 1908 and since that time it has been tested relentlessly in various ways, never failing a single challenge.[136]

In his final paper of 1905, *Does the Inertia of a Body Depend on its Energy Content?* Einstein combined the equations of Maxwell and Hertz along with his principle of relativity and concluded that mass, and energy were interchangeable:

> If a body gives off the energy L in the form of radiation, its mass diminishes by L/V^2.

Einstein was focused in his paper on the energy content of a body as related to its mass m, so naturally his equation sought the answer to that question:

$$m = L\,/\,V^2$$

where V was the velocity of light squared. But, of course, if this was true then the equation could also be solved for the energy it produced:

$$E = mc^2$$

as was suggested by Lorentz in a series of papers called *Einstein's Principle of Relativity*, published in 1914. This marked the first time the letter E was used for the loss of energy in the equation. The lower-case letter c was already in use for the constant velocity of light.

In the very next line of the paper, Einstein asserts "…that the energy withdrawn from the body becomes energy of radiation evidently makes no difference…The mass of a body is a measure of its energy content…"[137]

Essentially, what this last paper showed was that as far as nature was concerned, a paper clip and a beam of light were essentially the same thing, just in different forms. However, a paper clip represents an enormous amount of energy which, if it could be liberated, would match the bomb that destroyed Hiroshima. Fortunately, we cannot change a paper clip into pure energy by stepping on it or driving a train over it. The pressure and temperature needed to perform that feat are equivalent to those we find in the core of the Sun.

With the existence of atoms and energy quanta just being realized, this equation, coupled with the rest of Einstein's work, were benign enough and only caused a tremendous fervor in the science community, which set out to contrive ways to test the new theories for further insight into their implications. Several decades later, when the process of splitting the atom was discovered, Einstein's famous equation would be revealed to have far more menacing implications.

In the way that it fundamentally changed collective perceptions of the world we live in, Einstein's "Annus Mirabilis" (miracle year) of 1905, has been likened to those of Andreas Vesalius who published his study of anatomy in 1543, Nicolaus Copernicus who published his heliocentric theory in that same year,[138] and Isaac Newton who initiated a series of pioneering studies of physics in 1666.[139]

For several years, Einstein's papers lingered just outside the professional realm while he continued to work at the patent office. Finally, he received a boost from the most famous physicist of the day. Max Planck recognized the significance of special relativity and the mass-energy equivalence and began speaking about them to anyone who would listen. In 1907 Hermann Minkowski, Einstein's former professor at Zurich, improved upon some of the math in an effort to help it gain acceptance in the field.

By 1908, Einstein had gained a reputation as a leading physicist and was given a position as a lecturer at Bern. Sought after for his expertise in various fields he would move from Zurich to Prague and back to Zurich, taking one professorship after another until Planck was appointed dean at the University of Berlin and brought Einstein there as a professor in 1914. While there, Einstein and Planck became close friends and remained close even after Einstein felt obliged to leave the country after the Nazis came to power. He was known to have kept pictures of Faraday, Maxwell and Newton on the wall of his study.

In their personal lives, both men met with pain and pride during their lifetimes. After Planck's first wife Marie died in 1909 from a prolonged illness (perhaps tuberculosis), he married again two years later and his fifth child, Hermann, was born in 1911. During World War One his eldest son Karl was killed in action at Verdun in 1916 and his twin daughters died during child-birth two years apart. His fourth child, Erwin, led a most heroic life. After being taken prisoner by the French in 1914, he survived the ordeal and went on to become a prominent politician and a leader of the resistance movement against the Nazis. His luck ran out when he was sentenced to death and hanged in January of 1945 for his participation in an attempt to assassinate Adolf Hitler the previous year.[140]

By 1914, when the Einstein's moved to Berlin, the couple had grown apart, Albert having fallen in love with his cousin Elsa, who he would later marry. The relationship which had begun with such attraction between the two genius sweethearts had soured over the years as Mileva struggled with her lost professional opportunities and her husband's philandering and abnegation of spousal responsibilities. In contrast to historical couples such as Antoine and Marie-Anne Lavoisier, William and Margaret Huggins and Pierre and Marie Curie, the marriage was beset by early financial and personal setbacks and egoism that the two were unable to reconcile. Hans Einstein went on to become a widely respected engineer and a professor of hydraulic engineering at the University of California, Berkeley. At age 20, his younger brother Eduard was diagnosed schizophrenic and struggled to gain a foothold in life. He died of a stroke in 1965.

Issued in 1916, Einstein's General Theory of Relativity brought into focus Newtonian mechanics and electromagnetism with mass-energy equivalence and spacetime and is the third unification of modern physics.

By his treatise on the particle nature of light, a seed that had been planted by Planck, Einstein not only spawned the child that evolved into quantum mechanics but also fathered relativity in all its forms – an act of supreme scientific hermaphroditism.

Einstein continued to contribute to the creation of the field of quantum mechanics in the 1920s, linking Planck's law with the hydrogen atom theory of Niels Bohr in 1917 and co-creating with Indian physicist Satyendranath

Bose a statistical approach to Planck's law of radiation for bosons, the "carrier particles" of the fundamental forces of nature.

Attempts would be made by Einstein and others to unify Planck's atomic and subatomic realm with relativity, but as yet no one has succeeded. The highly unpredictable nature of quantum mechanics and the highly predictable nature of relativistic mechanics remain as the two central pillars of modern physics. Together, the new sciences would prove worthy in deciding the fate of the Steady State and Big Bang theories which were among the central conflicts in the story of Edwin Hubble and Milton Humason: relativity was the formulaic backdrop for expansion which the Mount Wilson astronomers explored in historic fashion, while quantum mechanics laid the groundwork for the discovery of the cosmic microwave background (CMB) radiation that ended the debate.

George Hale And The Modern Astronomical Observatory

While Albert Einstein was having his miraculous year in 1905, George Ellery Hale, the son of the Hale Elevator magnate William Hale who had made a fortune installing his patented hydraulic elevators in the new buildings that were erected in the wake of the Great Chicago Fire of 1871, was creating a solar observatory on Mount Wilson in the foothills north of Pasadena California.

He was born in Chicago on June 29th, 1868, into a family of English descent that had emigrated to colonial North America in the 17th century. A sickly child, young George was doted on by his father, who took every opportunity to inspire his growing appreciation for engineering and science.

At age 14, George made a telescope and set up an observatory at the corner of the plot of the family home in Hyde Park on Chicago's Southside. He was a brilliant boy with broad interests in the sciences, music, literature, and sports. He attended the finest schools in the city and learned about architectural developments and the art and application of social and political graces from Daniel Burnham, a friend of his father who would become the lead architect on the Chicago World Fair in 1893. The lessons instilled would serve him well in the years ahead as he spearheaded the creation and development of several observatories, research institutions and colleges, astrophysical publications, and national and international organizations.

From an early age, Hale decided to understand all he could about the evolution of a star. In 1885, aged 17, he enrolled at MIT where his interest in the work of William Huggins, astronomy and machinery led him to invent a spectroheliograph capable of photographing the Sun (our star) at various frequencies across its spectrum. In 1889 he began a two-year course at the Harvard College Observatory, studying under the great American astronomer Edward Pickering. By the time he left Harvard with his master's degree, Hale was already much sought after for professorships at various schools. Back at home he began work again at his own observatory, which had been officially named the Kenwood Observatory after his hometown village, and he later took a position teaching astronomy at Beloit College in Wisconsin.

While at Beloit, Hale began his transition from researcher and professor to state-of-the-art telescope and observatory founder. Under a contract with the University of Chicago and the donation of a sum of money from a wealthy hardware magnate by the name of Charles Yerkes, Hale designed and supervised the construction of the largest refracting telescope ever built, to be included in a newly founded observatory near Lake Geneva in Wisconsin dedicated in Yerkes' honor. The objective lens of the instrument was 40 inches in diameter and the light it refracted traveled 60 feet down the 20-ton tube to reach the eyepiece. Hale and his team rushed to get the instrument sufficiently finished to display in the Columbian Exposition (World's Fair) in Chicago, where it stood at the far end of the Manufacturer's and Liberal Arts Building on the shore of Lake Michigan.

For Hale, this was only the beginning. Not long after the new telescope was completed at the University of Chicago, he began looking for a site for a new observatory that would house a 60-inch reflecting telescope, ground and polished from a glass blank his father had bought for him. After some consultation with other members of the field, Hale decided on a former testing site of Harvard University on Wilson's Peak near Pasadena, California.

He first visited the mountain in 1902 and after some testing determined that it probably had the best "seeing" conditions of any site in the contiguous United States. In astronomical vernacular, seeing is related to the amount of turbulence caused by heat radiation and air flow in a given location. Good seeing indicates a lower amount of disturbance. Disturbed air makes stars appear to twinkle. A good observing location will have more nights of good seeing than a site with only average or poor seeing.

After spending several nights at the summit of Mount Wilson, Hale decided this was a prime location for his state-of-the-art observatory. The warm, dry climate would not only offer more clear nights, but it would also be good for his and his family's health.

By 1904, work at the site was underway in earnest with a grant from the newly created Carnegie Institution of Washington to finance it. The Mount Wilson Solar Observatory was born that year. Its first solar telescope was in operation the following year and several other solar and stellar instruments were under construction nearby.

The facility would become a "base camp" for astronomers studying stellar and universal evolution through the first half of the 20th century. It is here that the highly improbable duo of Hubble and Humason would reveal the practical evidence for expansion which left even Einstein scratching his head in wonderment.[141]

*   *   *

This introduction to some of the developments in science and those who produced them is necessarily brief. The lives of the individuals concerned were often very interesting. Those readers wishing further information are encouraged to seek it out. Furthermore, this section by no means covers all of the contributors to the advancement of science during this period. A great many people made incremental discoveries and additions that set the stage for the dynamic advances made by the likes of Einstein and Planck, Friedmann and Lemaître and Hubble and Humason. As Isaac Newton famously said, he saw so far across the landscape of science by being able to stand on the "shoulders of giants." A broader discussion of the development of the Mount Wilson Observatory and her sister facilities will follow in the coming pages.

The questions raised in the course of development in experimentation and theory leading to the turn of the 20th century made the situation ripe for someone with the right vision and a unique approach to coalesce the various departments of physics into one or two different overarching fields. That person, it turned out, was Einstein, whose four foundational papers in 1905 brought the fields of relativity and quantum mechanics to consciousness. At that very moment, Hale was opening what would become the world's leading research facility dedicated to understanding the implications of the findings of theoretical physics.

Similar developments with even greater implications for science were at hand in 1919, when Arthur Eddington's view a total solar eclipse confirmed that the Sun's gravitational influence bends starlight that passes by the Sun on its way to us. This proved a central tenet of Einstein's General Theory of Relativity, published several years earlier. In that year the 100-inch telescope on the summit of Mount Wilson went into regular operation at the same moment that Edwin Hubble and Milton Humason were starting their association with the institution.

References

1. Online Etymology Dictionary, http://etymonline.com, Accessed 7 August 2019
2. Kish, George (ed.), "A Source Book in Geography" (Harvard University Press, 1978), pgs. 51-54
3. Machamer, Peter, "Galileo Galilei," *The Stanford Encyclopedia of Philosophy* (Summer 2017 Edition), Edward N. Zalta (ed.), https://plato.stanford.edu/archives/sum2017/entries/galileo/
4. Museo Galileo, catalogue.museogalileo.it, Accessed 7 August 2019
5. Disraeli, Joseph, "Curiosities of Literature" (George Rutledge and Sons, 1893), pg. 523
6. Andriesse, C.D., "Huygens: The Man Behind the Principle," (Cambridge University Press, 2005), Pg. 6
7. Encyclopedia.com, https://www.encyclopedia.com/people/science-and-technology/physicsbiographies/christiaan-huygens#2830902105, Accessed 7 August 2019.
8. Devreese, Jozef T, Vanden Berghe, Guido. "Magic is No Magic: The Wonderful World of Simon Stevin." (WIT Press, 2008), pg. 275
9. Steele, Brett D., Dorland, Tamera. "The Heirs of Archimedes: The Science and the Art of War Through the Age of Enlightenment." (MIT Press 2005), pg. 20
10. Huerta, Robert D.; Vermeer van Delft; Jan, Vermeer, Johannes. "Vermeer and Plato: Painting the Ideal." (Bucknell University Press 2005), pg. 101
11. Bell, A.E., "Christiaan Huygens and the Development of Science in the Seventeenth Century," (Bell Press 2007), pg. 293-296
12. Wooton, David. "Galileo: Watcher of the Skies," (Yale University Press 2010), pg. 577
13. Wooton, David. "Galileo: Watcher of the Skies," (Yale University Press 2010), pgs. 291-294
14. Bell, A.E., "The Horologium Oscillatorium of Christiaan Huygens," (Nature 1941), pgs. 245-248
15. Bell, A.E., "Christiaan Huygens and the Development of Science in the Seventeenth Century," (Bell Press 2007), pg. 264, 473-74
16. Ackroyd, Peter, "Isaac Newton," (Vintage 2007), pg. 7
17. Ackroyd, Peter, "Isaac Newton," (Vintage 2007), pg. 1
18. Ackroyd, Peter, "Isaac Newton," (Vintage 2007), pgs. 4-6
19. James I. (2003). Singular scientists. *Journal of the Royal Society of Medicine*, *96*(1), 36-39. doi:https://doi.org/10.1258/jrsm.96.1.36
20. Ackroyd, Peter, "Isaac Newton," (Vintage 2007), pg. 6
21. Richard S. Westfall, *The Life of Isaac Newton*, (Cambridge University Press 1993), pgs. 17-18
22. Peter Ackroyd, *Isaac Newton*, (Vintage 2007), pg. 15
23. Peter Ackroyd, *Isaac Newton*, (Vintage 2007), pg. 26

24. Olivier Darrigol, *A History of Optics from Greek Antiquity to the Nineteenth Century,* (Oxford University Press 2012), pgs. 80-81

25. A Rupert Hall, *Isaac Newton: Adventurer in Thought,* (Cambridge University Press 1996), pgs. 79-80

26. Marcelo Dascal (ed.), *The Practice of Reason" Leibniz and His Controversies,* (John Benjamins Publishing Company 2010), pg. 45

27. Peter Louwman, *Christian Huygens and his telescopes,* ed. By Karen Fletcher. In: Proceedings of the International Conference "Titan – from discovery to encounter," April 2004 (ESA Publications 2004)

28. Richard S. Westfall, *Never at rest: a biography of Isaac Newton,* (Cambridge University Press 1983), pg. 270

29. Theresa Levitt, *A short, bright flash: Augustin Fresnel and the birth of the modern lighthouse,* (W.W. Norton & Company 2013), pgs. 29-32

30. Theresa Levitt, *A short, bright flash: Augustin Fresnel and the birth of the modern lighthouse,* (W.W. Norton & Company 2013), pg. 47

31. Bill Bryson, *A Short History of Nearly Everything,* (Broadway Books 2003), pgs. 106-110

32. The London Gazette, April 1, 1727, pg. 7

33. *Antoine Lavoisier: Scientist, Economist, Social Reformer,* Douglas McKie, (Henry Schuman 1952) pgs. 91-94

34. *Lavoisier in the year one: the birth of a new science in an age of revolution,* Madison Smart Bell, (W.W. Norton 2005), pgs. 5-6

35. *Lavoisier in the year one: the birth of a new science in an age of revolution,* Madison Smart Bell, (W.W. Norton 2005), pgs. 27-28

36. *Lavoisier in the year one: the birth of a new science in an age of revolution,* Madison Smart Bell, (W.W. Norton 2005), pg. 114

37. *Elements of Chemistry in New Systematic Order, Containing All Modern Discoveries,* Antoine Lavoisier Antoine (Edinburgh 1790), ed. William Creech, Translated by Robert Kerr, pgs. 175-176

38. *Charles Messier biography,* (http://messier.seds.org/xtra/history/CMessier.html). Students for the Exploration and Development of Space.

39. *Charles Messier, Napoleon, and Comet C/1769 P1,* Maik Meyer, (International Comet Quarterly 2007) www.icq.eps.harvard.edu/meyer_icq29_316.pdf

40. *The Herschel Chronicle,* Constance A. Lubbock, (University Press 1933), pg. 1

41. *A Popular History of Astronomy During the Nineteenth Century,* Agnes M. Clerke, (Adam and Charles Black 1908), pg. 11 *http://asterope.bajaobs.hu/cski/research/mktars/hegecikk/WilliamHerschel.htm,*archive.is

42. *William Herschel,* A. Armitage, (Doubleday 1963), pg. 21

43. *The History of the Telescope,* C.S. Hastings, The Sidereal Messenger, A Monthly Review of Astronomy, Volume X, (Carleton College 1891), pg. 342

44. *William Herschel,* A. Armitage, (Doubleday 1963), pg. 44

45. *A Giant of Astronomy: William Herschel had a conflicted relationship with his biggest creation,* Jacob Roberts, (www.sciencehistory.org/distillations/magazine/a-giant-of-astronomy2017)
46. *William Herschel,* A. Armitage, (Doubleday 1963), pg. 70
47. *Catalogue of Nebulae and Clusters of Stars,* William Herschel, (Philosophical Transactions of the Royal Society of London 1786), pgs. 457-499
48. *Catalogue of a Second Thousand of New Nebulae and Clusters of Stars; with a Few Introductory Remarks on the Construction of the Heavens,* William Herschel, (Philosophical Transactions of the Royal Society of London 1789), pgs. 212-252
49. *Catalogue of 500 New Nebulae, Nebulous Stars, Planetary Nebulae, and Clusters of Stars; with Remarks on the Construction of the Heavens,* William Herschel, (Philosophical Transactions of the Royal Society of London 1802), pgs. 477-528
50. *The Story of Astronomy,* Peter Aughton, (Quercus 2011), pg. 187
51. *Why the moons of Uranus are named after characters in Shakespeare,* Julia Franz, (Studio 360 2017), pri.org
52. *All About Mars,* https://mars.nasa.gov/allaboutmars/mystique/history/1700
53. *History of Astronomy: an encyclopedia,* John Lankford, (Garland 1997), pg. 258
54. *A Giant of Astronomy: William Herschel had a conflicted relationship with his biggest creation,* Jacob Roberts, (www.sciencehistory.org/distillations/magazine/a-giant-of-astronomy2017)
55. *Women in Science: Antiquity Through the Nineteenth Century: A Biographical Dictionary with Annotated Bibliography,* Marilyn Bailey Ogilvie, (MIT Press 1986), pg. 98
56. *The Comet Sweeper: Caroline Herschel's Astronomical Ambition,* Claire Brock, (Icon Books 2007)
57. *Satellites of Saturn,* Mr. Lassell, (MNRAS 1848), pg. 42
58. *Sir John Frederick William Herschel,* https://iphf.org/inductees/sir-john-frederick-williamherschel/
59. *The Cyanotype,* Richard T. Rosenthal, https://web.archive.org/web/20130330080304/http://vernacularphotography.com/vpm/vn1/the_cyanotype.htm
60. *Herschel at the Cape: Diaries and Correspondence of Sir John Herschel, 1834-1838,* David S. Evans, (Cape Town, Bakema 1969), pg. xxiii
61. *Burke's Genealogical and Heraldic History of Peerage, Baronetage and Knightage,* (Burke's Peerage Limited 1914), pgs. 1004 1005
62. *A Preliminary Discourse on the Study of Natural Philosophy,* John Frederick William Herschel, (London: Printed for Longman, Rees, Orme, Brown, and Green, and John Taylor 1831), Chapter VI
63. Charles Darwin, *The Autobiography of Charles Darwin,* ed. Nora Barlow, (Collins 1958), pgs. 67-68
64. *Ibid.* pg. 107

65. *William & John Herschel: Scientist, Musician, Mathematician and Astronomer,* https://www.westminster-abbey.org/abbeycommemorations/commemorations/williamjohn-herschel
66. Kitty Ferguson, *The Glassmaker Who Sparked Astrophysics,* (nautilus.us 2014), https://nautil.us/issue/11/light/the-glassmaker-who-sparked-astrophysics
67. *Ibid.*
68. Myles W. Jackson, *Spectrum of Belief: Joseph von Fraunhofer and the Craft of Precision Optics,* (MIT Press 2000), pg. 1
69. *Ibid.* pgs. 1-4
70. *Ibid.* pgs. 4-5
71. *Ibid.* pg. 53
72. *Ibid.* pg. 3
73. *Ibid.* pgs. 74-77
74. Kitty Ferguson, *The Glassmaker Who Sparked Astrophysics,* (nautilus.us 2014), https://nautil.us/issue/11/light/the-glassmaker-who-sparked-astrophysics
75. *Ibid.*
76. Myles W. Jackson, *Spectrum of Belief: Joseph von Fraunhofer and the Craft of Precision Optics* (MIT Press 2000), pg. 171
77. Wolfgang Jahn, Dr., Josef Kirmeier, Dr., Christoph Mewes, Carl R. Preyß, Leo Weber, Dr., *Fraunhofer in Benediktbeuern Glassworks and Workshop,* ed. Dr. Martin Thum, (Fraunhofer-Gesellschaft, München 2008)
78. Britannica, The Editors of Encyclopedia. "Robert Bunsen". *Encyclopedia Britannica*, 12 Aug. 2020, https://www.britannica.com/biography/Robert-Bunsen. Accessed 29 January 2021.
79. P. Graneau, A.K.T. Assis, *Kirchoff on the Motion of Electricity in Conductors,* https://www.ifi.unicamp.br/~assis/Apeiron-V19-p19-25(1994).pdf, Aperion, vol. 19, pg. 19-25
80. Gustav Kirchhoff, Robert Bunsen, *Chemical Analysis by Observation of Spectra,* (Annalen der Physik und der Chemie 1860), vol. 110, pgs. 161-189
81. Isaac Asimov, *The Secret of the Universe,* (Doubleday 1990), pgs. 106-109
82. Cornell University Astronomy, *Kirchhoff's Laws,* http://hosting.astro.cornell.edu/academics/courses/astro201/kirchhoff.htm
83. A.L. Cortie, *Sir Norman Lockyer, 1836-1920),* (The Astrophysical Journal 1921), vol. 53, pgs. 233-248
84. C. Robert O'Dell, *The Orion Nebula: Where Stars are Born,* (Belknap Press of Harvard University Press 2003), pgs. 25-26
85. *Henry Draper (1837-1882),* http://www.saburchill.com/border/about.html
86. William Francis Magie, *A Source Book in Physics,* (Harvard University Press 1963), pgs. 360-365
87. Alec Eden, *The Search for Christian Doppler,* (Springer-Verlag Wien 1992), pgs. 27-28
88. *Ibid.*
89. *Ibid.*

90. *Ibid.*
91. Alexander Hellemans, Bryan H Bunch, *The Timetables of Science: A Chronology of the Most Important People and Events in the History of Science,* (Simon & Schuster 1991), pg. 317
92. David M.F. Chapman, *Reflections: Comet Tales: Sir William Huggins and Jean Louis Pons,* (Journal of the Royal Astronomical Society of Canada), vol. 95, pgs. 107-108
93. Barbara J. Becker, *Eclecticism, Opportunism, and the Evolution of a New Research Agenda: William and Margaret Huggins and the Origins of Astrophysics,* (The Johns Hopkins University 1993), Chapter 1, http://faculty.humanIties.uci.edu/bjbecker/huggins/ch1.html
94. Joseph S. Tenn (Dr.), *The Hugginses, the Drapers, and the Rise of Astrophysics,* (The Griffith Observer 1985), pg. 3, http://www.physastro.sonoma.edu/brucemedalists/huggins/HugginsesDrapers.pdf
95. Barbara J. Becker, *Eclecticism, Opportunism, and the Evolution of a New Research Agenda: William and Margaret Huggins and the Origins of Astrophysics,* (The Johns Hopkins University 1993), Chapter 1, http://faculty.humanities.uci.edu/bjbecker/huggins/ch1.html
96. Barbara J. Becker, *Visionary Memories: William Huggins and the Origins of Astrophysics,* (Journal for the History of Astronomy 2001), pg. 44
97. Joseph S. Tenn (Dr.), *The Hugginses, the Drapers, and the Rise of Astrophysics,* (The Griffith Observer 1985), pg. 4-5, http://www.phys-astro.sonoma.edu/brucemedalists/huggins/HugginsesDrapers.pdf
98. Barbara J. Becker, *Visionary Memories: William Huggins and the Origins of Astrophysics,* (Journal for the History of Astronomy 2001), pg. 55-56
99. Joseph S. Tenn (Dr.), *The Hugginses, the Drapers, and the Rise of Astrophysics,* (The Griffith Observer 1985), pg. 4-5, http://www.phys-astro.sonoma.edu/brucemedalists/huggins/HugginsesDrapers.pdf
100. Simon Singh, *Big Bang: The Origin of the Universe,* (Harper Perennial 2005), pgs. 244-246
101. Marilyn Bailey Ogilvie, *Women in Science: Antiquity Through the Nineteenth Century: A Biographical Dictionary with Annotated Bibliography,* (MIT Press 1986), pgs. 101-102
102. Jim Lucas, *What is Parallax?* https://www.space.com/30417-parallax.html Dec. 12, 2018
103. Rolf Willach, *The Heliometer: Instrument for Gauging Distances in Space,* Journal of the Antique Telescope Society, Issue 26 2004, pg. 5
104. Jim Lucas, *What is Parallax?* https://www.space.com/30417-parallax.html Dec. 12, 2018
105. *Ibid.* pg. 11
106. Doris A. Simons, Caroline Hertzenberg, *Scientists, Mathematicians, and Inventors: Lives and Legacies: An Encyclopedia of People Who Changed the World,* (Onyx Press 1999), pg. 19

107. Friedrich Bessel, *A letter from Professor Bessel to Sir J. Herschel, Bart., dated Konigsberg, Oct. 23, 1838,* (RAS), pgs. 152-161

108. Rolf Willach, *The Heliometer: Instrument for Gauging Distances in Space,* Journal of the Antique Telescope Society, Issue 26 2004, pg. 11

109. Friedrich Bessel, *A letter from Professor Bessel to Sir J. Herschel, Bart., dated Konigsberg, Oct. 23, 1838,* (RAS), pgs. 152-161

110. Isaac Asimov, David Wool III, *How Did We Find Out About the Speed of Light?,* (Walker 1986), pgs. 13-16

111. IMAGE, *Hippolyte-Louis Fizeau,* (Journal of Photography of the George Eastman House May, 1952, vol. 1 no. 5), pgs. 3-4

112. Isaac Asimov, David Wool III, *How Did We Find Out About the Speed of Light?* (Walker 1986), pgs. 22-25

113. Alexander Hellemans, Bryan H Bunch, *The Timetables of Science: A Chronology of the Most Important People and Events in the History of Science,* (Simon & Schuster 1991), pg. 319

114. Isaac Asimov, David Wool III, *How Did We Find Out About the Speed of Light?* (Walker 1986), pgs. 27-30

115. *Ibid.* pgs. 30-31

116. Michael Guillen, *Five Equations That Changed the World: The Power and Poetry of Mathematics,* (Hyperion 1995), pgs. 123-129

117. *Interactive timeline: Michael Faraday,* https://www.rigb.org/our-history/michael-faraday/about, Retrieved Feb. 2nd, 2021

118. Brian Clegg, *Ten Physicists Who Transformed Our Understanding of Reality,* (Little, Brown Book Group 2015), pgs. 76-77

119. James Clerk Maxwell Foundation, Facts Page, http://www.clerkmaxwellfoundation.org/html/key_facts_about_maxwell.html

120. Charles Paul May, *James Clerk Maxwell and Electromagnetism,* (Franklin Watts, Inc. 1962), pg. 89

121. James Clerk Maxwell Foundation, Facts Page, http://www.clerkmaxwellfoundation.org/html/key_facts_about_maxwell.html

122. *Schleswig-Holstein,* Britannica.com, retrieved Feb. 2nd, 2021

123. Max Born, *Max Karl Ernst Ludwig Planck, 1858-1947,* (Biographical Memoirs of Fellows of the Royal Society 1958), pgs. 161-162

124. Brian Clegg, *Ten Physicists Who Transformed Our Understanding of Reality,* (Little, Brown Book Group 2015), pgs. 160-161

125. Banesh Hoffman, *Albert Einstein,* (Paladin 1975), pgs. 36-39

126. Jimena Canales, *The Physicist & the Philosopher: Einstein, Bergson, and the Debate That Change Our Understanding of Time,* (Princeton University Press 2015), pgs. 107-110

127. John S. Rigden, *Einstein 1905: The Standard of Greatness,* (Harvard University Press 2006), pgs. 19-20

128. *Ibid.* pgs. 33-34

129. *Ibid.* pgs. 38-39

130. Elisabetta Canetta, Dr., *Physics and Beyond: "God does not play dice." What did Einstein mean?*, https://www.stmarys.ac.uk/news/2014/09/physics-beyond-god-play-dice-einsteinmean/ , 2014, Retrieved Feb. 5th, 2021

131. Albert Einstein, *Concerning an Heuristic Point of View Toward the Emission and Transformation of Light*, (American Journal of Physics, 1965), pgs. 1-2

132. John S. Rigden, *Einstein 1905: The Standard of Greatness*, (Harvard University Press 2006), Pg. 38

133. *Ibid.* pgs. 68-71

134. Albert Einstein, *On the Electrodynamics of Moving Bodies*, translation by Methuen and Company, Ltd. Of London 1923, http://www.fourmilab.ch, retrieved Feb. 5th, 2021

135. John S. Rigden, *Einstein 1905: The Standard of Greatness*, (Harvard University Press 2006), pgs. 91-95

136. *Ibid.* pg. 97

137. Albert Einstein, *Does the Inertia of a Body Depend on its Energy Content?*, translation by Methuen and Company, Ltd. Of London 1923, http://www.fourmilab.ch, retrieved Feb. 5th, 2021

138. Gerard Rempel, Professor, *The Scientific Revolution*, https://mars.wnec.edu/~grempel/courses/wc2/lectures/scientificrev.html, Retrieved Feb. 5th, 2021

139. Glenn Elert, *Universal Gravitation*, The Physics Hypertextbook, https://physics.info/gravitation/, Retrieved Feb. 5th, 2021

140. Shareen Blair Brysac, *Resisting Hitler: Mildred Harnack and the Red Orchestra*, (Oxford University Press 2000), pg. 393

141. Helen Wright, *Explorer of the Universe: A Biography of George Ellery Hale*, (E.P. Dutton & Co., Inc. 1966)

2

A Portrait (1928)

A fictionalized account of the meeting between Edwin Hubble and Milton Humason as they embark on the historic partnership which will uncover the expanding universe and reveal for the first time the impossibly great vastness of its depths. By this point they have known of each other largely in passing, each of them having made his own unique impression on the men, women and research at the Mount Wilson Observatory and the science world at large. Despite working in the same department of research, they have worked little with one another before this 1928 meeting in Hubble's office. This account is intended to give the reader a sense of the character and the position each man held at this moment in their respective careers, as well as a sense of time and place, offering a glimpse into the world at large and the life in Los Angeles, California, in particular, toward the end of the 1920s.

Thirty-seven-year-old Milton Humason closed the door to his small one-story rented house on 1034 North Hudson Avenue in Pasadena, California,[1] and walked out to the car port where his bicycle stood awaiting him. He toed the kickstand back and slung his right leg over the bike. Fixing his shoulder bag, he pushed off and glided down the driveway waving goodbye to his wife Helen, standing in the doorway, and waving back.

The offices of the Mount Wilson Observatory (MWO) were a short flight directly south as the birds fluttering about in the trees could fly, and a five-minute ride for a mere mortal riding a bicycle at a leisurely pace. "Mile-a-minute Milt,"[2] as friends and family affectionately called him, would make it in about two and a half minutes, flat.

© Springer Nature Switzerland AG 2021
R. Voller, *Hubble, Humason and the Big Bang*, Springer Praxis Books,
https://doi.org/10.1007/978-3-030-82181-4_2

After just a few yards he was at full speed, careening around corners and smiling back at the disgruntled drivers as they honked their horns in disapproving scorn. It seemed everyone was waving at him this morning, some with open hand, others with clenched fists.

It was another perfect summer morning in Los Angeles and the day held with it the promise of new mysteries and adventures for the veteran astrophotographer. He didn't talk about it much, but he felt like he had to pinch himself from time to time to be sure he wasn't dreaming all this. Just 10 years earlier, he wouldn't in his wildest dreams have expected to be living the life he was living now.

Since his arrival in the area from Minnesota with his family in 1902,[3] Milton had witnessed firsthand the incredible expansion of his adopted hometown both in population and development. From his vantage point atop Mount Wilson, he had seen the city grow tenfold from its population of roughly a hundred thousand in the year of his family's arrival to over a million strong now, fueled in large part by the Hollywood film industry, which churned out hundreds of short pictures each year to eager audiences in the east. The Los Angeles of the day was very much at the forefront of popular culture with modern cars, paved streets, museums, stadiums and amphitheaters, all catering to crowds of visitors waiting to hear the latest from Duke Ellington and his Cotton Club Orchestra or witness the skeleton of a giant saber-toothed tiger or cheer for their favorite college football team at the Coliseum otherwise known as the Rose Bowl. Victorious in the Great War, the United States had rebounded from the field of battle and a raging flu pandemic in roaring fashion, and Los Angeles had become the place where dreams could come true.

George Ellery Hale (See chapter one) was among those whose imaginative genius and immutable spirit had spearheaded much of the development in science and education in Los Angeles since his arrival in 1904 with a handful of men, a pocketful of cash and the determination to build the world's greatest astronomical observatory dedicated to studying the evolution of the stars. It would be hard to imagine how the legendary director could've been more successful in this task.

In the 25 years since its inception, Mount Wilson had contributed no fewer than three major breakthroughs in astronomy: the discovery of the Sun's magnetism in 1908, the proper location of the solar system far from the center of the galaxy in 1917, and proof of the existence of external galaxies in 1923.[4] From one vantage point or another, Milton had seen it all unfold.

It was hard to figure how he could've gone from a teller at the nearby hotel, to a cowboy muleskinner working the trails up the mountain, to a citrus rancher in the valley below to the position he held now. By some stroke of mad luck, Milton had become an integral part of the observatory itself, even playing a pivotal role in the research of nearly everyone on the staff of the stellar research department.

One of the few that he had yet to work with extensively was Hubble. He was probably the most popular astronomer in the world at that moment and Milton was headed in for a meeting with him this very morning to discuss a possible partnership.

Around Mount Wilson, Hubble's reputation was very different from the shining beacon of the human spirit often conveyed in the press. Here he was seen as more of an outsider, introverted, intransigent, self-promoting and arrogant or all of the above depending on who Milton talked to.

Milton's own dealings with Hubble had been limited to a few encounters while a night assistant in the early part of the decade and, more recently, the weekly discussions between the staffs at Mount Wilson Observatory and Caltech. As a result, he had had few one-on-one discussions with Hubble over the years save for the nights he worked as his assistant, which didn't amount to much more than a brief discussion of the desired coordinates and the set-up of the telescope. As he remembered it, Hubble knew how to lead a team. He was forthright and well informed but never condescending or controlling, and he let the night assistants do their job. If they were in error, he could be stern, but Milton thought he knew where he stood when he worked with Hubble. The astronomer even sat in on a poker game that broke out one night when the weather wasn't cooperating.

Still, there was something about Hubble that didn't sit right, Milton thought as he sped along the street past the beautiful sunlit homes, their yards festooned with colorful flora. The affected British accent was part of it, sure.[5] After all, Hubble was from Missouri like Milton's friend and mentor Harlow Shapley. But unlike Shapley, who was effusive, funny, and fun-loving, Hubble was much more reserved. He seemed like a guy who wasn't comfortable in his own skin, as though he was presenting himself to the world as someone else, some alter ego. It was a quality in a man that tended to make Milton leery of his intentions and undermined his trust in him.

A warm Southern California breeze teased Milton's wavy brown hair as he swung his bike around the corner from El Molino onto Santa Barbara Street and glided up onto the sidewalk past the optical and instruments shops with their long green lawns running to the entrance of the observatory's two-story office building. The building was designed in the Spanish Colonial Revival style that was popular in California at the time, with its sparsely landscaped lawn, columned entrance, and terracotta tile roof.

Milton hoisted the front wheel into a slot in the bike rack and walked down the sidewalk toward the front steps. As he spat the wad of tobacco he was chewing into the bushes, he was joined by his friend Seth Nicholson who had arrived at the same time. Nicholson was one of the most respected and liked researchers at Mount Wilson and one of Milton's best friends.

"Humason!" It was customary for astronomers to use last names only, both in person and in professional writings. "I thought that blur of color might be you riding up. I'm pretty sure you were approaching light speed, old pal."

"You know, it's funny you mention it, Nicholson," Milton replied as they walked. "The other day I got pulled over for running a red light. I tried to explain to the officer that as I was approaching the intersection the light must have been yellow. I was going so fast that the yellow light was actually blue shifted, so the yellow and blue mixed together making the light appear to be green."

"Yeah, and what did he say to that?"

"He tore the ticket up and wrote me one for speeding, instead!"

This was met with a subtle guffaw as Nicholson begrudgingly acknowledged Milton's ingenuity as they walked in through the front entrance and headed toward the stairs to their second-floor offices.

"With the elections coming up I'm guessing you'll have the Republican boys on the solar staff tied up at work on the mountain election day, as usual?"

"All's fair in love and war, Milt."

"Well, you're going to need all the help you can get, Nicholson. Hoover's going to crush Smith!"

"Step into the light, my friend."

"Never!" Milt said with a laugh.

"Say, Alma wants to invite you and Helen over for dinner Friday night."

"Well, I'll have to check with Helen, but I don't see why not."

"Great! We'll look forward to it. Oh, say, I heard you're meeting with the Major today," Nicholson said sarcastically. "What's that about?"

"Don't know, exactly, but Walter thought I'd be perfect for the job, so he asked me to hear him out." Walter Adams was the observatory director and also a friend of Milton's.

"Well, if anyone can deal with him, you can. Good luck with that one, pal!" Seth smiled mischievously as he walked away down the hall to his office.

Standing over his desk sipping a cup of coffee, 39-year-old Edwin Hubble examined a large print photograph of the sky that he held gently between thumb and forefinger. Had a stranger entered at that moment, he could not have noticed his heart pounding in his chest or the renewed feeling of excited anticipation at having yet another chance to experience the glorious, life-affirming joy of discovery he had felt several years earlier. Any minute, though, a relative stranger with the potential to help him deliver on that opportunity would do just that. When he did, Edwin thought, he would be ready to state his case on behalf of his cause.

As the faint rays of morning sunlight cast scant traces of dust in a Brownian flying circus near the window, Edwin allowed himself a moment to reflect on the momentous changes in his life.

When he arrived in Pasadena from England after the war, he was a man on the run, in search of a life that had long eluded him and the mountain had welcomed him, immersing him in its wild embrace. There he had carved his new life right into the granite walls of the summit, just as Hale had done years before. A buzz of excitement coursed through his veins as he eagerly hiked the mountain trail, his military knapsack on his shoulder, "Major Edwin P. Hubble, 343rd Inf" proudly emblazoned on its side.[6]

Then there was the Cepheid in M31, the great nebula in Andromeda, a moment of pure discovery so joyous it still made the hair on his arms and neck stand at attention. And not long after, the woman of his dreams, who for years had been close enough to touch but, married as she was to another was as distant as the starlit sky, fell into his arms after being widowed by her young husband's untimely death. The image of his father passed through Edwin's mind like a silken cloud. He had at long last been gifted with both Grace and the heavens. He glanced at the picture of Grace on his desk. Astronomy was his job now, and Grace was his universe. Finally, he could leave God and his legal training behind.

The recent five-month trip to Europe with Grace had been a time of joy for the couple and personal pride for Edwin. He had recently been selected as the youngest member of the National Academy of Sciences (NAS) in Washington, D.C., had been made a Foreign Associate to the Royal Astronomical Society (RAS) and was set to chair the Commission on Nebulae and Star Clusters (CNSC) at the third General Assembly of the International Astronomical Union (IAU) from July 5th through the 13th in Holland. Grace and he had taken time to visit friends and explore England and France between Edwin's scheduled lectures before arriving in the Dutch city of Leiden, five miles northeast of The Hague, for the conference.[7]

It was there that he learned of the mysterious link between Einstein's general relativity and the possibility spacetime was evolving. There were differing opinions on the subject, with some conjecturing the universe might be contracting and others that it was expanding, or perhaps oscillating by expanding and contracting at seemingly timeless intervals. After consulting with astronomers (because astronomy was not his field) Einstein (See chapter one) himself had opted for a static universe, neither contracting nor expanding. If he could uncover the truth one way or another, Edwin reasoned, it would mean a second even more fundamental discovery than the Cepheid variable in Andromeda and, he dared to dream, virtual immortality.

Testing the hypothesis would require understanding three intrinsic characteristics of the nebulae: their distance, their velocity (if any), and their motion relative to Earth (approach or recession). Measuring the velocity of a nebula required the exposure of certain elements of its spectrum, which could then be placed on a slide opposite an existing spectrum of the same elements made in the lab at Mount Wilson for comparison of the Fraunhofer lines (See chapter one) of the two spectra. If matching lines in the nebula's spectrum were offset toward the blue end of the spectrum, this indicated the nebula had a velocity of approach; if the offset was toward the red end of the spectrum, it was a velocity of recession (Fig. 2.1). If he could gather data on a significant number of nebulae, he might be able to ascertain whether the universe was static, as Einstein believed, or dynamic in a state of expansion or contraction.

But the devil was in the details, of course. Nabbing a useful spectrum from M31 was no problem at all for the average astronomer, but to go after an object 10 or a 100 times fainter than anything that could be seen with the naked eye was another matter altogether. For that he would need an expert, the finest spectroscopist he could find. That man, Walter Adams had assured him, was Milton Humason.

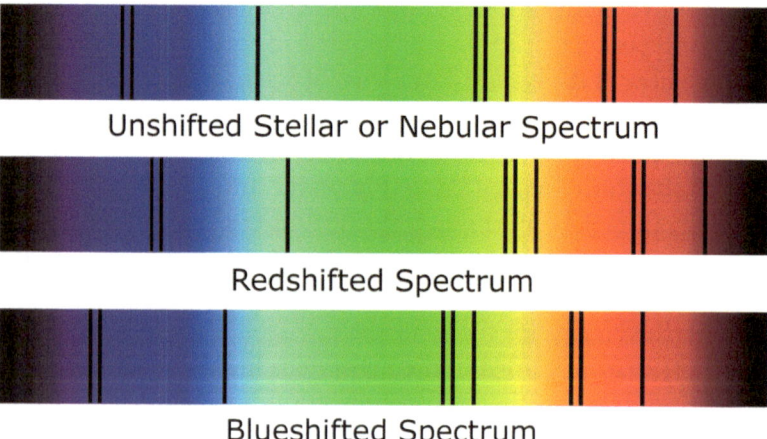

Fig. 2.1 Diagram showing comparison lines in three spectra. The lines in the redshifted spectrum indicate a velocity of recession compared with the unshifted spectrum. The lines in the blueshifted spectrum indicate a velocity of approach. The double line in the violet end of these spectra represents the H and K lines of calcium often used by spectroscopists for comparison.

Who was this guy and how did he get here? Edwin, like everyone else, had heard some of the stories of Humason's exploits as a cowboy and muleskinner in the early days of the development of the observatory and he was not immune to the rapture the mountain and its local lore often provoked.

For this project, though, he needed a serious researcher with incredible skill. His limited exposure to Humason had left him with the notion that he was something of a simpleton, quick to laugh, not predisposed to the attention this task might require. He had listened to Adams's suggestion, but he didn't trust the observatory director any more than he trusted most men in the field. He would take Humason's measure and see for himself.

A knock at the door interrupted his train of thought and Edwin cleared his throat.

"Come," he said in a voice loud enough to be heard through the partially opened door, and Humason stepped in.

"Good morning, Major." Milton knew it was Hubble's preference to be referred to by his military rank and thought they should start off on the right foot.

"What-ho, Humason?" This was an expression that Edwin knew made even his beloved wife's eyes roll but he couldn't resist his adopted English tone and mannerisms. "Thank you for popping by. Please have a seat, won't you?"

Milton sat in the wooden chair in front of Hubble's desk, sizing up both the man and the situation. The office was nondescript, the walls were bare and the desk tidy with a single picture of his wife Grace in a simple frame. Hubble cut a formidable figure, standing six feet two inches tall, sturdy, and athletic with a chiseled chin. The mole on his right cheek only seemed to give more character to his leading-man appearances. The accent was more pronounced in this setting, spotty and uneven, more detectable as a fraud. Milton had little idea what the former Oxford man wanted, but he was the best poker player on the mountain for a reason. He would sit patiently and wait for Hubble to show his hand.

Edwin watched as Humason glanced around the room. He was above average height, maybe five feet ten inches with clear blue eyes and a short crop of wavy brown hair. He seemed calm and cool as if he didn't have a care in the world. Did he even wonder why he had been asked to meet? Of that he couldn't be sure, but one thing was clear, educated or not, he didn't look a fool. Edwin was searching for a way to break the ice when he noticed Humason's eyes fall briefly upon the picture of Grace. He saw his opening.

"How is your wife, Helen, is it?" Edwin struck a match as he sat down behind his desk and lit his pipe, drawing several short puffs of smoke from the smoldering tobacco. Better to start off easy than to rush into the discussion.

"She's fine, thanks. How is your wife, Grace? Okay, I hope?"

"She is well, thank you."

Another pause brought silence back to the room. Milton shifted his weight and crossed one leg over the other, hoping Hubble would get to the point. But Hubble continued in the same vein.

"I hear you are something of an expert fly fisherman. Adams tells me you know the back country here as well as any man and that if I should desire a guide for a tour of some of the prime fishing holes, why, I couldn't do better than you."

"I've been known to load the rod once in a while," Milton replied with a wry smile. Any self-respecting fly fisherman should've understood what he meant by that phrase and the slight grin forming on the other's face told him he did. After all the apprehension leading into the meeting that morning, this was a welcome distraction. But what was Hubble up to? Milton still wasn't sure, so he continued.

"What sort of fly do you like to use?"

"I'm a dry fly man, myself, a drifter you might say."

"Oh well, you'll do well on the West Fork, then. There are so many trout in there, in some places you could walk across the river without getting your feet wet."

Edwin sat back and relaxed in his chair. A man who could fish, especially an expert, might be trustworthy after all. Humason wasn't being pushy or trying to impress him with his brilliance either. Edwin had the sense that he was an unassuming fellow. This may have accounted for his earlier impression of Humason as being of a simpler mind. He was a woodsy chap, no doubt, but he might just be up to the task.

"Perhaps we can arrange an outing. I fancy an outdoor excursion when the occasion and the mood strike. A chance to commune with nature is good for the constitution, for the spirit, don't you think? We could try out a few of your favorite fishing holes?"

Milton's poker face was expressionless. The great Edwin Hubble wanted to "commune" in nature with him, a mere mortal? What was so important that he couldn't just get to it?

"Well, I don't know about my favorite fishing holes. Giving up ones best kept secrets makes for a hungry fisherman," he pointed out. "But if there's fishing to be done, you can count me in."

Humason's mischievous grin was warm and genuine. Smiling at the quip, Edwin nodded his head. It was clear to him now why the man was so well liked. Still not sure how to lead into his intended topic he continued with small talk.

"I hear you're a big game hunter as well, is that right?" He had heard the story of the man's exploits in shooting and killing a mountain lion in 1919 prior to Edwin's return from England after the war. It was a chance for Humason to puff out his feathers a bit if he chose, but instead his expression grew cool, almost grim.

"I don't hunt big game, no. Unless, of course, the big game is hunting my son. Hoge and some of the guys went on about it but I don't like to talk about it much."

Get on with it already, Milton thought. "Love fishing, though. Stars or streams, it's all the same to me. If I may, though, Major, you didn't ask me here to talk about fishing and hunting, did you?"

Edwin leaned forward in his chair. It was a reasonable question. They had gone on long enough and it was time to get to the point.

"Quite right. Jolly good, Humason. Allow me, then, to walk you through it."

Almost immediately Milton sensed a mood change, an energy in Hubble's speech that had been hidden until now. He had the tone and body language of a schoolboy who was holding a secret he was dying to disclose.

"I have recently returned from the IAU conference in Leiden wherein I had a most auspicious conversation with a couple of colleagues that piqued my interest, considerably."

Milton dropped his other foot and sat up in his chair, the fingers of his hands woven together as he listened to Hubble's brief description. He had heard about Einstein's theory of relativity, but no one could explain it so that he could understand it. Thankfully, Hubble wasn't going that far, either because he couldn't or because it wasn't necessary; he didn't know which and it didn't matter. What was clear was that Hubble had a most serious mind and a solid grasp of his topic.

Edwin finished his summary of the problem and was confident Humason was with him so far, a reassuring sign that his instincts were correct. With that he decided to delve into the experiment and see if Humason would go for it.

"You're familiar with the spectroscopic work of Slipher at the Lowell Observatory, I take it?" Now the ball was firmly in Humason's court. Vesto Slipher's work between 1909 and 1926 had been the first extensive research on nebular spectra and any self-respecting spectroscopist should know of it. If he was worth his salt, Humason should've been well abreast of it even if he wasn't aware that Hubble had been corresponding with Slipher since the early 1920s. To his delight, Humason's answer was swift, thorough, and concise.

"He published the first known velocity of a nebula for Andromeda in 1913 and 14 more the following year. The lion's share showed recessional velocities. And then in 1921, I think it was, he announced that the radial velocity for

NGC 584 in Cetus was 1,800 kps, and that's the highest velocity ever recorded. Oh yeah, I wouldn't be sitting here today if I didn't know Doctor Slipher's work."

In the field of astronomical spectroscopy Slipher was a living legend. Harlow Shapley, Seth Nicholson, and Walter Adams had all corresponded with the Lowell man for years, although he had had little contact with Slipher to that point himself.[8] Adams, who was elected to the NAS in 1917, had written to congratulate Slipher on his election to that group in 1921, the year Slipher published the great velocity in Cetus.[9] Slipher's nebular work had ended around that time mainly because he had pushed the 24-inch refractor at Lowell to its limits – but not before he had nabbed over 40 nebular velocities. By that time, Slipher had gained the directorship of the observatory in the wake of the death in 1916 of its founder and inaugural director, Percival Lowell. What Milton knew that Hubble probably didn't, was that in obtaining the spectra for his deepest targets Slipher sometimes spent as much as 60 hours at the eyepiece, which was an unheard-of amount of time for a single spectrum.[10]

"Bully," Edwin continued, content that his audience was sufficiently up to speed. "What I propose is the following. Slipher's work includes only the radial velocities for the nebulae in his studies, and those were taken at the 24-inch refractor using his specifically designed spectrographic camera. What we don't yet have, and what we will surely need, are the *magnitudes* for these nebulae in order to get their distances. Now if we assume that nebulae are of comparable intrinsic luminosity, the fainter the nebula appears the more distant it is. If these greater magnitudes correspond to increasing velocities, then I just may be able to decipher a relationship between velocity and distance."

"It's a sound plan, I follow you so far." Milton had a feeling he knew what was coming next.

"Excellent, I'm glad you agree," Hubble continued. "Now, the part of the plan I was hoping to enlist your expertise for is two-fold. One, I would like to go back and remeasure the nebulae in Slipher's work with the 100-inch to verify or update their velocities. Given the overwhelming gain in observing power of that telescope, I think it wise to revisit these for corroboration. And two, I shall need a velocity for another nebula that is presumably a lot farther off due to its faintness. The goal is to determine whether the velocity-to-distance relationship exists in reality."

Now Milton knew for sure why he had been summoned to Hubble's office. He started running numbers in his head, trying to remember the exposure times in Slipher's samples and relating those to the 100-inch. Most, if not all, would be well within safe distances and easy targets, he thought.

"What did you have in mind, Major?" He knew it was a loaded question.

Sensing Humason's tacit acceptance of his plan, Edwin stood up, pipe gripped firmly between his teeth, and lifted the corner of the print of the sky he had been examining prior to Humason's arrival. The spectroscopist stood and leaned over the front of the desk to get a glimpse of the print for himself.

"This is a wide-angle of the sky around Pegasus. I thought maybe we could start here." Edwin thought he would give his counterpart the opportunity to chart his own course.

Milton studied the image. He knew the sky very well but for the most part his work to date had been centered on stars. The notion that the nebulae were distant star centers similar to the Milky Way had only been confirmed practically three years before, thanks to Hubble. But that was too painful a memory to revisit now.

"There are one or two targets located between Pegasus and Pisces that would probably work. Even the brightest of these would be fainter than anything Slipher would've seen at Lowell. I could maybe start there."

Was Humason saying he was in? It all seemed too easy. Maybe this wasn't such a good idea, Edwin thought. A better educated man, perhaps, someone with a lifetime commitment to science might jump at the chance to make history. This Humason character was a curious sort. But dealing with better educated, ambitious men inevitably led to conflict when they couldn't or wouldn't accept his findings or wanted to claim priority for a discovery. Adams had been clear, though, saying Humason was the right man, perhaps the only man for this work.

Hubble decided to take the chance. "So, you'll do it?"

"Yeah, why not. I'll give it a try." Humason was looking at him with a hint of mischief emanating from razor sharp blue eyes that peered at Edwin through wire rimmed glasses. After a pause, Humason continued. "You should know I have several other projects I will need to continue working on as well. I don't want to give the wrong impression."

Even as he agreed to this assignment, Milton wasn't certain he liked the idea. His attack on the Wolf star had been difficult enough. He couldn't be sure, but this new target might require two or three times the exposure time to get a measurable spectrum. On the other hand, he was always up for a challenge.

"Yes, of course," Hubble interrupted. "Adams informed me of your already prodigious observing schedule. Your exploits at the 100-inch are becoming legend, Humason. Good show, old man!"

"Thank you, Major." Milton was mustering all the pleasantry he could, but he was eager to get out of there and dig deeper into Hubble's proposal. He had only considered meeting Hubble at Adams's request. He really did have an already full schedule and taking a crack at a target like NGC 7619 could require several days if not more just to get one plate with a usable spectrum. And then there was the fact that Hubble did not exactly have a reputation as a guy that people wanted to work with.

But Adams was responsible for hiring Milton and there was almost nothing he wouldn't do for the observatory's director. If things didn't work out with Hubble, he could back out of the deal with the director's blessing.

"I'll need some time to organize my thoughts," Humason continued. "What do you say we meet again in a week, and in the meantime, I'll talk to Joy and try to work this into the schedule?" Alfred Joy was a respected astronomer and the secretary of the observatory who managed the stellar observing schedule on the mountain.

"Sounds like a plan." Edwin was elated. As Humason shook his hand and turned to leave Hubble could hardly contain his enthusiasm. He couldn't wait to return home and fill Grace in on the discussion. She didn't understand the science but her insights on human behavior were always welcome and profound.

As the door to Hubble's office closed behind him Milton reflected on the meeting. There was an eagerness to Hubble's mood that inspired him. He bore the look of someone who was in possession of a great secret. Humason wasn't going to get too excited, though. He'd been on the verge of discovery before only to have the chance slip away for one reason or another. He was honored to be thought of for the task, of course, but in the end, it was all gravy. He was living a dream he hadn't sought, providing for himself and his family as part of a community of brilliant scientists. Facing an opportunity to give something back to that world in the shape of a discovery of groundbreaking significance, he'd surely give it a try.

* * *

This account of the meeting between Edwin Hubble and Milton Humason as they embarked on their historic pursuit of what is known as the Big Bang theory of universal evolution is meant to offer the reader some perspective of the mindset, status, and disposition of the two men as they first considered partnering on the project.

The exact date of the meeting in 1928 isn't known but must have occurred sometime between the end of the Leiden meeting of the IAU on July 13th and Humason's first assault on NGC 7619 and its neighbor NGC 7626 on the 6th of August.[11]

The dialogue has been fictionalized to move the plot forward and so is not a real account of their discussion. It is merely intended to offer a snapshot of one of the most momentous meetings in scientific history. This incongruous pairing between two of the oddest, most overachieving, and highly charismatic characters who were by varying degrees misfit to the time, their respective fields and each other, nevertheless placed astronomical research on a momentous new course.

From the start the two were a study in contrasts: Hubble the preening, self-promoting, Old South conservative from Missouri and Humason, the publicity shy former muleskinner and rock-ribbed old-school republican from Minnesota. Hubble had come to Mount Wilson via a troubled childhood and law school at Oxford in England before starting his education in astronomy in earnest at Yerkes Observatory under Edwin Frost. Humason, from a tight-knit family, first arrived in the summer of 1902 before there was an observatory. He ended his education after his first day of high school and was hired as a bellhop by the small hotel on the mountain, later becoming a cowboy and citrus rancher. How he managed his rise to the pinnacle of the astronomical world carrying Slipher's spectroscopic baton into the outer reaches of space has become the stuff of legend.

Despite their many differences, Hubble and Humason shared a love of nature, a sense of personal conviction and a passion for their chosen field. Each entered into his career in astronomy relatively late in life, by different paths: Hubble from a distinguished academic career and Humason down a path more suitable to a character in a Jonathan Swift novel. Each had ultimately shunned his parents' guidance in choosing his life path: Hubble in law and Humason in banking, finance, and real estate.

They were both avid fishermen, enjoyed the outdoors, hiking and camping, and were devoted husbands. Above all, they shared a deep interest, purpose, and ambition in excelling at their own disciplines. This drive, more than anything, would propel them to the success that awaited them.

At the time of their pairing, each of them was entering a unique time in his development as a scientist. Hubble, the now famous discoverer of external galaxies whose large physical frame and leading man looks made him a media darling, was nevertheless starting to feel the strain of imposition from other scientists eager to claim partial credit for his discovery. As a result, the youthful exuberance he first brought to the field had taken on a patina of disillusionment and introspection. The theoretical possibility of universal expansion had revitalized the dormant eagerness that had driven him in his early days at the observatory. However, he remained wary of those around him and was interested foremost in finding a colleague he could trust with his results – either because he didn't seek notoriety or out of personal integrity that prevented him stealing Hubble's prodigious thunder. In Humason, he hoped, he had found both qualities.

Humason, on the other hand, was also at a milestone in his development as an observer. He had left behind his earlier ambitions regarding research and fallen into the craft of deep space photography and spectroscopy in which he was excelling at an unprecedented level. In contrast to Hubble, he had few, if any, adverse relations with the men and women at the observatory. Humason's deep inhibitions and anxieties around expressing himself to others in the field stemmed from the choice that he made early on to forego his education in favor of creating a life on Mount Wilson. His amiable personality and simple manner of speech endeared him to those he worked alongside, but also betrayed his lack of confidence, not in his ability but his intelligence, despite the contrary opinion of many of his colleagues.

These personal traits and experiences shaped the characters of Hubble and Humason, forming the backdrop for a partnership that would carry them forward into history. These same characteristics will also help to define the disparity in their individual popularity and legacies: Hubble to world acclaim but wavering respect within the field, and Humason to the margins of history, although romanticized as a cult hero and well respected both for his work and his character.

Oddly enough, as will be shown in the coming pages, they would later come to a point of role reversal, Hubble's usually brimming confidence eroding as Humason's remained steady, as they and other members of the practical and theoretical fields of astronomy and physics scratched their heads trying to understand and clarify the results of their work.

The details of these two highly sympathetic, incredibly dynamic, and altogether human lives and the impact they had on science and our collective understanding of the universe we inhabit are as remarkable as any in the annals of human endeavor. It is both a story of the ages and for the ages.

References

1. Address of Milton and Helen Humason, Pasadena City Directory, 1928, Retrieved 2010
2. Interview with Ann Humason Bernt, 2006, from family photo album
3. Letter to Thomas A. Humason jr., Sales at Harcourt, Brace and Company, Inc., New York from Milton Humason, October 24th, 1950

4. Mount Wilson Observatory, *Timeline: Historical Highlights from Mount Wilson Observatory's First 100 Years,* (www.mtwilson.edu), Retrieved Feb. 6th, 2021
5. Gale E. Christianson, *Edwin Hubble: Mariner of the Nebulae,* (University of Chicago Press 1995), pg. 134
6. *Ibid.* pg. 105
7. *Ibid.* pgs. 187-188
8. V.M. Slipher from Shapley, Nicholson, various dates., http://library.lowell.edu/Research/library/pub/search.php?archive=correspondence
9. V.M. Slipher from Dr. Walter S. Adams, Mount Wilson Obs., Pasadena, Calif., June 14th, 1921 http://library.lowell.edu/Research/library/pub/search.php?archive=correspondence
10. William Graves Hoyt, *Vesto Melvin Slipher 19875-1969: A Biographical Memoir,* (National Academy of Sciences 1980)
11. Logbook entry for the 100-inch telescope at Mount Wilson Observatory from August 6th, 1928, (Huntington Library), Retrieved May 10th, 2019

3

Childhood and Other Mysteries
(1889-1909)

Midwest natives, Edwin Hubble's and Milton Humason's young lives unfold in very different manners, both stemming from issues regarding their fathers. Living under the watchful eye and stern hand of his father, Edwin Hubble must navigate between his desire to pursue a career in astronomy and his pious father's wish for him to enter the field of law. His quiet rebellion against John Hubble's religious dogma and domineering persistence fuels an inner competitiveness that follows Edwin into his college years. And a Huck Finn early childhood living on the banks of the Mississippi River ends for Milton Humason when his family, suffering tragedy and threatened by his father's ill health, moves from Winona, Minnesota to Los Angeles seeking to reverse its fortunes. A naturalist at heart, young Milton follows his bliss to the summit of Mount Wilson where he begins his loose association with the observatory 15 years before the arrival of Edwin Hubble.

On a crisp fall morning, November 14, 1899, as steam hissed from under the massive iron horse at the front of the train waiting to transport him and his mother and siblings north to Chicago, 10-year-old Edwin Hubble's attentions were divided evenly between his family and the scene unfolding before him. The engine's whistle wailed urgently as the train sat beside the short platform at the depot in Marshfield, Missouri.[1] The familiar dance of conductors and passengers arriving and departing began as the short steps were placed on the platform near the open doors of the train. Passengers stepped down and greeted loved ones while those departing said their goodbyes. It was a setting he had seen many times as his father had come and gone on business in the

© Springer Nature Switzerland AG 2021
R. Voller, *Hubble, Humason and the Big Bang*, Springer Praxis Books,
https://doi.org/10.1007/978-3-030-82181-4_3

years leading up to their departure. Now finally, the day had come, and it was his turn to board the train to some as yet unexplored distant land.

He could recall vaguely walking alongside his father and mother as a young boy in St. Louis where the family had lived for a year or two, owing to his father's ever-changing employment. Marshfield was really the only home he had ever known, and he was eager to see what else the world had to offer.

On this day though, Edwin's feelings were mixed. Standing on the platform he reviewed a family whose arrival in the New World predated the Revolution, who had fought on both sides of that war as well as in the war of Northern Aggression of just a few decades prior.[2] He could still see the ravages of conflict and the toll it had taken in the faces of those he was leaving behind.

Although it wasn't part of the deep south, Missouri, like the rest of the nation at the time was deeply segregated. In its ruling on Plessy vs. Ferguson in 1896, the Supreme Court of the land had sanctioned the segregation of the races and this ruling had sent shockwaves through the country. The most prominent African American leader of the age, Booker T. Washington, protested the extension of the so-called Jim Crow laws while cautioning his followers to resist passively and peacefully but the situation only seemed to be worsening. In the decade since Edwin's birth there had been more than 20 public lynchings of African American men in Missouri, in most cases incorrectly or baselessly accused of offenses which ranged from annoying a white woman to murder, although in many cases the actual murderers were still at large. The most recent victim had met his untimely and unjustified doom a fortnight earlier in Fayette about 150 miles north of his hometown. Things were worse in the deep south, where thousands were losing their lives to angry mobs of white people, men, women, and children.[3]

The inner torment that the nation felt, past and present, could be felt directly within the Hubble family, descended from Old South Virginia slaveholders on both his mother's and father's side. Hailing from Missouri and Tennessee, the generation of fighting age young men were nearly torn apart at the onset of war between North and South.

Marshfield was a smaller township of Springfield a few miles west of the family home and below the 38th Parallel, the proverbial line drawn in the sand below which point slavery was still legal. After Missouri voted not to secede when the war broke out, a declaration of martial law and the freeing of the slaves in that state was quickly overturned by President Abraham Lincoln. This was meant to assure Missourians, who were sharply divided on the issue of slavery, that their livelihoods would not be altered without a plan, but Lincoln was pilloried on one side for waffling and on the other for being just another abolitionist.

Tennessee had been the last state to secede from the Union and the state remained torn for years after the war. About two-thirds of the able-bodied fighting men enlisted in the Confederate army and the others were recruited by the Union army. Edwin's grandfather Martin Jones Hubble had lived with his maternal grandfather in Tennessee after the age of 12, owing to his father's untimely death. His grandfather, a slaveholder, released his slaves when the war broke out, perhaps realizing the inevitable, but Martin and his brothers were divided over the war. His older brother had died fighting for the Southern cause while his younger brother, George Washington Hubble became a standard bearer and was wounded in the leg carrying the Stars and Bars into battle. Still hobbled by his wound, uncle George stood beside Edwin's grandfather on the platform to see the family off.

Martin Jones Hubble, perhaps taking a page from his Grandfather Jones in Tennessee, finally joined the Union army and, ironically, spent the first year or so of the war guarding against any rebellion by the local slave population until Lincoln issued the Emancipation Proclamation in September of 1862. After the war he used his local celebrity to become a leading businessman, politician, and trustee, and established himself as a man of intellect, ambition, and civic mindedness. He still bore the features of his impressive frame standing six feet in height and brawny of build even though to Edwin's young eyes the 54-year-old appeared very old. Although he had seen little or no action during the war Martin was still known as Captain Hubble around town, a charade his grandson would play out on his return from Europe after the Great War.[4]

As the disruption in the country continued after the fighting, Martin Hubble had put the family on stable footing among the middle class, capitalizing on the growth around the area to establish a commercial insurance firm in nearby Springfield in the 1880s. Later, attempting to capitalize on his successes, he bought a 640-acre plot of usable land in Marshfield and established the Hubble Land and Fruit Company. The operation of the "Hubble Ranch" was a family affair. Edwin's grandmother, Mary Jane (Powell) Hubble, acted as president, his uncle Joel as vice-president and his father, John Powell Hubble, as secretary, and all were listed among the company's board of directors.[5]

What his grandson was less aware of was the fact that Martin Hubble was an obsessively puritanical and equally ambitious man who ruled over his family with an iron fist. Vulgar language, drinking alcohol, billiards and consorting with young ladies outside the public sphere were strictly forbidden in the Hubble household. Growing up, John Powell Hubble and his siblings received regular instruction in the gospel. John was pressed into studying law, a profession his father saw as bearing similar, albeit secular features to the reading of scripture. Although John never graduated law school at Washington University, he passed scrutiny before a judge and gained membership to the bar. His fortunes as an attorney were short lived and his practice shut down after a few years.

Brimming with confidence but still seeking his fortune, Martin Hubble had speculated heavily in real estate with John Hubble, who had knowledge of the markets and could read the law serving as the middleman for deals made for and with his parents. These holdings failed to yield the promised profits however, and the overextended elder Hubble had to file for bankruptcy protection.[6]

Next to Edwin stood his mother Virginia (Jennie) Lee Hubble, whose focus was first and foremost on her parents, William and Virginia James. William Henderson James had been a successful doctor and drugstore owner who had a reputation as a specialist in treating typhoid fever. The James's were descended from Scottish Irish and English ancestry dating back to the Mayflower and the early Plymouth settlers. Like the Hubble's, their families were heavily steeped in the traditions and values of Southern slaveholders. Before settling back in Missouri, they had traveled to California during the Gold Rush while William sought his fortune in precious metals and merchandising, failing in his quest to join the ranks of the mining barons. It was nevertheless during this period that Edwin's mother had been born in May of 1864 in Virginia City, Nevada.

It was a twist of fate, a farming accident, which had brought the two families together. While working on the ranch on his summer break from law school, a team of plow horses was spooked and dragged John Hubble, entangled in the harness, for hundreds of yards across the fields. He was driven by wagon to town to the home of his future father-in-law, who called his daughter to help him with the mangled patient. The youngsters fell in love during John's recovery and were wed several years later.[7]

His father's work as a traveling insurance sales manager kept him away from home for weeks at a time, leaving the raising of their children to Virginia Hubble, who sought help from her parents. Their declining health had precipitated the move north to be nearer to the regional sales office where John Hubble spent most of his time.

A consequence of his father's long absences was that Edwin felt a greater bond with his grandfathers than his own father. He loved to hear his grandfather Martin tell stories of the Old South and Hubble family lore. He dreamed of visiting the Confederate Museum to see the bloodstained flag that had been wrapped around the wounded leg of his great-uncle George as he was carried off the field of battle.

After struggling with his attention early on, Edwin had become a star student and an avid reader of various genres from *Alice in Wonderland* to Kipling's *Jungle Book*. Two years earlier, on his eighth birthday, Edwin's grandfather William invited him to spend an evening looking at the stars through a telescope he had built from materials he had acquired through the local general store. The experience sparked something in Edwin, who began to read and talk about astronomy with both of his grandfathers. They were duly impressed by the boy's passion and his grasp of the topic. One of the highlights of Edwin's young life had come on June 23rd, just months before he and his family were set to embark on their move north, when he stayed up all night with a friend to behold a lunar eclipse for the first time.[8]

The whistle from the big 4-4-0 interrupted the family's tearful goodbye. The conductor, pocket watch in hand, shouted the "All aboard" warning as passengers rushed to mount the waiting train cars. Virginia Hubble, holding Edwin's infant sister Helen in her arms, gave her mother a hug and a kiss and hustled the children on the train. Edwin followed his older brother Henry, sister Lucy and younger brother William as they climbed up the steps and wandered down the aisle of the Pullman with its plush seats and table service. It was all very fancy to a young country boy. He could hear the bell on top of the engine's boiler ring as the giant steel drive wheels grabbed at the tracks and the train slowly lurched forward.

By now, Edwin, a planner and list maker by nature, knew the route to his new home by heart. Hours after their trip began on the St. Louis and San Francisco line, nicknamed The Frisco, they would be in St. Louis. From there they would transfer to the Chicago & Alton line to head north to Chicago. The advertisements for that line promised the finest comforts in train travel of the day: reclining seats, luxurious dining, and sleeper cars. It was a hotel on wheels and the Chicago & Alton Railroad Company was building the biggest and fastest trains in the country.[9]

Train travel offered time for reflection and as the train clicked and clacked its rhythmic beat along the tracks between stops at the dozens of sleepy Midwestern towns on its route, Edwin thought of those left behind and of the adventures that lie in wait for him. He would celebrate his 10th birthday in a new home in Evanston, Illinois on November 20th, in just six days' time, far removed from friends and family.

Although he was mostly unaware of his grandfather Martin's ambitions and imperious puritanical views, Edwin was fully aware of these traits in his father, who seemed obsessed with becoming a member of the elite society. At six feet three inches, John Hubble was an imposing figure whose religious views went beyond even those of his father. He was an ardent purveyor of scripture who insisted on strict adherence to the Bible and its teachings. Just as his father had set out for his family, John Hubble didn't tolerate vulgar language or impious behavior from his children. But his prolonged absences from home disrupted the continuity and consistency of family life and caused friction during his brief returns. In the year since the birth of the sixth child, Helen, John's presence at home, where presumably a religiously adherent man would desire to be, had grown increasingly rare. No doubt this accounted for at least some of the discord between Edwin's parents.[10]

Unlike his father, Edwin was fascinated by recent developments in America's industrial complex and the astonishing opportunities and advances they were making possible. The transcontinental railroad built between Omaha, Nebraska and San Francisco, California was an international marvel since its completion in 1869, enabling travel from coast to coast in North America. The Brooklyn Bridge, construction of which started the year the great cross-country railway was completed, now carried thousands of people across the east fork of the Hudson River from Brooklyn to New York City every day. The increased manufacturing muscle was being fueled by a modern steel industry led by the steel magnate Andrew Carnegie and wealthy oil men like John D. Rockefeller and supported by a robust banking industry led by John Pierpont "J.P." Morgan. Great modern steel-framed buildings were rising above urban streets in New York City and Chicago and in just the past 10 years the Eiffel Tower in Paris and the first Ferris Wheel in Chicago had been opened to hordes of sightseers.

Favorite among these wonders for young Edwin was the 40-inch refracting telescope (Fig. 3.1) at the recently created Yerkes Observatory at Williams Bay, Wisconsin. The brainchild of its founder George Ellery Hale, Yerkes opened in 1897 as part of the University of Chicago. Hale's refractor, the largest in the world, was the key attraction at the new observatory with its Alvan Clark & Sons lens and mounting by Warner & Swasey Company. Edwin eagerly read the news reports of the activities at Yerkes and the new instruments that were being developed by Hale and his chief optician and instrument maker George Ritchey.

As he made known his affinity for astronomy and science, Edwin noticed a growing displeasure in his father over the pursuit of such passions. To the strict Calvinist, the idea of understanding the heavens was an insult to God's grace as the Almighty Creator. God's laws were simple for John Hubble. The fate of a brilliant young son was preordained. Just as the laws of the Creator governed men's souls, so did the laws of the land govern their hearts and

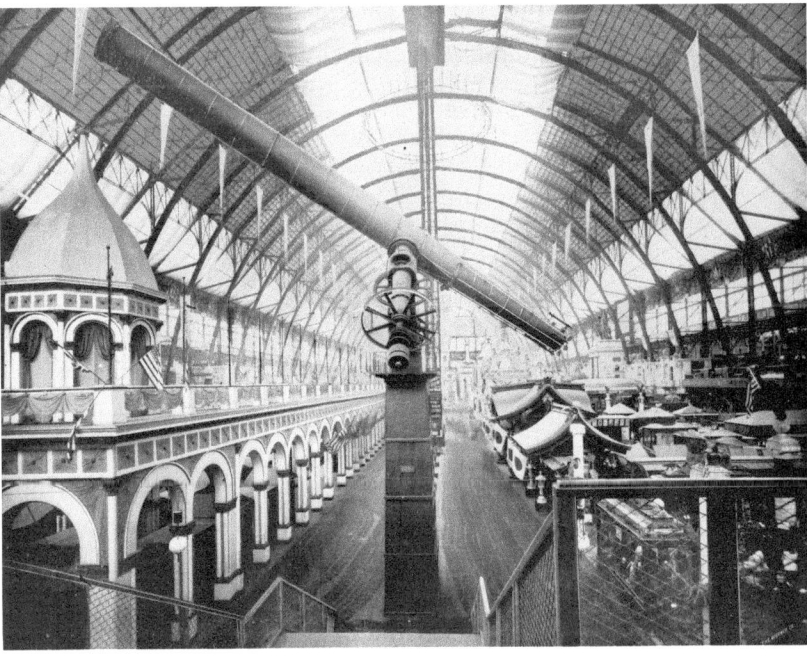

Fig. 3.1 The Yerkes 40-inch refracting telescope inside the Manufacturers and Liberal Arts Building at the 1893 Chicago World's Fair. (Courtesy, Field Museum. GN90799d_CG_166w)

minds. Hence there could be no greater pursuit among earthly beings than the study and practice of law. In John Hubble's opinion, a life in astronomy for his son was completely out of the question.

Despite his father's religious earnestness, Edwin had begun to dream of a life studying the stars. But as the train sped away from his childhood home and his beloved grandparents, he worried that his dream might be starting to slip away.

Little is known about the Hubble's brief stay in Evanston, Illinois. It is likely the move to the southwestern suburbs of Chicago was precipitated by the noxious odors from rotting entrails of slaughtered livestock at the huge Union Stockyards 15 miles to the south. John Hubble's disgust with the relatively poor living conditions and rampant hedonism within the city limits was probably also a contributing factor.

The township of Wheaton, established after Andrew Jackson's Indian Removal Act of 1837 which sent the regional tribes marching west on the so-called "trail of tears," was one of the wealthiest and most conservative cities in all of Illinois. It was established by the Wheaton brothers in 1890. Like the Hubble patriarch, they were Calvinists who hailed from Connecticut and moved to the area in the 1830s in search of fertile soil for farming. Warren Wheaton, a farmer and schoolteacher who was the first of the brothers to arrive in the area, acted as the first president of Wheaton and led the development of the town into a city as a businessman, legislator and philanthropist. The population was nearly three times that of Marshfield and the town was a veritable fortress of religious rectitude with a standing ban on alcohol. With the steeples of the many churches standing like watchtowers above the herd, the community seemed the perfect choice for the 36-year-old insurance salesman who brought his family to the quiet southern suburb a year after their arrival in the Chicago area.[11]

For a boy arriving at the threshold of the city at the dawn of the 20th century, the city's sprawling industrial and intellectual complex held all the promise of the age. Chicago had been the center for growth in the United States for half a century and its streets and alleys were teeming with nearly two million people. This remarkable rise as the center of intellectual, technological, and industrial growth had culminated in the World's Fair of 1893 (its original 1892 opening was set back by difficulties in construction) in commemoration of the 400th anniversary of Columbus's discovery of the Americas with its beautiful "White City" lined with canals and all the wonders of the modern world.[12] At the foot of the old exhibition grounds, just 30 miles or so east of the town where Edwin and his family lived, was the university where George Hale worked with its newly founded observatory in the countryside 60 miles to the north.

Edwin was the third member of the Hubble family: Henry (1886), Lucy (1888), Edwin (1889), William (1892) and Helen (1898). The death of Virginia (1894) of an unknown illness at 14 months had plunged a five-year-old Edwin into a depression, thinking he might have caused his baby sister's death.[13]

Still in grade school, Edwin was already starting to show the athleticism and scholastic aptitude that would characterize his academic career. Sports may have been one of the few sanctioned escapes available to the aspiring young student to spend time outside the specter of moral tyranny he and his siblings lived under at home. Although he did his best to please his father and abide by his wishes Edwin was already weary of stern rule. Like a canary in a cage, he looked out at the world around him and wondered if he might ever be set free.

***

In 1902, a year after Edwin Hubble and family had relocated to Wheaton, Illinois, Milton Humason stood at the head of the old Indian trail at the foot of Mount Wilson on the western slope of the Sierra Madre mountains in Pasadena, California. The dusty rustic scene before him was a far cry from the gilded refinement he encountered at the home of his uncle Henry Clayton Witmer and his wife Alice Petterson Witmer, with whom the Humasons had been living since their arrival in Los Angeles earlier that year.

Of the two, this was far more to Milton's liking. People of various stripes were preparing to hike or ride the 10-mile-long trail to the summit on whatever they could afford, horse or mule, while a man barked orders at a small group of young cowboys saddling one beast or another for service in the endeavor. Dust kicked up by a breeze or a passing horse and cart swirled and rose into the morning sky to reveal the narrow mountain path weaving its way into the nearby hillside only to disappear into the mountain.[14]

Standing at his side was Milton's father, William Grant Humason, an eighth generation American whose distant ancestor arrived from England around 1640 and settled in the area around New Haven, Connecticut. Milton's great-grandfather, Lawrence Stern Humason, was the first to move to Dodge County, Minnesota shortly after his marriage to Ruth Tyler in 1832. Lawrence died in 1898 leaving behind his widow who still lived in the area.[15] Humasons had fought in every war the young republic had waged since their arrival. Milton's grandfather, Lewis Abel Humason married his cousin Ellen Amelia Humason in a small ceremony on July 5th, 1862, in Winona and their first child was born a year later. Lewis enlisted in Company C of the 3rd Minnesota Volunteers in January 1864 and served as warden of the prison at Little Rock, Arkansas until his regiment was mustered out in September 1865. Milton's father, born on September 30th, 1868, was the second and only surviving child of four.

A miller by trade, Lewis A. Humason moved his young family west to California in the 1870s, probably drawn by the influx of jobs and business opportunities after the Gold Rush. It was here that Milton La Salle Humason gained his distinctive middle name years prior to his birth in Dodge Center, Minnesota on August 19th, 1891. As family lore has it, young William Humason was exploring the hillside near his home when he inadvertently fell into an abandoned mineshaft, injuring himself in the process. Unable to move he called for help and was rescued after an exhausting length of time by a French-Canadian man bearing the surname La Salle. In memory of the kindness of his rescuer, William promised to name his first-born son after the stranger.[16]

William Humason later met and fell in love with the vivacious Laura Grace Petterson, and the couple were married near her home in San Francisco in 1889. They relocated to Minnesota near William's family shortly thereafter. After Milton's birth, the family moved to nearby Winona, where his brother Lewis Howard (1893) and sister Virginia (1899) were born. Then came the move west.[17]

A group of boys stood with Milton and his father waiting to head up the Mount Wilson trail to the newly reopened Strain's Camp. A notice in the Los Angeles Herald on Milton's 11th birthday, August 19th, may have moved his uncle Clate, as Henry Witmer was known to friends and family, to suggest the idea.[18]

The cowboys finally finished readying the horses and mules, carts and wagons for the morning's journey and William paid for a mule to serve as his young son's mount for the trip to the peak. Soon the mule train, guided by cowboys at its front center and rear, slowly moved off and up the steep and narrow trail (Fig. 3.2).

When he wasn't peering helplessly over the edge of the steep canyon to his right or arguing with the cantankerous beast which he had been literally and figuratively saddled with, Milton might've recalled aspects of the world's he had known.

Fig. 3.2   Pack train to Wilson Peak on the Sierra Madre Trail in 1900

Winona in the 1890s had been in the throes of a battle for supremacy in the grain and lumber industry that it would eventually lose to its larger neighbor Minneapolis, upstream on the Mississippi River. William Humason worked at the Grain & Lumber Exchange near the waterfront and Levee Park.[19]

Milton and his younger brother Lewis spent their formative years playing by the river with friends. Winona was typical of 19th century towns that sprang up near large waterways in the United States, being a quiet but growing residential community set beside a roaring riverside industrial and economic complex where smoke spewed from the mill chimneys, and riverboats lined the shoreline providing entertainment for the men seeking a diversion from everyday life. This is probably the place where a young Milton Humason got his first taste of gambling and poker, two pastimes he would later be known as an expert in. Left to his own devices and desires while his mother tended to his infant sister and his father struggled with his ailing health there was no limit to the mischief Milton and friends could get into.

The riverboats of the Mississippi River were used for a multitude of purposes. Milton had grown up watching them navigate their way past the rotating train bridge with a huge lumber barge of floating logs hundreds of yards long. One boat would sit sideways at the front of the barge while another pushed from the rear. A toot from the pusher would signal to the pilot of the boat at the front to go forward or backward, to steer the barge to port or starboard and avoid trouble. Still other boats were used for ferrying businessmen and other travelers from one town to the next up and down the river. The owners of these boats had seized on the opportunity to create a little additional income by installing pianos and poker bars and tables for men to gamble and enjoy a dance from the girls hired to entertain them. The parlors of these boats could be magnificent with huge chandeliers hanging from their ceilings, carpeted floors and tables and plush chairs where men and women sat reading or talking. The casinos were often frequented by rugged, well-dressed, mustachioed men with fancy coats and hats who stood or sat around large tables or at the bars drinking whiskey and trying their luck at cards.[20]

For a kid whose favorite books included *Treasure Island* and *Huckleberry Finn*, Winona was an ideal place to grow up. The Chicago and Northwestern line arrived at the central depot several times a day loading and offloading passengers and luggage, and ships ran up and down river with passengers and cargo. There was a large park at the edge town where people played baseball and fairs were a regular occurrence most of the year. At the last fair commemorating the new century, a pair of tightrope walkers in an elephant suit shimmied across a line between two buildings to the applause of spectators. The bustling city had all the hallmarks of the modern era, electricity, streetcars and telegraph stations and phone lines. It was about as modern as a city could get at the turn of the 20th century in the United States. Attractions ranged from Benedict's Pool Hall to Laird's Drugstore, where Milton and friends would sometimes go for a toffee, to climbing the high butte of Sugar Loaf west of town. Memories of these early childhood adventures would stay with Milton for the rest of his life.[21]

Milton's mother, Laura Humason and her sister Alice were the third and fourth daughters born to George Howard Petterson and his wife Mary Catherine (Shirkey) Petterson. George Petterson was a native of Turin, New York, where Milton's paternal grandmother Ellen originated, raising the possibility that William and Laura had been introduced through some family or friendly connection. The older Petterson sisters were born in New York while Laura and Alice were born in San Francisco, a move that may have been caused by their parents' desire to remove the family from the ongoing civil war.

George Petterson first found employment in a sash and blind factory and later became a shipping clerk. Being born less than two years apart and a decade younger than their sisters, Laura and Alice were very close growing up. Alice, who appeared in the 1870 census as "not named," subsequently made a name for herself by marrying the wealthy California banker Henry Witmer.[22]

It was in 1898 that the Humason clan was dealt a blow by the hand of cruel fate. William Humason contracted typhus while enlisted in the 12th Minnesota Volunteers during the Spanish-American War. The deadly disease kept him quarantined in a Chicago hospital for six months.[23] During this time Laura Humason gave birth to their daughter, Virginia. Eventually, William was able to return home to his family and regain his foothold in the area in the spring of 1899.

Meanwhile, in 1898 Alice Petterson married the millionaire banker and entrepreneur Henry Clayton Witmer in Los Angeles. But the occasion had been veiled in sadness owing to the sudden death of Witmer's younger brother and brother-in-law within months of the wedding. Alice gave birth to the couple's only child, Joseph, a year later but the appearance of a healthy baby boy barely assuaged the pall that had descended on the family.[24]

No record of it survives, but it is likely that while Laura Humason was pregnant with Virginia and still awaiting the arrival of her husband home from his sick bed in Chicago, she received a telegram from her sister in California. Although the Witmers may have had telephone service in their home, the device was still not in large scale use at the turn of the century. The death of his brother and brother-in-law, who were also his business partners, left Henry Witmer to carry the load of running the family's many business holdings, which included banking, real estate, oil production and a cable car company. Alice worried that the strain of the operation would take her husband's life prematurely. They were in dire need of another able-bodied man to help out. On top of that the warmer, drier climate might have a beneficial effect on William's constitution. By uniting the Witmers and Humasons in Los Angeles, the sisters might just preserve the lives of both of their husbands.

Milton would later relate the year of the family's arrival in California in 1902 in a letter to a relative.[25] Not far from the western border of his home state of Minnesota the country was largely barren, with no sign of civilization day after day for 1,000 miles along the railway. This was truly frontier country, but Los Angeles, where their journey ended, was a striding, glistening exception with over 100,000 residents.

The Witmer mansion on 3rd Street in Crown Hill was grandiose by the standards the Humasons had known previously. The surrounding lakes and parks offered scope for adventure for a growing boy and his brother and the Witmer's business and land holdings offered a bright future. But his uncle's business interests were of little or no interest to Milton. A bright and curious boy, he was more attuned to nature than to the trappings of the modern world. Of his uncle's land holdings, the citrus ranch in Lordsburg (La Verne) was by far the most exciting place to visit. In the backwoods and orange groves he could resume the pursuits he had so adored in his old childhood home on the banks of the Mississippi. There were animals to attend to and fruit crops to harvest and it was there that he probably learned to shoot a rifle and ride horseback.[26]

And now he was riding a mule up to the summit of Mount Wilson like an old mountain man making his way into the wilderness. Strain's Camp was named for Ashbel Greene Strain, a Pennsylvania-born former real estate owner and rancher who had built a cabin on a homestead that he purchased in a shaded glen of cedars and firs in 1889. This was during a population boom that occurred in the Los Angeles area shortly after the completion of the transcontinental railroad there in 1885 by the Atchison, Topeka, and Santa Fe Railroad Company. Capitalizing on his investment, Strain next opened a camp on his land north of the summit of Mount Wilson[27] (Fig. 3.3).

Wilson Peak was named for the former mountain man, entrepreneur and politician, Benjamin Davis "Don Benito" Wilson. A colorful character who among other things was the grandfather of General George S. Patton jr., a senior US Army commander in Europe in World War II, Wilson had purchased the land in the 1860s and built a toll road (now the Little Santa Anita Canyon Trail) from Sierra Madre to the summit of the mountain in 1864.[28]

Strain abandoned the camp after seven years of operation and the land lay idle for nearly five years before a deep well of water was discovered in the

Fig. 3.3 A.G. Strain's Camp, Mount Wilson ca 1914 (Image courtesy of the Observatories of the Carnegie Institution for Science Collection at the Huntington Library, San Marino)

center of the site early in 1902. This convinced the Mount Wilson Toll Road Company, landowners of Wilson's Peak Park, to reopen the camp with plans to create a tent city for hikers and campers. The well would later be expanded, and its underground water reserves used to feed both the Mount Wilson Hotel, operated by the Toll Road Company, and the Mount Wilson Observatory.[29]

At the foot of the camp stood a small wooden shack shaded by trees where Milton and his fellow summer campers were checked in to the campsite and issued a tent for their stay. Instantly captivated by the location, he had found a new home away from home, a place of wonder and adventure.

***

On a pleasant early spring day in March 1910, Edwin Hubble smiled as he stepped out of the English Gothic building that housed the department of mathematics at the University of Chicago and across the courtyard toward East 59th Street. There he would turn left and walk the half mile along the Midway Plaisance to Jackson Park for a stroll through the old Japanese garden that still stood almost as it had during the Chicago World Fair 17 years earlier.

The Midway itself had been very popular during the Fair, offering side shows, music, educational and entertainment venues, as well as one of the park's main attractions, George Ferris's 264-foot-tall original Ferris Wheel. The wheel was a smashing success, boasting more than 1.5 million riders by

Fair's end. The ride set several firsts on its completion and could carry 2,160 passengers at a time to a height 26 stories above the grounds of the park in 36 well-appointed cars. From there the entire park and most of the city were visible all the way out to Lake Michigan to the east. The massive marvel had stood at the center of the Midway at just about the exact point where Edwin made his turn toward the lakefront. The Ferris Wheel was disassembled after the park closed and moved to the city's North Side for several years prior to being taken to St. Louis for that city's World Fair in 1904, after which it was finally demolished.[30]

The principal buildings of the "White City," as the grounds of the World's Columbian Exposition of 1893 were called, had burned to the ground several months after the Fair's closing and not many of its original attractions remained intact. One of these was the replica of Columbus's ship, the Santa Maria, which still sat at harbor near the new yacht club. The Palace of Fine Arts was the only fireproof building built for the Fair and consequently was spared during the fire. For the moment, the 1,500-square-foot building was abandoned and would remain so until 1930 when it would be rehabilitated and expanded to accommodate the Museum of Science and Industry.[31]

Apart from the history itself these features held little interest to Edwin, who preferred to amble through the peaceful setting inside the Japanese Garden on an island between the East and West Lagoons. He was not the only student from the university to visit the park by far, but on this day, he bore one unique quality above all others. He was the only Rhodes Scholar to be making his way through its pleasant surroundings. On what might be his final walk around the grounds of the campus and its adjoining parks, he could relax and celebrate a job well done.

The past decade had been one in which Edwin had waged a quiet rebellion against his father, whose insistence that he pursue a career in law had simply not aligned with his own ambition of becoming an astronomer. Out of love and respect toward the family patriarch he had played the role of the dutiful and highly capable son right to this very day. He had excelled as a student and an athlete at the Central School in his adopted hometown of Wheaton, Illinois. He lettered in track and basketball, becoming one of the most celebrated high school athletes in the area while averaging in the mid-90s in his schoolwork. He even managed to pull up his grades in deportment and spelling, a deficiency that would remain with him. To many of his classmates, Edwin appeared aloof and acted like he had all the answers. But he was focused and determined and despite the lack of respect from his peers he had managed a scholarship to the prestigious college where he began his undergraduate studies in October of 1906.[32]

It was the first step in a long-range plan to prepare himself for a life in astronomy while complying with his father's wishes, hoping he would eventually come around to Edwin's point of view.

If he wanted to be an astronomer the Yerkes Observatory was the place to be, and that facility was part of the University of Chicago. The observatory's director George Hale had by this time moved on to California, but the institution had the largest operational refracting telescope in the world and its second director was the highly respected Edwin Brant Frost, who also served as editor of Hale's Astrophysical Journal.

But Edwin was still under the watchful gaze of his father and knew he had to give every appearance of preparing for a law degree. With that in mind he planned to take the courses he would need in science, to be used later for the study of astronomy, while also satisfying the requirements for advancement in law.

If he could accomplish this much, his next goal was to win a Rhodes Scholarship and study at Oxford where he would no doubt have to carry on the charade barring any change in his father's attitude to the sciences. Edwin had had this goal in mind since his freshman year in high school. To earn the Rhodes, one needed not only to excel at school but also to display the characteristics of leadership, a devotion to duty and kindness toward others, as well as athleticism. He took on these challenges in full view of the school's administration, carefully planning the moves that he hoped would lock him in for the award.

His former athletic prowess was overshadowed by others at the collegiate level, but he had nevertheless managed to letter in track and field and in basketball. He had joined the Kappa Sigma fraternity in his first year and progressed through the core of mathematics, science, and language courses[33] (Fig. 3.4).

One of his favorite teachers in those years was Forest Ray Moulton, associate professor of astronomy, whose textbook, *An Introduction to Celestial Mechanics*, Edwin would hold onto well after it had outlived its usefulness.[34]

It was high season in the physics department at the time. The head of the department, Albert Michelson, was awarded the Nobel Prize in Physics in 1907 for his measurements of the velocity of light and the development of scientific gadgets for the study of physical phenomena. He was the first American to receive the award and Edwin hoped he wouldn't be the last. Such were the ambitions of the young man from Missouri.[35]

In his final three years he took additional language courses in French, Latin and Greek and added political science too. All were calculated, however

Fig. 3.4 Studio portrait of Edwin Powell Hubble as a young man ca 1907. (Image from Edwin Hubble Papers, Huntington Library, San Marino, California)

carefully, with an eye toward earning an extended trip to England. Coupled with his commitments in athletics it had been an exhausting schedule.

With an eye firmly fixed on the finish line, Edwin made a few decisive final moves. As the test neared in October 1909 he put in extra time on the study of Greek and ran for class vice-president, a job that apparently no one wanted since he won the title unopposed. His obsessively large course schedule had earned him early graduation in March, which meant he could serve on the class executive committee. This gave him not one but two leadership roles to point to on his application to the Rhodes selection panel.

All his hard work evidently paid off as he was the only member of his class who passed the qualification exam that fall. With the remaining competition being at other schools in the state, Edwin began seeking a letter of recommendation from a professor of stature that would give him the edge. In the end it was Robert A. Millikan, the assistant director of the department of physics, who delivered the persuasive arguments in Edwin's favor, speaking of him as "a man of magnificent physique, admirable scholarship, and worthy and loveable character," and concluding, "Seldom, have I known a man who seemed to me to be better qualified to meet the conditions imposed by the founder of the Rhodes scholarships than is Mr. Hubble."[36]

Edwin had worked as Millikan's lab assistant as part of a Junior College Scholarship in Physics award he had earned his sophomore year, starting in the fall of 1908. At that time, Millikan was involved in the famous oil-drop experiment that enabled him to calculate the charge and mass of an electron and led to his receipt of the Nobel Prize in 1923. His stature, as much or more than any other, had aided Edwin in the final tally but the eager graduate had done his part to ensure victory. Millikan would serve as the inaugural president of the California Institute of Technology (Caltech) from 1920 to 1946 at the behest of Hale, who had been instrumental in transforming that institution from its guise as the Throop College of Technology. The close relationship between Caltech and the Mount Wilson and Palomar Observatories that ended in the administration of the observatories being absorbed by the college would bring Edwin and his former professor and champion together in the future.[37]

The news of his selection as the Rhodes scholar from Illinois for that year almost seemed too good to be true. All the hard work had paid off and he was heading to England to study at Oxford. The prize allotted $1,500 per annum ($41,000 in 2021) for his graduate work in the field of his choosing at Queen's College. This was the only remaining sticking point in his plan. Edwin knew he would either have to study law or stand up to his father and pursue his dream of astronomy. For a 20-year-old who was still living very much under his father's influence, revolt seemed impossible.[38]

There were times when even the Edenic setting of Wheaton township seemed not to live up to the standards of John Hubble. When a train carrying gamblers came through the area to bet on the horseraces at the local track, his father had forbidden Edwin and his siblings from leaving the backyard for fear they might get a whiff of the alcohol sold on the trains to entice would-be betters to make the trip...and descend into the depths of depravity. This was the way of world according to John Hubble.[39] But exclusion from exciting events only made the youngsters more curious, especially Edwin and his younger brother Bill, who the family regarded as Edwin's academic and athletic equal.

As time went on Edwin derived ever diminishing solace from his mother who, although she was sympathetic to her children's desires for freedom, had nothing to offer in the way of a solution. The deaths of her father and mother in the years immediately following John Hubble moving the family to Chicago had left Jennie Hubble without a support network in the south apart from her brother Jefferson Beauregard James, Edwin's favorite uncle. The sudden death of his beloved grandfather, who had done so much to pique his interest in the stars, from an apparent heart attack in the fall of 1900 had been a blow to Edwin and upset the balance in the family.[40]

Around 1906 the Hubble's moved to a large clapboard home at 606 North Main Street, the last home they would have prior to their move back south. For the children this was the setting for some of their fondest memories. The big two-story home now housed the entire clan, including two new sisters, Emily Jane, born in 1902 and Elizabeth in 1905.

Life maintained its usual course during Edwin's visits home from college. He sang in the church choir on Sundays and attended mass at First Baptist Church. Apart from church, weekends at the Hubble home were often filled with laughter and games. From checkers to charades to spelling matches there was always something going on. As they grew older, young adult entertainment became available in the form of hayrides and fairs, and the nearby nickelodeon was a favorite. When science fairs and demonstrations were offered at Wheaton College, Edwin and his brothers liked to go and see what new wonders were on display.

The inner drive that was seeded in his high school years remained in Edwin through his undergraduate years. He still paid little or no attention to girls except for the occasional flirtation. He was on a personal mission and prolonged distractions were unwelcome. His seclusion from daily family life while at school had loosened these restraints to a degree, but an inherent arrogance and imperiousness remained. Bill Hubble was said to be more emotionally available and intellectually approachable than his overachieving older brother. But this mattered little to Edwin, who remained close to his brother despite his preferential treatment from their classmates. In the summer the brothers worked cutting and hauling ice blocks to the homes of Wheaton. The money was good and the gratitude of residents whose ice boxes were filled with the giant frozen blocks was paid in cookies and other treats.

In the summer between his junior and senior years of high school, Edwin's father set up a job for him on a surveying crew in Wisconsin, a memory he would cherish. After a course called "Introduction to Surveying" in his sophomore year at Chicago, Edwin found well-paying and stimulating work during the summers of 1908 and 1909 making surveys for the Chicago, Burlington and Quincy Railroad Company.

While an undergraduate, Edwin still managed to find time for a quick trip south to visit his paternal grandparents. He exchanged correspondence with Martin Hubble throughout his college years and delighted in extolling to the aging family patriarch the knowledge he was gaining of the stars.[41]

By 1910, as he prepared for his departure to England, Edwin Hubble was starting to form the social and political opinions and convictions that would inform his rhetoric and actions throughout his adult life. At the time the United States was beginning to show its influence on the world stage. Capitalizing upon its naval victory in the Spanish-American War, the country had just opened a naval base at Pearl Harbor on the Hawaiian island of Oahu. The Great White Fleet of 16 new battleships and a convoy of escorts had finished its 14-month round-the-world trip in February 1909 on a mission to introduce American naval power to various "friendly" nations. It had been authorized by President Theodore Roosevelt, whose penchant for action Edwin could relate to even though Roosevelt was a dreaded republican.[42]

Other American technological progress was evident at that time. The Queensboro and Manhattan Bridges became the third and fourth suspension bridges to span the East River into New York City from the boroughs of Queens and Brooklyn, respectively. The city of Detroit welcomed the Hudson Motor Car Company, which officially opened for business. The fledgling automaker would grapple for market share with Ford Motor Company which had recently introduced the first low-cost automobile, the Model T, and was turning them out in record numbers on Henry Ford's new assembly line manufacturing system.[43] The burgeoning scientific scene brought the renowned neurologist Sigmund Freud to America that year, where he gave the only lecture on psychoanalysis in the country to students and faculty at Clark University in Worcester, Massachusetts.[44]

For Edwin Hubble, heading to England as a representative of his country was an honor the earnest collegian welcomed with energy, purpose, and seriousness. But while he openly committed to focusing on the study of law while at Oxford, he was eager to maintain his connection to astronomy, still clinging to his dream of embarking on a career in the field. He was keenly aware of developments in science and technology in general and astronomy.

By this time, George Hale's new state-of-the-art observatory in the Sierra Madres was home to some of the most technologically advanced equipment of the age, including a solar telescope set on a 60-foot tower for atmospheric stability. Hale had discovered evidence of the Sun's magnetism two years earlier by his research with the new instrument. The same year, in 1908, the 60-inch reflector saw its first light. It was the largest research telescope in the world and there were reports of an even larger reflector with a 100-inch mirror under construction at the observatory.[45]

The furious development underway by Hale and his associates fueled Edwin's dream of joining their ranks and simultaneously sank his spirits as he reflected on his father's wishes. All he could do now though was bide his time and wait for his opportunity, should it come.

As if he needed further incentive to stay his course, the year 1910 ushered in the arrival of a much-anticipated visitor and the unanticipated arrival of another. The Great Daylight Comet, so-called because its brightness enabled it to be seen in the twilight sky, burst into visible range of the Earth over the Southern Hemisphere in January and made its way into the Northern Hemisphere toward the end of the month.[46]

Then in April there was Halley's Comet. Edwin noted the death on April 21st of Mark Twain, one of his favorite authors in childhood. Twain had been born in 1835, the year that this comet had last flown by the Earth, prompting the great American author and satirist to observe in his autobiography, "The Almighty has said, no doubt: 'Now here are these two unaccountable freaks; they came in together, they must go out together.'"[47]

An article in the New York Times on February 7th mentioned a report from Edwin Frost at the Yerkes Observatory that cyanogen bands had been found in the more famous comet's tail by spectroscopic observations taken at the 40-inch refractor. With the Earth on a course to pass through the comet's tail in May and cyanogen a lethal toxin, Camille Flammarion, a French astronomer, and author of dubious scientific and cult-like literary repute, suggested that the cyanogen would be absorbed in the atmosphere and extinguish all life on the planet. The speculation set off a wave of paranoia spurring runs on gas masks, so-called anti-comet pills and umbrellas, and various other scams. Reassurances from more respected members of the astronomical community that the cyanogen would be distributed so thinly across the globe as to have little or no impact on life soon quelled the hysteria.[48]

In May, the Hubble family, like many others across the nation, enjoyed the magnificent light show of the comet, the broad tail extending far behind its bright white head as it made its way through the night sky.

The summer of 1910 was the last the Hubble family would spend together. John Hubble, increasingly laid up by chronic malaria, had moved the family south to the quiet community of Shelbyville, Kentucky, a half hour commute from Louisville by streetcar. Edwin's oldest brother Henry had dropped out of school after a year at Wheaton College and was working with his father in the insurance business. Bill was home from his first year at the University of Missouri where he was studying agriculture, intending to get into farming after school. Lucy was home helping her mother with the younger children and housework. The family and guests spent Sunday afternoons celebrating with food and song. By September, Edwin would be on his way to England with his spirits and hopes held high.[49]

* * *

Staring at that same twilight sky in May of 1910 under which Edwin Hubble was dreaming of a future in astronomy, Milton Humason sat at the edge of the granite cliffside ridge known as Echo Rock several hundred yards from the dome of the 60-inch telescope on Wilson's Peak. The cool breeze bore the anticipation of the coming darkness and the spectacular tail of Halley's Comet. This might have been the climactic event in the life he had lived for the past seven years on the mountain, whose every waking day brought a tingling sensation and universal presence of being ensconced within nature's constant embrace. It was a life that began not long after his family's arrival.

When the Humasons arrived at the Arcade Depot in the heart of downtown Los Angeles in 1902, beleaguered from their weeklong 2,000-mile journey, they were almost certainly met by Laura Humason's sister, Alice Witmer, her husband Henry, and their three-year-old son Joseph. The five-story-tall wooden depot was 500 feet long and 90 feet wide and was considered the finest train depot on the Pacific Coast. The waiting area was brightly lit with sunlight pouring through the skylights that lined the roof.[50]

Luggage, personal effects and whatever else they could transport on the journey to their new life, was likely loaded onto one of the waiting dray wagons and sent to the Witmer's address on 1422 West Third Street in the beautiful Crown Hill section of the city, two miles west of the station. From there the two families may have climbed onto one of the Witmer's Second Street Cable Cars, which could accommodate them all comfortably and would offer a brief tour of the city on the way to their new home. Henry Witmer sold the company that year, anticipating the arrival of the more efficient electric street-car system that was being used in San Francisco to the north.

The Witmer mansion, one of three large family homes built on the hill by Henry Witmer, his brother Joseph, and their brother-in-law Samuel Lewis, had more than enough bedrooms to hold both families until William Humason managed to find a suitable home of his own. The large family home had servants' quarters, a kitchen, reading, dining, living and family rooms and a huge porch area front and back that overlooking an expansive park and a pond nestled in the rolling hills.

Uncle Clate was a tall, confident, and savvy businessman in his mid-forties whose gentle, unassuming spirit won him the respect and admiration of the entire community. Since his arrival in the area with his brother in the 1890s he had slowly built on his family's estate, starting one of the first state banks in the city, buying and developing land and investing in the burgeoning oil industry. He was a trustee of the local chamber of commerce and a positive, progressive voice for the development of the city. With such a diverse portfolio of holdings and his business and civic responsibilities, Witmer was doubtless very happy to have his brother-in-law's help keeping the books and managing various aspects of the operation of the family business.[51]

The favorite weekend escape for the Witmer and Humason clans was to visit the Witmer ranch in Lordsburg, near the city fairgrounds. The pride and joy of the Witmer family, the ranch was lined with orange and apple groves and yielded smaller crops of berries and other fruits and nuts. The farm had a stable, corral and trails running through the forested foothills of the San Gabriel Mountains which were part of a confluence of mountain ranges that includes the Sierra Madres extending northwest from Mexico, the Sierra Nevada range of coastal California and the Rocky Mountains to the east. From the ranch, Mount Wilson could be seen 25 miles northwest, as could Mount Baldy 15 miles northeast.[52]

With his finger on the pulse of the goings-on within the community, Henry Witmer was in an excellent position to suggest interesting adventures for Milton and his younger brother Lewis. Lewis was the more studious and engaged in his schoolwork. Although Milton was an avid reader and appeared more than capable, he proved disinterested in schoolwork from an early age.

So it was that in the summer of 1902, at his uncle's suggestion, Milton had made his way up the mountain trail to Strain's camp on the back of one of old man Barrett's infamous mules. The mules and their handlers had a local reputation for the ranker of the beasts and the colorful language of the cowboys as they handled them. Despite the tongue they often used in addressing their herd, the cowboys were known to take great care of the animals and understood their individual capabilities. On one occasion, one of the strongest of the mules in the corral took a 225-pound bathtub up the old trail to the halfway house.[53]

Particularly tuned to the natural world, Milton took to the skills and activities involved in mountaineering and outdoorsmanship with enthusiasm. The rugged wilderness evoked images and stories of the Old West, even as the last of the notorious gunfighters of the era had just met his end. Harry Tracy, perhaps the most murderous of all the gunfighters, had his name in the headlines of newspapers around the country that year. Having escaped an Oregon prison with an accomplice in June, Tracy reportedly killed as many as a dozen men in fleeing one posse after another, finally meeting his doom from a self-inflicted gunshot wound as a posse closed in on the 6th of August in Creston, Washington. The more famous Butch Cassidy and his friend and accomplice, the Sundance Kid, who Tracy claimed to have run with for a time in the 1890s, committed their last robbery two years prior and were reported to have fled to South America to escape capture by Pinkerton Detectives.[54]

Such stories captivated the boys and young men that combed the wilderness on Mount Wilson in those years and were the source of some outdoor theater as Milton and friends play-acted the roles of the gunmen and those who brought them to justice. These elaborate sketches lasted well into his teenage years.[55]

During those first few years, Milton learned that to survive life on the mountain he had to understand his surroundings. Black bear, rattlesnakes and cougars had to be treated with respect to avoid an early exit from the world. No record of his time at the camp survives but his love of Mount Wilson is apparent in the fact that he returned there each year during summer breaks from school. When the hotel opened in 1905, he dropped out of high school after just one day to join the hotel staff as a clerk.[56] From a young age he was headstrong and decisive and the only future he wanted any part of was awaiting him on his new mountain home.

His parents were not necessarily willing participants in Milton's decision regarding his education. "Like his future wife Helen Dowd, the granddaughter of a published author of English grammar, both of Milton's parents had received thorough educations."[57]

Initially, Milton spent most of his days off from the hotel at the Humason family home on 1345 Carroll Avenue, where he played big brother to Lewis and "Ginny." Family visits to the Lordsburg ranch with the Witmers were a favorite excursion, as were visits to their home on Crown Hill and the occasional picnic at Echo Park or the sequoia forests north of the city. Despite likely frustration at the decisions their oldest son was making, the adults in both families recognized and extolled the virtues of family life in their children. Milton could never have been confused with a person having a deep understanding of Taoism, but he seems to have had a most existential outlook on life and lived very much in the moment. He was not given to dreaming of the future any more than he reminisced about the past. He was possessed of a sharp, earthy wit and a quiet disposition that when combined with his love of storytelling and practical jokes won him friends easily.

Joining the muleskinners had always been the target for Milton, a group he ascended to in 1907 by lying about his age to start months before his 16th birthday. His hiring coincided with the point at which the construction of George Hale's state-of-the-art observatory was nearing its high point (Fig. 3.5).

Hale and his chief instrument designer George Willis Ritchey recognized the superior light gathering capability and cost effectiveness of reflecting telescopes vis-à-vis refractors using the photographic results from the 24-inch reflector Ritchey had built at Yerkes years earlier. With the small reflector outpacing the quality of its larger and much more expensive cousin, Hale and his team decided the 60-inch glass disk that his father had made available years before would be best utilized as the primary mirror in a modern reflector. A reflecting telescope of this aperture would offer many times the light gathering ability of the Yerkes refractor in a dome just two-thirds the size.[58]

It was a daunting challenge to build a major facility atop a mountain in a remote area of the country accessible only by a steep and narrow 10-mile native trail and having no tools, supplies or utilities except those you carry up the mountain yourself.

Into this furious frenzy of unprecedented industry, Milton Humason was thrown in the summer of 1907, the youngest member of a team of cowboys whose job was of the utmost importance in the transport of men, materials and supplies to the observatory grounds. For one thing the old trail had to be widened to support the largest pieces of the telescope tube and other assemblies. An alternative route to the summit was soon revealed to offer more suitable and more stable ground for expansion and work was begun immediately to widen it. Teams of workers were brought in with shovels, picks, pry bars, axes, wheelbarrows, and dynamite to dig, blast and smooth the mountain floor into a surface suitable for the transport of Hale's precious cargo. Sometimes torrential rains caused landslides which stopped work and created extra work because the roads had to be cleared and regraded. But the road was complete by May of 1907 and its opening was celebrated by driving a Franklin automobile along its twisting path to the top of the mountain.

At the summit, the dome was being constructed by a team of welders and steelworkers assembling the steel beams and cables using a DC electric

Fig. 3.5 Milton La Salle Humason wearing his favorite Mountie hat ca 1909. (Courtesy, Ann Humason Bernt family)

generator to operate all manner of welders, drills, grinders, and other tools. The almost 60-foot-diameter building would be faced with sheet metal around its base and dome, then a steel tube armature was added two feet from the metal surface of the dome for the application of its white canvas cover. The air space between the cover and the metal beneath would be essential for maintaining cooler air temperatures in the dome. The massive cement pier that would hold the telescope mount had to be poured and finished before the first parts of the assembly could be hauled up the mountain.

The telescope itself, being built at the Union Iron Works in San Francisco, was nearly destroyed the year before during the great earthquake of 1906 that burned much of the city, destroying buildings, and killing over 3,000 people. Somehow, work was continued, and by late 1907 the first elements of the base were heading to the summit. Among the first major pieces to arrive at the dome were the seven-ton triangular base (so large it had to be brought to the observatory in two pieces and welded together on the pier), the 15-foot, four-and-a-half-ton polar axis on which the telescope would track the stars, and the five-ton cast iron fork to grip the telescope itself. Next was the four-ton steel float 10 feet in diameter (along with the 650 pounds of mercury which would support 95 percent of the giant's 22-tons of moving parts) and the two-ton drive gear. And finally, there was the mirror, with its surface shaped and polished to an accuracy of millionths of an inch by Ritchey in a specially fitted room over a period of three years.

In all, the telescope and dome equaled 150 tons of weight, nearly all of which had to be hauled up by mule teams, donkeys, and horses. A special motorized truck with an onboard DC generator powering an electric motor at each wheel was created to help carry the largest pieces of the telescope, including the six-and-a-half-foot diameter, 18-foot-long steel tube. The vehicle failed to live up to expectations when hauling the largest loads and had to be harnessed to mule teams to aid in its ascent[59] (Fig. 3.6). Hailed as a marvel of modern machinery, it seemed like little more than a kind of noisy, gas-drinking, oil spitting beast. As far as "Milt" and the other cowboys were concerned, it may have eaten differently, but it was still just another stubborn mule.

The cowboys piloting the innumerable expeditions of mule teams in those days were in high demand and regarded with sincere respect by the men who worked at the observatory, Walter Adams among their most spirited and vocal champions. Not just the materials that formed the dome, but all the manpower and supplies to sustain them had to make their way up the mountainside. Most of the men were too big and heavy for a mule to carry and were given horses to ride up the trail.[60]

The 60-foot tower solar telescope was finished in 1908, the same year the big reflector saw its first light, and it wasn't long after the solar telescope went into regular use that Hale made his now famous discovery of the magnetism in sunspots. Reporting in the yearbook of the Carnegie Institution that year

Fig. 3.6 Metal tube of the 60-inch telescope being transported along a road on Mount Wilson ca. 1908 (Image courtesy of the Observatories of the Carnegie Institution for Science Collection at the Huntington Library, San Marino, California)

he proclaimed the "existence of magnetic fields in sun-spots…has been placed beyond doubt" and explained that "sun-spots are electric vortices."[61]

The early results from the 60-inch reflector provided insights into objects that had long fascinated astronomers, in particular "a remarkable spiral structure near the center" of M31, the Andromeda Nebula.[62]

A new means of photographing the spectral lines made by the spark from various metals was being used for comparison with "the spectrum of the Sun's limb" and other objects. Spectacular new images of the towering plumes of gas shooting out from the solar surface were displayed in the yearbook. Although the scientific revelations were compelling, the dangers of the equipment were evident as well. A stout 35-year-old University of Chicago physics professor named Henry Gale was put in charge of comparing field studies made by Adams of the Sun's limbs with spark spectra in the laboratory. While experimenting, Gale was severely burned when he somehow "came into contact with the high-potential wires of the transformer used to produce the spark." He took several months to recover and was back the following year to pursue the experiments further.[63]

Living and working on the mountain brought Milton into close contact with the men at the observatory. As part of his public outreach program, Hale offered guests regular night viewings at the 60-inch, where a member of the observatory would guide visitors through a beginner's crash course in observing methods and the capabilities of the telescope. Other smaller telescopes were also set up near the hotel to enable guests to view the stars.

Adversity struck when the "monastery," as the observatory dorms were known, burned to the ground in December of 1909, only months before Andrew Carnegie, the observatory's chief benefactor, was to arrive for a tour of the grounds and buildings. With the proficiency that was a hallmark of the organization, a new, larger, and much more modern facility was constructed in its place with the capacity to accommodate both solar and stellar observers as well as a library and reading room.[64]

Milton was regularly exposed to the men, the machines, and the objects of their attention in the skies above California at Mount Wilson Solar Observatory (MWSO) in those years. He was present when the cornerstones were laid and was a proud member of the roughneck cowboy clan responsible for getting anything and everything relevant to the building of the facility to the top of the mountain safely. His superior ability as a horseman, friendly nature and swift decision-making helped him rise quickly through the ranks of the cowboys. His long association with the mountain had made him intimately aware of every foot of it from base to top and east to west, earning him a measure of appreciation from the astronomers as well.

When not working the trailhead, his days were often spent riding and hiking its many trails, fishing its streams, and swimming in the nearby water hole down the hill on the West Fork River. A conservationist by nature he had rescued a fox kit whose mother had fallen from a ledge, perhaps trying to elude a predator, or hunting prey and he nursed the youngster to adulthood before releasing it to fend for itself.[65]

There were bonfires and cookouts, jamborees, and dances at the Mount Wilson Hotel throughout the year. It was at one of these dances that Milton had met the woman of his dreams. She was now sitting beside him at Echo Rock with the same look of wonder in her eyes as she gazed at the strange comet in the evening sky.

Helen Dowd, the daughter of Merritt C. "Jerry" Dowd, was the same age as Milton. She was probably the only woman the young muleskinner had ever really known outside of his family. Whether it was the sight of the young man on horseback looking every bit the 19th century cowboy or his boyish charm, something about him captured the young girl's heart and they became sweethearts. It wasn't long before Milton, always one to follow his bliss, began contemplating marriage. Once again things appeared to be falling into place for the young muleskinner until suddenly the decisions, he had made earlier in life came back to bite him (Fig. 3.7).

Fig. 3.7 Photograph of Helen Dowd prior to her marriage to Milton Humason ca. 1910 (Courtesy, Ann Humason Bernt family)

Jerry Dowd was Mount Wilson's chief electrician and engineer. Tall, highly intelligent, rangy, mustachioed, and ornery as a twice stuck hog, he had worked hard to provide a life for his family, and he intended to see his daughter married to a man who could do likewise. He perceived the impending dead end in Milton's future, even if the love-struck youngsters could not.

Although construction on the mountain was still in full swing with the 100-inch Hooker reflecting telescope under development and the 150-foot solar tower telescope completed that year, Dowd knew the end of the line was approaching for the cowboys. The widening of the road, necessary for hauling the massive new reflector's tube and other parts to the summit, was underway, and auto and truck technology was improving. In fact, in the not-too-distant future, Dowd figured, the road from base to summit would be paved, an event that wouldn't occur until 1935. But you get the idea. His daughter's intended was destined to be jobless soon and there was no way he was going to allow her to marry a young man with no prospects. He let them know his opinion in no uncertain terms.

This presented a problem whose solution was not immediately apparent. Milton had no formal training in a trade, no education and really had never even entertained the notion of attaining either. Because of that, he was being denied the hand of his beloved by his would-be father-in-law. He did have one thing going for him though: family.

Tragedy appeared to come in 10-year cycles to the Humason and Witmer families, and 1909 was particularly devastating. The death of Henry Clayton Witmer, Milton's uncle, in March of 1909 at age 52 left behind his widow, Alice, and their 10-year-old son, Joseph. His passing put the family's future in doubt.[66] The only able-bodied male left with intimate knowledge of the workings of the vast holdings of the Witmer Brother's Company was William Humason. The families needed to rally together and once again they did. By the time the 1910 census was being taken, William, Laura, Lewis, and Virginia Humason were once again living at the Witmer home on Third Street with Alice and Joseph Witmer. William was recorded as the head of the household, as was customary at the time.[67] On learning of her brother-in-law's death Josephine Witmer, the widow of Henry's younger brother Joseph, crossed the country by train to visit and introduce her college-aged sons William and David to the family. The infusion of youth along with the addition of one or two of the Witmer women would eventually help to fill the void left by Henry Witmer's death.

A photograph taken around the time of Henry Witmer's death was included in Virginia Comer's book *In Victorian Los Angeles: The Witmers of Crown Hill*. A copy of the same picture also found its way into the Humason family photo archives. In it, members of both families stand around a picnic area on the farm in Lordsburg with a large tent in the background and an outdoor stove in the foreground. It recorded the inflection point where the fortunes of the Witmer empire were handed to the heirs of Joseph Witmer. The two families would avail each other for the next generation as they banded together to help propel the Witmer family businesses forward.[68]

That year, Milton was living at the base of the mountain near the trailhead. After some discussion he had decided to take work with his cousin Letha Lewis at the citrus farm she ran with her husband Samuel Lewis, a Massachusetts engineer. The plan was that he would learn the ropes of the Witmer and Lewis family farms and, in time, move up to the position of foreman in charge of operations. A future.

Now, as he gazed at Helen on that cool spring night at Echo Rock on Mount Wilson, in his favorite spot next to his favorite girl, Milton likely gave little thought to the struggle he faced gaining a foothold in the valley below. He had been wired from birth to roll with the changes, he had every confidence in his abilities, and he was determined to win Helen's hand.

References

1. Gale E. Christianson, *Edwin Hubble: Mariner of the Nebulae,* (University of Chicago Press 1995), pgs. 19-20
2. *Ibid.* pgs. 5-6
3. Quinn Malloy, *New Data on an Old Disgrace: Missouri had second highest number of lynchings outside Deep South,* (Missourian July 2nd, 2017), Retrieved Aug. 18th, 2019
4. Gale E. Christianson, *Edwin Hubble: Mariner of the Nebulae,* (University of Chicago Press 1995), pgs. 4-6
5. *Ibid.* pg. 10
6. *Ibid.* pgs. 7-9
7. *Ibid.* pgs. 10-12
8. *Ibid.* pg. 16-18
9. *History of the Frisco,* (thelibrary.org 2020), Retrieved April 12th, 2020, https://thelibrary.org/lochist/frisco/history/1962history.cfm
10. Gale E. Christianson, *Edwin Hubble: Mariner of the Nebulae,* (University of Chicago Press 1995), pgs. 14-5
11. *The History of Wheaton, Illinois,* (www.oldplaces.org), Retrieved Feb. 8th, 2021, http://www.oldplaces.org/dupage_county/wheaton.htm
12. *Bird's-Eye View of the World's Columbian Exposition, Chicago, 1893,* (World Digital Library 2021), Retrieved Feb. 8th, 2021, https://www.wdl.org/en/item/11369/

13. Gale E. Christianson, *Edwin Hubble: Mariner of the Nebulae,* (University of Chicago Press 1995), pg. 14
14. Walter S. Adams, *Early Days at Mount Wilson,* (Publications of the Astronomical Society of the Pacific October 1947), pgs. 213-214
15. Interview with Ann Humason Bernt, 2006
16. *Ibid.*
17. *Ibid.*
18. Paul R. Spitzzeri, *At Our Leisure: Strain's Camp, Mount Wilson, 1904-1907,* (The Homestead Blog 2017), Retrieved April 30th, 2020, https://homesteadmuseum. blog/2017/07/16/atourleisure-strains-camp-mount-wilson1904 1907/
19. Humason, W G, bkpr Western Elevator Co, res 77 E. Wabasha, Winona City Directory:1902, Retrieved 2007
20. *Steamboat Times: A Pictorial History of the Mississippi Steamboat Era,* (steamboattimes.com), Retrieved Feb. 8th, 2021, https://steamboattimes.com/rafts.html
21. *Places of the past: 57 historical photos showcasing Winona history,* (Winona Daily NewsOnline), Retrieved Jan. 9th, 2019, https://www.winonadailynews.com/ news/local/placesof-the-past-57-historical-photosshowcasing-winona-history/ collection_69ecd0ec-2bcb-573f-a2bc-173767dc24e6.html#1
22. U.S. Census Reports for 1870 and 1880, City and County of San Francisco, California, Enumerated July 18th, 1870; June 5th, 1880, (ancestry.com), Retrieved, October 16th, 2007
23. *Archive for the 12th Minnesota, Company C: Humason, William G.,* Minnesota American History and Genealogy Project, (sites.rootsweb.com/-mnahgp 2005), Retrieved September 2nd, 2006
24. Virginia Linden Comer, *In Victorian Los Angeles, The Witmers of Crown Hill,* (Talbot Press 1988), pg. 68
25. Letter to Thomas A. Humason jr., Sales at Harcourt, Brace and Company, Inc., New York from Milton Humason, October 24th, 1950
26. Virginia Linden Comer, *In Victorian Los Angeles, The Witmers of Crown Hill,* (Talbot Press 1988)
27. Paul R. Spitzzeri, *At Our Leisure: Strain's Camp, Mount Wilson, 1904-1907,* (The Homestead Blog 2017), Retrieved April 30th, 2020, https://homesteadmuseum. blog/2017/07/16/atourleisure-strains-camp-mount-wilson1904 1907/
28. Nat B. Read, *Don Benito Wilson: From Mountain Man to Mayor,* (Angel City Press 2008)
29. Paul R. Spitzzeri, *At Our Leisure: Strain's Camp, Mount Wilson, 1904-1907,* (The Homestead Blog 2017), Retrieved April 30th, 2020, https://homesteadmuseum. blog/2017/07/16/atourleisure-strains-camp-mount-wilson1904 1907/
30. Julie K. Rose, *The World's Colombian Exposition: Idea, Experience, Aftermath,* (University of Virginia 1996), http://xroads.virginia.edu/~MA96/WCE/ legacy.html
31. Museum Facts page, Museum of Science and Industry, Chicago, https://www. msichicago.org/explore/about-us/museum-facts/

32. Gale E. Christianson, *Edwin Hubble: Mariner of the Nebulae,* (University of Chicago Press 1995), pgs. 30-35
33. *Ibid.* pg. 38
34. *Ibid.* pg. 41
35. *Ibid.* pg. 27
36. Robert Millikan letter to Edmund James, Jan. 8th, 1910, HUB, Box 18
37. Robert A. Millikan, *The Autobiography of Robert A. Millikan,* (Prentice-Hall, Inc. 1950)
38. Gale E. Christianson, *Edwin Hubble: Mariner of the Nebulae,* (University of Chicago Press 1995), pg. 58
39. *Ibid.* pg. 21
40. *Ibid.* pgs. 28-45
41. *Ibid*
42. Mike McKinley (Journalist Second Class), *Cruise of the Great White Fleet,* (Naval History and Heritage Command Online 2017), Retrieved April 17th, 2020, https://www.history.navy.mil/research/library/online-reading-room/title-listalphabetically/c/cruise-great-white-fleet-mckinley.html
43. *1900 to 1909 Important News, Significant Events, Key Technology,* (The People's History Online), Retrieved January 23rd, 2019, http://www.thepeoplehistory.com/1900to1909.html
44. Kendra Cherry, *Important Dates in the Life of Sigmund Freud,* (verywellmind.com 2020), Retrieved February 10th, 2021, https://www.verywellmind.com/sigmund-freud-timeline2795846
45. *Timeline: Historical Highlights From Mount Wilson Observatory's First 100 Years,* (mtwilson.edu), Retrieved Feb. 10th, 2021, https://www.mtwilson.edu/timeline/
46. *Not Much is Known of Daylight Comet,* (New York Times, Jan. 30th, 1910), Retrieved April 18th, 2020
47. Justin Kaplan, *Mr. Clemens and Mark Twain: a biography,* (Simon and Schuster 1966), pg. 386
48. *Comet's Poisonous Tail: Yerkes Observatory Finds Cyanogen in Spectrum of Halley's Comet,* (NY Times Tuesday, Feb. 8th, 1910), Retrieved Aug. 7th, 2019,
49. Gale E. Christianson, *Edwin Hubble: Mariner of the Nebulae,* (University of Chicago Press 1995), pgs. 60–61
50. Nathan Masters, *Lost Train Depots of Los Angeles History,* (kcet.org Jan. 17th, 2013), Retrieved August 20th, 2019, https://www.kcet.org/shows/lost-la/lost-train-depots-of-los-angelesHistory
51. Virginia Linden Comer, *In Victorian Los Angeles, The Witmers of Crown Hill,* (Talbot Press 1988), pgs. 29-41
52. *Ibid.* pgs. 61-64
53. Walter S. Adams, *Early Days at Mount Wilson,* (Publications of the Astronomical Society of the Pacific October 1947), pg. 216

54. J. Mark Powell, *Meet the Old West's Last Gunfighter,* (jmarkpowell.com 2015), Retrieved April 19th, 2020, http://www.jmarkpowell.com/meet-the-old-wests-last-gunfighter/

55. Interview with Ann Bernt Humason, family photo album, Retrieved 2006

56. Letter to Jocelyn Jurkovich of Sacramento, California from Bill Humason, October 23rd, 1985

57. J.N. Dowd, *English Grammar: Exhibiting A New and Improved System of Teaching Grammar in the Form of Questions & Answers with Explanatory Examples,* (Edwin T. Greenfield 1830), Archives of Ann Humason Bernt, Retrieved 2006

58. G.W. Ritchey, *The Two-Foot Reflecting Telescope of the Yerkes Observatory,* (The Astrophysical Journal Nov. 1901), pgs. 217-235

59. George E. Hale (Director), *Carnegie Institution of Washington Yearbook, 1907: Mount Wilson Solar Observatory,* (CIW 1907), pgs. 146-152

60. Walter S. Adams conversation about the early days at Mt. Wilson [sound recording], circa 1918. by Adams, Walter S. (Walter Sydney) 1876-1956, Call number: AV 5-70-3; 2013-1223

61. George E. Hale (Director), *Carnegie Institution of Washington Yearbook, 1909: Solar Observatory,* (CIW Feb. 1910), pg. 159

62. *Ibid.* pg. 160

63. *Ibid.* pgs. 165-167

64. *Ibid.* pg. 176

65. Interview with Ann Humason Bernt, family photo album, Retrieved 2006

66. Virginia Linden Comer, *In Victorian Los Angeles, The Witmers of Crown Hill,* Talbot Press 1988), pg. 75

67. Thirteenth Census of the United States: 1910, State of California, County of Los Angeles, Los Angeles Township, Precinct 62, City of Los Angeles, Enumerated April 20th, 1910, (ancestry.com), Retrieved July 16th, 2005

68. Interview with Ann Humason Bernt, family photo archive, 2006

4

Per Aspera Ad Astra (1910–1917)

Between 1910 and 1917, Edwin Hubble and Milton Humason face events that will shape the remainder of their lives. A bit of strategic gamesmanship has secured Edwin a Rhodes Scholarship and a trip to Oxford, where he continues to pursue his dream of becoming an astronomer in the shadow of his father's indomitable spirit. John Hubble's unexpected death in 1913 opens the possibility that Edwin might pursue a graduate degree in astronomy. But he must weigh personal ambition against the obligation of providing for his widowed mother and school-aged sisters. In a decision that will set the course for the rest of his life he takes the selfish route, but his plans are again put on hold when war breaks out in Europe. After years of living and working on Mount Wilson as a hotel clerk and muleskinner, Milton Humason falls in love with the daughter of the establishment's head electrician and leaves the mountain to pursue a new life as a citrus rancher. Seven years later he returns to the mountain as janitor and groundskeeper just as the 100-inch telescope sees its first light. From there he begins his astonishing reinvention as a stellar photographer and astronomer.

Sitting at his desk in Queen's College at Oxford in England, Edwin Hubble studied with grave intensity the telegram in his hand. It was from his mother back home in Louisville, Kentucky, where the family had lived for the past two years. His father, 52-year-old John Powell Hubble, had died in his bed on Sunday, January 19th, 1913, of kidney failure brought on by chronic malaria and possibly Bright's Disease or acute nephritis. The grieving son's mind was spinning as he processed his mother's words.

The 17th century astronomer Edmond Halley had once walked the very corridors he had called home since his arrival in the fall of 1910. It was Halley's

© Springer Nature Switzerland AG 2021
R. Voller, *Hubble, Humason and the Big Bang*, Springer Praxis Books,
https://doi.org/10.1007/978-3-030-82181-4_4

namesake comet that Edwin and family had watched from outside their home the last time they were all together. He remembered the happy hours they had spent singing songs accompanied by his sister Lucy at the family piano or meeting with friends and family during those final summer months.

As John Hubble's condition worsened, Edwin had written home asking permission to return to his father's deathbed but the ailing patriarch, wanting to be sure his son stayed in England to finish his degree, denied the request. This exchange strangely illustrated again the disparity between a father's earnest desire to see his son succeed as he had been unable to do, versus a son's desire for autonomy and self-determination. Swirling around amid the cauldron of misery and dread in the child's mind at learning of the loss of his father were the negative emotions left unassuaged, the arguments not pursued, the anger and frustration held in reserve. This intermingled with a very real sense that his father's heart was usually in the right place, even when it ran afoul of Edwin's personal ambitions.

Whatever his shortcomings, John Hubble had been a provider, a mentor and an able steward to his wife and children and his passing carried with it the awesome specter and gravity of life and death. As childhood propriety was overcome by self-reproach, and as unwavering hope turned to momentary despair, Edwin took pause to contemplate his next course of action. Would he honor his father's wishes or would be follow his own dream?[1]

He had been brimming with confidence when he boarded the steam ship Canada bound from Montreal for England two and a half years earlier. Even then, as a Rhodes Scholar in command of his collegiate career, he could hear his father's voice in his head warning him not to fall prey to the temptations of the flesh and drink awaiting him down life's many darkened alleyways.[2]

Queen's College was the school chosen by Edwin out of the more than two dozen at Oxford at that time (another dozen or so have since been added). Named in honor of Queen Philippa, wife of King Edward III, the school was founded by chaplain Robert de Eglesfield in 1341 and boasted both front and back quadrangles, a library and a chapel located at the center of the campus between the quads.[3] It owed its architecture to one of the great English architects, the 17th century polymath Christopher Wren who designed St. Paul's Cathedral in London. The customary trumpet call signaling dinner was a favorite daily occurrence for the young postgraduate, as were other traditions at the school.[4]

Eager for acceptance, Edwin immediately began to adopt an English accent in an attempt to fit in with members of his newfound community. Letters home were soon peppered with bits of English expression. Instead of being a guy, someone he met was a "chap" or a "splendid fellow" and they were off on weekends having a "ripping good time." Although he didn't immediately take to every fashion trend that the country had to offer, by his departure he was often dressed in billowy legged, calf length, plus fours and a tweed Norfolk jacket, which he topped off with a flat cap, black cape, and cane in the cooler months.

But despite his attempts to endear himself to the young Englishman at the college, it was a while before he gained their acceptance and respect. This was formed out of Edwin's physical and intellectual prowess. Apparently, he was one of the few at the university who could read the old texts in German, Greek, and Latin. The men around the observatory there were likewise amazed that Hubble, ostensibly on campus to study law, was so familiar with past and present astronomy.

A natural athlete, by his 23rd year he cut quite a physical form at 190 pounds, and it was just a matter of time before the handsome young American won favor with the home crowd. He joined the rowing and track teams and even captained an intramural baseball team that captured the attention of many of the cricket playing Brits on campus.

By the beginning of his second year at the college a large crowd of like-minded young men from several countries had formed a lasting kinship and Edwin was one of those at the head of the group. They traveled extensively throughout Europe during the many extended breaks within the school year, engaging in impromptu singalongs with Edwin occasionally accompanying them on a mandolin that he had acquired in Germany and taught himself to play. He briefly attempted to earn letters, first in rowing and later in track but was largely unsuccessful, achieving a half a blue (letter) in track in his first year. Although his native athleticism would remain an asset for years to come, he had been surpassed by many of his peers and no longer assigned a high priority to competition at the highest levels of sports. Whether or not he returned home after his studies at Oxford, Edwin had fallen in love with England and his adopted country had responded in kind.[5]

–oOo –

At this moment in history, Europe was the center of Western civilization and England was the capital of its commerce and culture. Upon his coronation in 1910, King George V, the grandson of Queen Victoria, ruled over an empire that included the United Kingdom, the Indian Empire, East and West Indies, South Africa, Ceylon (Sri Lanka), Malaya and other nations.[6]

Like many European countries, the heritage of England was steeped in monarchic rule tempered by the arts, including some of the legends of literature of the day with Rudyard Kipling, H.G. Wells and Thomas Hardy being among Edwin's favorite authors.

The disgrace at the tragic loss of the RMS Titanic on its maiden crossing of the Atlantic in 1912 with the loss of more than fifteen hundred passengers and crew had not dampened the unshakable spirit of the English people.[7]

England's reach was being felt back home as well as in Europe. It had introduced the world to the sport of golf with the Open Championship in 1860 and tennis with the opening of the Wimbledon Championships in 1877, two of the world's oldest and most prestigious sporting events.

America was for many a land of great interest, intrigue, and hope. Citizens in eastern and southern Europe were experiencing great hardship due to the fallout from the depression that had persisted since the mid-1870s as well as drought, famine, and armed conflict. To many millions in these countries, the promise of America was too powerful to resist. Its fast-growing industrial complex, built on a foundation of innovation, was brimming with opportunity. Consequently, the office of admissions at Ellis Island in New York City was filled to capacity from December 17th, 1900, when the new Main Building was opened and 1910. Of the nearly nine million people who had passed through the gates during that decade, 1.25 million in 1907 alone, more than two-thirds came from southeastern Europe.[8]

−oOo −

At Oxford, Edwin had a chance to explore his thoughts, read and listen to those of others, and experience aspects of life denied him during childhood. Along the way, he developed a strong sense of the man he wished to become and gained confidence in his ability to seek it out upon returning home. But even from more than four thousand miles away, the spirit of John Hubble loomed large, cajoling, preaching, shaming his son into submitting to his wishes. A spirit of compliance would have required little taming and hence might have led a less remarkable life. Edwin's spirit was rebellious if not resolute at that moment. No one among the student body that knew him, at least, was under any illusion as to where his true passions lay.

Whenever conversation allowed, Edwin would expound on the subject of astronomy to his friends at Oxford. Seeking out those of like mind and spirit, he had met and befriended Herbert Hall Turner, the Savilian Professor of Astronomy and Director of the Radcliffe Observatory at Oxford University. Not celebrated for his contributions to the field, Turner nevertheless deserves some mention within the story of Hubble and Humason. Prim and scholarly with a thick mustache and greying hair parted on the left, the 50-year-old former secretary of the Royal Astronomical Society (RAS) was a popular lecturer and author of a six-part lecture series *A Voyage in Space,* published as a children's book in 1913. With his research using astrophotography, Turner had improved on methods of obtaining positions and magnitudes for stars and other objects from photographic plates and was credited for having coined the term "parsec" as a distance measure to objects outside the solar system. Years later, after the discovery of the ninth planet it would be Turner who would forward the suggestion of an 11-year-old Oxford girl that it be named Pluto. In recognition of his distinguished career, which also included contributions in the field of seismology, he was awarded a Bruce Medal in 1927.[9]

Edwin's passion for astronomy was a match for Turner's enthusiasm for passing on his knowledge of the field to the next generation. The young scholar was often an invited guest in the Turner home and made a good impression on his hosts.[10]

Although he continued to pursue astronomy in the background of his study of the law, Edwin shouldered the silent burden of his father's righteous dogma. Throughout his time at Oxford the ebb and flow of his emotions were evident in letters home, mainly to his mother, who was the recipient of his more candid thoughts and ideals. In an early letter he confided that labor for labor's sake in effect "becomes an aversion," alluding to the practice of law as routine, adding that work ought to be designed toward some greater good or goal and that without it "work is hardly satisfactory." Although he was unsure exactly where he would find his footing, Edwin felt he was meant for a higher purpose, "for whose sake [he] could leave everything else and devote [his] life."[11]

In truth, it seemed, Edwin detested the law and in moments of frustration wished that he could "chuck everything and just wander in mind as well as body," the guilty confession of a youth whose course plagued his mental state. An attempt to divulge his hopelessness at his prospects for a career in law came in a cryptic letter to both his parents in December of 1911 where he stated that he "rather doubt[ed] the outcome" of such a life.[12]

His defiance of his father's priggish persuasions also extended to vice. He had not just tasted alcohol; he was now a regular at the local brewery where he enjoyed the apparently strong ale that was served on tap and had developed a taste for wine as well. Despite his attention to the work at hand, he had also found time to mingle in the company of women who were introduced to him by way of one circle or another. The carefree abandon and youthful joy of public intoxication among friends and the hair-raising allure of the scent and touch of a young lady were welcome additions to Edwin's daily life in those years, although he would never have admitted as much to his father. He mentioned only in passing "one or two excursions into the realms of astronomy" in a letter home just weeks before his father's death. Whether this was the last-minute guilty expression of inner desire from a son to his dying father or the precursor to a stand that he would later have to take against the ailing patriarch in the event he miraculously recovered, Edwin nevertheless was seizing on an opportunity to broach the topic, however delicately, to the man who had for so long suppressed his youthful curiosity and desires in science and in life.[13]

The death of John Hubble would seem to have removed the only real obstacle to the fulfilment of Edwin's dream of becoming an astronomer. He had held the field within reach during his time at Oxford, he was well regarded by Turner and others within the astronomy department, and he had already attained the credits he needed to pursue an advanced degree in the subject during his time at the University of Chicago.

In an ironic twist, the loss of his father created another impediment that was seeded in the lessons of family values to which Edwin was now duty bound. His older brother Henry and his sister Lucy were doing their best to hold things together, but they were ill-equipped to sustain the family long term. Another breadwinner was needed, and Edwin was not only next in line he was also in the best position to help. He knew he had to return home to tend to his family's needs. To do otherwise would be irresponsible. Thus, even from beyond the grave, John Hubble's influence weighed heavily on the ambitions of his son whose dream of becoming a man of the world, an explorer, stargazer, and adventurer, were once again placed on hold.

–oOo –

Standing on his porch, Milton Humason struck a match and lit the strands of tobacco jutting haphazardly from his freshly packed pipe. A red glow formed in the pipe's well and short staccato puffs of smoke rose into the air. The reddish yellow glow of the Sun was sinking into the horizon, fading the wisps of clouded sky from red to violet. From this vantage point he could only imagine the waves of the Pacific Ocean crashing into the sandy California shoreline. In minutes the veil of night would fall upon hill and valley.

For as long as he could remember, water had occupied his thoughts. He was awed by its mysterious depths and compelled by its currents. He was obsessed with the whir of the river, lost in the vastness of the sea. In recent years he had made a connection between the sea and the sky, the timeless perspective that seemed a part of both as if they were one and the same. How subtle, the hand of fate? One minute you were heading somewhere and the next you were headed somewhere else. The key was to roll with the changes.

Three years earlier in April of 1910 he was the only boarder at the house of the 27-year-old stable manager Chester Heston at 215 Mira Monte Avenue, not far from the Mount Wilson trailhead. A three-year veteran of the cowboys that roamed the Pasadena hillsides, Milton was listed as "Packer Burro Train" on the census taken at the end of that month. An expert horseman, he had become an almost indispensable member of the muleskinner team and a friend and ally of his manager, seven years his senior, whose parents also hailed from the Midwest.[14]

Up to that time, Milton's life seemed to be on a steady course and sure footing. He had no plans for the future and didn't need any, preferring to live his life in the moment, in the spirit of the old mountain men of the 19th century. He felt a special kinship to his friends among the cowboys, the ragtag group of foul-mouthed but well-meaning young men who could cuss out a misbehaving mule while doffing their hat for a lady. There was a cook by the name of "Mrs. Moyer" who "weighed over three hundred pounds." She arrived one day while Milton was at the trailhead. According to his later account, "she waddled up the street and asked for a burro," which was against the rules. Mules and burros were not capable of handling that kind of load and Milt and crew gently suggested to Mrs. Moyer that she pay a bit extra for a horse to take her to the summit. To his lasting surprise she decided to walk it. "She went up to the halfway house the first day," he recalled, "and stayed overnight and went up the next day." When Mrs. Moyer turned out to be "one of the best cooks" on the mountain "a wagon" was secured for her to ride up and down from the summit.[15] In the years to come a motor stage would be used to ferry employees of the hotel and observatory up and down the mountain.

The winter and early spring of 1910 had been busy with construction crews working overtime to rebuild the monastery in time for the arrival of the Carnegies and a large group of astronomers for the fourth meeting of the International Union for Cooperation in Solar Research (IUCSR), the brainchild of George Hale and an inspiration for the creation of the International Astronomical Union (IAU).[16]

Carnegie and his wife and daughter stayed at the hotel on the mountain and Milton and a fellow clerk were sent down to the valley "to get Mr. Carnegie some fish," which he liked "freshly broiled." For their efforts, Carnegie gave each of them "a five-dollar gold piece," which those who knew of Carnegie's miserly ways told them "was a pretty generous tip." Later, he recalled watching Carnegie pace back and forth at the foot of the large fire in the hotel lobby as he and Hale talked for many long hours.[17]

Days on the trail could be long, mules and burros could be cranky, and the road could be treacherous where frequent mudslides and rock falls made reaching the summit by the old trail impossible. There were also bears and mountain lions and rattlesnakes sunning themselves on the path that might spook a mule. A mule in a foul mood could dislodge a load, sometimes throwing both the load and the animal over the side of the mountain and down the steep slope, where it would have to be rescued. His stories of life on the mountain made him something of a family celebrity and he sometimes took members of the Humason and Witmer families up the trail and into the rolling hillside.

Milton was at the height of his stature as a muleskinner when he met Helen Dowd the year before. Almost as quickly as the two fell for each other his life started to take a turn, and as spring turned to summer, as Halley's famous comet rose "over the ridge leading to Mt. Baldy and came up like a great search light," the familiar life which he had known for five years was about to be altered inexorably[18] (Fig. 4.1).

On the morning of his departure Milton filled his saddle bags with the few belongings he had with him and walked to the stable where his muscular black stallion, Blackie, all of 17 hands at the withers, stood with its head out of the stall door in expectation of his arrival. The horse was soon saddled, and Milton was on the dusty road toward the Storrow ranch at 534 Palmetto Drive, eight miles east southeast.[19]

He would head directly east to the Arroyo Seco a mile and a half north of the ranch and from there head south. The 25-mile-long watershed began at Red Box Saddle near Mount Wilson and ran downhill west of Altadena, through Devil's Gate – the rock formation that resembled, for some, the demon's profile – and straight through Sam Storrow's citrus groves. It was a perfect water source for the farm, which was located two miles west of Tournament Park where the Tournament East-West football game was played. It was two miles south of the eventual site of the Rose Bowl that would become the host of the game beginning in 1923. He may have heard the wail of the Santa Fe rail line roaring over the bridge across the waterway on its way from downtown Los Angeles to Pasadena where well-to-do passengers from the east would disembark for a stay at the impressive Raymond Hotel near the ranch.[20]

Milton's cousin Ralph Witmer was Letha Lewis's younger brother and 10 years Milton's senior, but the two had become close in the years since the Humasons' arrival in California. Ralph was set to take over at the Lordsburg ranch and would help teach Milton the ropes at the Storrow ranch as well.

At the time, Pasadena consisted of a small but vibrant downtown surrounded by a rural community of citrus ranchers and traditional farmers of whom Letha Lewis and her husband were a part. Out of all the members of the Witmer and Lewis families of Crown Hill, Letha Lewis was probably the

Fig. 4.1 The cowboy muleskinner, Milton "Milt" Humason near the foot of the trail to the peak of Mount Wilson ca 1910. (Courtesy, Ann Humason Bernt family)

closest to understanding Milton's way of life. The 40-year-old daughter of Henry Witmer's older sister Agnes, widow of Colonel Samuel Lewis, a Union Army civil war veteran, was a real Renaissance woman well before it was socially acceptable to be one. A maven of the theater with a flair for the dramatic who rubbed elbows with Charlie Chaplin, Letha was a liberal socialite who was as adept at skeet shooting and 'motoring' in knickers and boots as at hosting black tie dinner parties. Although fancy by his standards, Milton, who enjoyed play-acting, was likely fond of his older cousin's zest for life. Whether or not it was Letha who convinced him to come to work at her husband's citrus ranch, the two would probably have gotten along well.[21]

Within a year, Milton had secured his place on the farm by turning out crop yields of oranges, apples, almonds and other assorted nuts and berries. The speed with which he grasped the responsibilities involved in managing the operation at the orchard must have impressed Jerry Dowd because when the young man called again to ask the Mount Wilson electrical engineer for his daughter's hand, he gladly consented to the union.

A short time later, Milton took his beloved Helen to the summit of Mount Wilson where they had met two years earlier and proposed to her at an outcropping of tree roots that they had dubbed "the perch," near Echo Rock. The roots offered a nice seating area and views of Mount Baldy 20 miles to the east. It was their favorite place on the mountain and seemed a fitting place to make his marriage proposal, which Helen accepted after what must have seemed an exhaustively long wait.[22]

Milton and Helen were wed at the Dowd family home at the base of the mountain on October 14, 1911, with Helen's friend Nellie Campbell and Milton's cousin Ralph Witmer serving as witnesses. The couple moved into a small house on the grounds of the Storrow Ranch where they would live for the next several years.[23]

The arrival of the next generation of Witmers in 1912 gave the family a welcome boost of youthful energy, enthusiasm, and well-educated manpower. William Witmer graduated Harvard with an engineering degree that year and slowly took over the reins of the Witmer Brother's Company operations. William later married and had three children with his wife Mary, two of whom survived childhood. William's older brother, David Witmer, graduated Harvard a year later with an advanced degree in architecture and soon joined William and family in Los Angeles, where he became prominent in that profession. He served as a pilot in both world wars, earning a Bronze Star and a Legion of Merit in Europe after 1943. But he is best known for his service during 1941 and 1942 as chief architect of the Pentagon in Washington D.C.[24]

The early years of Milton and Helen's marriage were lived in peaceful country comfort amid the orange groves overlooking the foothills. Cradled in the warmth and security of the family, their new life also included several of the trappings of the well-to-do Witmers, turning up at fairs and festivals finely groomed and dressed in all white from hat to heels, always brimming with joy in candid photos from the period.

Now, after a little more than two years in this new life, the couple had just brought a baby boy, William Dowd Humason, into the world. As he stood drawing on his pipe on the porch of their small cottage, Milton could hear the stirrings of his newborn son. Mother and son were doing fine and were well looked after by Helen's mother, Katherine Dowd. In his hand he clutched a copy of the English folk tale, *Robin Hood*. This epic of heroism and rebellion obviously resonated with Milton and Helen, and they were eager to instill its lessons on their newborn son when he was old enough to appreciate them. On the inside cover of the book Helen had written: "To William Humason – from his Mother and Father – October 30th, 1913."[25]

Still grieving the loss of her husband, Alice Petterson Witmer had left for a lengthy tour of Europe with her son Joseph in June of 1912. They sailed for London on the 21st and Joseph was enrolled in boarding school in Montreux, Switzerland.[26] In November of 1913, shortly after the birth of her grandson, Laura Humason boarded a train for New York with Milton's sister, Ginny, and from there they sailed for London.[27] Ginny was enrolled in the same boarding school as her cousin Joseph, and the Petterson sisters then met for a tour of the continent.[28]

As William "Billy" Humason began his life, the United States was beginning to flex its innovative muscle.

Thanks to his new assembly line, Henry Ford was rolling out thousands of Model T's from his plant in Detroit. Not to be outdone, General Motors Corporation had introduced the first electric self-starter in its Cadillac models, eliminating the need for a driver to wind a crank handle at the front of the vehicle. American automobile manufacturers had begun facing off on the racetrack. The first Indianapolis 500 was run in 1911 with Ray Harrounat winning with an average speed of just under 75 mph.[29]

The city of Los Angeles opened its first gas station that year as the automobile continued to impose itself on the landscape of American cities. With a population of more than three hundred thousand it was growing quickly and becoming increasingly modern. The nascent movie industry was producing short films at a rapid pace and Milton and Helen were among the thousands of avid moviegoers eagerly paying for entry into movie theaters. Legends of early cinema such as D.W. Griffith, Lionel Barrymore and Mary Pickford were getting their starts and the director Cecil B. de Mille had just made *Squaw Man* as the first feature-length Hollywood movie.[30]

Basking in the glow of the moment and the smoldering pipe clenched between his teeth, Milton stole a glance at the peak of Mount Wilson 20 miles to the northeast. He could see the large white dome of the 60-inch reflector shining proudly from the dense pine forests near the summit, while the smaller domes of the 60-foot and the 150-foot solar tower telescopes stood just above the tree line nearby. Just to the north was the area designated for George Hale's new 100-inch reflector that was experiencing delays owing to problems pouring and annealing the glass disk for its primary mirror. Despite these delays, Hale and his team were confident on-site construction would start the following year.

As the growing city encroached on their vast citrus groves, the Humason, Witmer and Dowd families continued to work and play in the shadows of Mount Wilson. While doting on their infant son, Milton and Helen made plans and saved their pennies to buy a ranch of their own.

–oOo–

On the 28th of June 1914 in Sarajevo the young Bosnian Serb nationalist Gavrilo Princip assassinated Archduke Franz Ferdinand of Austria-Hungary along with his wife, Duchess Sophie. As a Yugoslav nationalist, Princip sought reunification of Serbia with Bosnia and Herzegovina, Croatia, Montenegro, Kosovo, Slovenia, and Macedonia. He and colleagues in the secret society known as the Black Hand were arrested, tried, and jailed. Princip died of tuberculosis in a Bohemian prison in 1918.[31]

The Triple Alliance between Germany, Austria-Hungary and Italy had been formed in 1882 as a means of establishing a hold on the center of Europe. The Triple Entente between Great Britain, France and Russia was intended to counter it. This standoff between central European nations and their neighbors to the north, east and west began to intensify, drawing in smaller constituents of the whole. On the heels of the Archduke's assassination, Austria-Hungary effectively declared war on Serbia, which reciprocated. As the alliances defended their constituents, July 28th, 1914, marked the slide into a conflict that would be named the Great War.[32]

The emerging specter of all-out war caught Laura Humason and Alice Witmer by surprise. They removed their children from boarding school in Switzerland and sailed for England, where they were admitted to a school in Eastbourne. After further discussion they decided that Laura should head home, and Alice would return with Ginny and Joseph after they had finished boarding school. Laura Humason sailed for the Port of New York from Glasgow, Scotland on August 15th, 1914.[33] As the conflict boiled over into full-scale warfare the following year, Alice Witmer decided she ought to get the children out while the getting was good. She applied for emergency passports for them on April 29th and booked passage to home on the RMS Lusitania.[34] The liner left New York on May 1st and as it neared the coast of Ireland it was sunk by a German submarine, killing almost twelve hundred people. On its return to America, the ship was to have carried Alice and the children. News of their close call sent a chill through both families.[35]

For several months Woodrow Wilson attempted to appeal to the German government through one tactic after another to halt attacks on merchant shipping from neutral nations, with limited success. On August 19th, the White Star liner SS Arabic was sunk by a German submarine with two Americans among the dead and Wilson feared the slide to war. But for the moment anyway, the German government began to soften its position.[36]

As Germany pulled back its attacks on non-combatant nations, Alice Witmer felt the coast was finally literally and figuratively clear. She set sail with Ginny and Joseph, now both aged 16, on the S.S. St. Louis from Liverpool on September 4th, 1915. One can only imagine the collective sigh of relief once the ship cleared dangerous waters. In a couple of weeks, they were at last safely home in Los Angeles.[37]

–oOo –

In the year since his return home Edwin Hubble had put on a brave face, trying to live up to his promise to his mother that the family would hold together. A central occupation was paying off his father's debts to creditors and the medical bills from dealing with his illness. After all was said and done, John Hubble's estate left his wife and seven children around $7,000 (a little over $180k in 2021), approximately $500 of which would go to each child. With school-age children still at home and their mother's long-term care in the balance, the adult Hubble children were eager to sustain the small fortune their father had left.

Henry, still living at home, was an insurance analyst at the Kentucky Actuarial Bureau while Edwin's older sister Lucy taught piano. Younger brother Bill spent the summer of 1913 working on his dissertation and graduated in the spring of 1914 from the University of Wisconsin with a degree in agriculture. His long-term ambition was to buy farmland to be cultivated in support of the family, but he enlisted in the army out of college and spent the war at Fort Leavenworth in Kansas in charge of the post's creamery. While there, Bill sent home his extra cash to help pay the bills, beginning a lifelong commitment of sustaining the health and well-being of his mother and sisters, making him a family hero.

On his arrival home, the boy that the family had known as "Ed" Hubble had transformed himself into an English fop in knickers, cape and cane, his school ring boldly displayed on his finger for all to see. He was Edwin Hubble now and his transformation in appearance matched his English accent, which he maintained after his return either out of nostalgia for his adopted motherland or to remind him he was no longer intended for the world in which the family lived, or both. He sat at the head of the table, demanded tea served in the English style and displayed his collection of pipes in the library where he spent most of his downtime.[38]

If his mother had any objections to Edwin's new persona, she rarely said anything about it, doubtless happy simply to have him home again. As the addressee of several of her son's more candid letters, Jennie Hubble could have been under no illusions concerning his inner desires. The only question was whether he would leave home to pursue them or remain and honor his promise to his father. The answer to this question began to show in the spring of the following year. Edwin had taken a position teaching Spanish and mathematics at New Albany High School in Indiana, across the Ohio River from the family home in Louisville. His reputation was somewhat mixed among the alumni of the school but just as he always had Edwin made his presence felt in form and in action, coaching the school basketball team to an undefeated record and third place in the state championships. By the end of his first year at the school he had earned the adoration of many of his graduating students and athletes.

Despite his successes as a teacher, coach and mentor, Edwin must have felt trapped by his new situation. The familiar fear, anger, and self-loathing he so often felt in his childhood years still haunted him. He had no use and no passion for the law as a career, having spent much of his time at college learning it solely to appease his father.

According to his biographer Gale Christianson, Edwin Hubble never passed the bar and did not present himself to a circuit court judge for examination, although he later said he had been practicing part time and had earned the healthy sum of $10,000 for a year's work.[39]

This prevarication, one of many to come, was to be part of the revival of Edwin Hubble, the young, ambitious scholar eager to share tall tales with the world. Having long felt that his destiny lay in the celestial realm he decided to escape the pull of his father's spirit and pursue his love of astronomy once and for all (Fig. 4.2).

Fig. 4.2 Edwin Hubble looking with a small telescope ca 1914. (Image from Edwin Hubble Papers, Huntington Library, San Marino, California)

He wrote to his former professor, Forest Ray Moulton at the University of Chicago in May of 1914 asking for his assistance in joining the astronomy department in pursuit of his doctorate. Moulton, in turn, reached out to Edwin Frost at the Yerkes Observatory who was more than happy to have a star in the making come to the observatory as his assistant. The class was full for the next semester, but Frost found a little extra cash and invited Edwin to join the school term in the fall. By August of 1914, with war breaking out in Europe, Edwin Hubble found himself in Williams Bay in Wisconsin enrolled in the only course he would require for his doctoral work at the University of Chicago, research at Yerkes Observatory.[40]

Edwin's feelings about the war were absolute. He knew intimately the history behind the struggle in Europe having spent many hours reading and talking with his friends and schoolmates at Oxford. In the face of Germany's mounting military presence, he felt it was of the utmost importance to match that force in order to neutralize it and keep the peace. With the US outside the conflict, he could only watch with concern as it unfolded, worried for the health and safety of his friends who likely now found themselves engaged in battle. He made no secret of his intention to enlist in the service of his own country on behalf of England and her allies should America get into the war, an outcome he hoped would come swiftly.

But Hubble would spend the next three years as an observing assistant to Frost who, as it happened, was losing his eyesight, the worst possible fate for an astronomer. This left Edwin and the other assistants more observing time to carry on investigations on behalf of their beloved director as well as the establishment. It was believed at the time that the best way to obtain an education in astronomy was to train in observing the night sky and subject one's results to rigorous mathematical analysis.

From the day he joined the campus at Williams Bay, Hubble and Frost were in almost daily contact. Frost was keen for Hubble to read as much as possible on astronomy and the latter was an eager and highly capable student.[41]

He already felt like a member of the team, having stopped in Evanston, Illinois in August at Frost's suggestion to attend the 17th meeting of the American Astronomical Society (AAS) at Northwestern University. It was there that Vesto Slipher read his groundbreaking paper detailing the extraordinary radial velocities of a selection of nebulae.[42]

–oOo –

Vesto Melvin Slipher was born November 11th, 1875, in Mulberry, Indiana on a farm owned by his father Daniel Clark and his mother Hannah App Slipher. He and his brother, Earl, five years younger, would work together for their entire careers at the Lowell Observatory where Vesto was hired in 1901 by its founder Percival Lowell on a "short term" basis. His association with the observatory would last 53 years.

By the time Slipher achieved adulthood, life on the farm had made him into a vigorous specimen and he was possessed of a mind that was equally adept at solving both mechanical and observational problems. On graduating from Indiana University at Bloomington with a B.A. in mechanics and astronomy in 1901 he remained and received his master's degree two years later.

He married Emma Rosalie Munger in Frankfort, Indiana on January 1st, 1904, and they set up home on Mars Hill at the Lowell Observatory, where they would raise two children. With his family's affairs in order, Slipher returned to Indiana University where he received his doctorate in 1909. His close association as assistant to the Mars-fixated Lowell, starting just after his undergraduate days, no doubt influenced the title and topic of his dissertation: *The Spectrum of Mars* (Fig. 4.3).

In constitution, manner, and ambition, Slipher shared much in common with his younger counterpart Milton Humason, who would bear the torch set alight by Slipher in the autumn of 1914.

In the 15 years from 1901 until Lowell's death in 1916, Slipher tended to the director's cow "Venus" and his vegetable garden in addition to his observing duties. He was reserved and reticent, shying away from the public sphere and often arranging for colleagues to read his papers at astronomical meetings. Most of his contributions were published only in the Lowell publications and seldom saw the light of day outside of that journal.

In his work as a photographer and spectroscopist he was careful, detailed and patient and never published results he was unsure of or made assertions in his reporting he wasn't clear about. He was probably more ambitious than his natural demeanor would permit him to be, which plagued him privately in his later years, and he was a calm, forthcoming and reassuring leader in science, business and in his community.

As an observer he displayed more of the open-minded and insightful qualities that his younger counterpart Edwin Hubble showed. In the course of his research, he discovered that spiral nebulae rotated and made significant contributions to the understanding of their relative motion and distribution, confirmed the existence of interstellar gas and dust and demonstrated that some nebulae shine by reflecting light from nearby stars – all before 1916, the year he took over the reins at the Lowell Observatory.

Slipher's education in engineering differentiates him from both Hubble and Humason. In addition to his great skill and incisiveness as an observer he was an extremely competent mechanical technician. The new 40-inch reflector was to be used primarily for studying the planets, so Slipher had to settle for the older 24-inch refractor to which he attached a three-prism spectrograph in 1909. His goal was to collect a useful spectrum of a galaxy, choosing M31 in Andromeda as his target. But the smaller telescope equipped with the spectrograph and camera on hand were insufficient to the task. As he began to work through the problem, he soon discovered that focal length, aperture, and the degree of dispersion that the prisms in the spectrograph produced were of little importance in accomplishing his goal. The key ingredient was a

Fig. 4.3 Vesto Melvin Slipher using his spectrograph at the 24-inch refractor at Lowell Observatory (Image courtesy of the Lowell Observatory Archives)

"faster" spectrographic camera. A year later he had created a single-prism spectrograph from parts lying around the observatory shop. This camera reduced the light lost in refraction between the three prisms of the old instrument and reduced dispersion or the spread of the light on the image. Overall, he increased the "speed" of the instrument 100 times.

Working like a mad scientist in a movie with sparks flying from a high-voltage induction coil beside the telescope (the Leyden jar capacitor array he used to vaporize traces of iron and vanadium inside the new spectrograph for his comparison spectrum) Slipher tried again to get a useful spectrum of the light from Andromeda. What he found intrigued and excited him. The Fraunhofer lines of the nebula appeared to be offset from those of his comparison spectrum toward the blue end of the spectrum.

True to form, Slipher decided he needed to be sure of his findings and that would require a faster camera. He worked for two years on and off with various lenses until finally opting for a commercial Voigtlander f/2.5 lens that increased the spectrograph's speed 200 times over the original. It was this camera that on January 1st, 1913, yielded the definitive result. Slipher had made history, becoming the first person to measure the radial velocity of a galaxy, with the result showing M31 to be moving toward us at the incredible rate of about 300 km/sec.

With this information in hand and spurred on by his exuberant and delighted observatory director, Slipher continued this work so that by the time the AAS convened at Northwestern on that August day in 1914 he had confirmed spectra for 15 galaxies with three surprising results.

First, all of the galaxies he examined had measurable velocities.

Second, the majority of the velocities were recessional, some as large as 1,100 km/sec. This led Willem de Sitter three years later to suggest that the universe might be expanding after Einstein published his general theory of relativity.

Third, inclined lines on the spectrum of at least one of the galaxies indicated rotation at a high rate of speed.

When he finished presenting this extraordinary paper the Lowell Observatory stalwart received a standing ovation, which was virtually unheard of in scientific circles at the time.

The ability of the 24-inch refractor and spectrographic camera to delve deeper into space would be exhausted by 1921, but by then Slipher had published more than forty velocities with the largest rate of recession of approximately 1,800 km/sec for NGC 584 in Cetus. He received both a Gold Medal from the RAS and a Henry Draper Medal from the NAS in 1933 for his work in establishing the field of nebular spectroscopy.[43]

–oOo–

Slipher's paper to the AAS in August 1914 provided an exciting and inspiring beginning to Edwin Hubble's career in astronomy and may have compelled him to pursue the study of faint nebulae upon his arrival at Williams Bay. From the outset he appears to have been interested in a holistic approach to the subject that would evolve in later years into a search for a classification system and greater understanding of galactic evolution and dynamics.

Prior to delving into his own research, time was needed to understand and operate the instruments and their associated equipment. The prize at Yerkes was the 40-inch refractor in the large dome at the west end of the observatory building which, in the words of George Hale, was shaped in the form of "a Roman Cross."[44]

The largest working telescope in existence when it was completed in 1897, the 40-inch was quite a sight to behold nestled in the 90-foot-diameter dome that rose six stories above the hilltop. The 60-foot, six-ton steel tube balanced effortlessly on the 40-foot-tall cast iron pier supplied by Warner & Swasey Co., with the enormous glass lens made by Alvan Clark & Sons housed in the leading end, near the dome slit. Perhaps the most intriguing element was the 75-foot circular platform floor that could be raised or lowered using an electrically powered winch system to bring the astronomer to the viewfinder. Altogether telescope and dome weighed in at a massive 240 tons, making it a spectacle both in size and stature.[45]

Throughout his young life Edwin had dreamed of guiding the giant Yerkes refractor across the night sky in search of never-before-seen phenomena of the cosmos. By now however, his philosophy more seasoned and his eye more mature, he preferred to work with George Ritchey's original 24-inch reflector, affectionately named the "two-foot reflector," in the southeast dome, some 90 feet from where the large refractor was housed. The mighty little instrument with a clear aperture of 23.5 inches had been designed and built by Ritchey and his assistant Francis Pease before the turn of the century and installed temporarily in the heliostat room for testing. The base, mounting and frame were built in the observatory's own instrument shop while Ritchey and Pease ground, shaped and polished the mirrors at Ritchey's lab in Chicago. The blank disk for the mirror had been poured at the St. Gobain Glassworks near Paris in France, which was now trying to produce one for the 100-inch at Mount Wilson. More on that story later.

The two-foot reflector had showed itself to be more powerful and clearer of focus than its 40-inch cousin. What is more, its focal length of 93 inches meant that the telescope was no more than seven feet in length.

The dome in the southeast tower had been designed to house a 16-inch refractor that required a comparatively long focal length. This meant that the pier and the floor had to be raised 12 feet in height to bring the shorter reflector up to a height where its axis of rotation was near enough to the dome slit to be effective, with steps added to gain access to the observing floor (Fig. 4.4).

This "experimental" telescope had convinced Ritchey and Hale of the effectiveness and efficiency of reflecting telescopes over refractors. Ritchey laid out the entire scope of the development, along with his plans for what would

Fig. 4.4 George Willis Ritchey's 24-inch reflecting telescope in its dome at Yerkes Observatory in 1901. (Image courtesy of the Observatories of the Carnegie Institution for Science Collection at the Huntington Library, San Marino, California)

become the 60-inch reflector at Mount Wilson (originally to have been built at Yerkes) in a treatise on the two-foot reflector for the ApJ in November of 1901.[46]

Although the Bruce 10-inch doublet was already in operation in its own dome on the Yerkes campus and Edward E. Barnard was using it to capture images for his famous 1919 *Catalogue of Dark Markings in the Sky*, the two-foot reflector was the instrument of choice for young Hubble. He settled into regular research with the instrument as soon as he was cleared to operate the telescopes on his own. The eyepiece was never more than 11 feet from the observing floor, so a chair or observing ladder was all that was necessary to reach the more difficult targets in the night sky.

Of the roughly one hundred fifty thousand estimated nebulae "within reach of existing instruments," only around seventeen thousand had been catalogued by the time he started working.[47]

Although there was much speculation as to their structure and relationship to the Milky Way, there was no clear evidence as to their true nature. In particular, what accounted for the spiral nature of some nebulae? Or their brightness? Were they star fields like the Milky Way as the supporters of the "island universe" theory predicted, or simply clouds of dust? What was the nature of their evolution? Could they be categorized according to the stages of their development? No system for nebular classification had been established. And what caused the mysterious redshifts that Slipher's spectrographic camera had revealed?

No one at that time could've predicted the extent to which the five increasingly powerful telescopes that George Hale was to develop between 1895 and 1950 would fundamentally alter our perceptions and understanding of the inner workings of the nebulae and hence the universe itself. No one could've foreseen that Edwin Hubble would become one of the few astronomers to work with all of them, and with such profound impact on the evolution of scientific knowledge.

In 1915 Edwin Hubble was just a young man from Missouri with a chip on his shoulder and a dream in his heart. These two opposing forces would remain with him throughout his life and career, the dream leading him to glory and the chip damaging his reputation.

The beckoning frontier in the field of nebular research must have attracted him. Telescope and photographic technologies were just becoming advanced enough to begin to unravel some of the questions put forth by the greatest astronomers of the past two centuries from Messier to Slipher. Solving these mysteries would involve years of research to even begin to cover such a large field. The appeal, the wonder and the potential for objective discovery formed a center of gravity between the nebulae and the ego of Edwin Hubble. He believed he was the right man for the job.

For his initial target he selected NGC 2261, a small cometary nebula in the constellation Monoceros whose "nucleus [wa]s known as the variable star R Monocerotis, said to range from magnitude 9 to 13.5, with an irregular period." William Lassell in England had earlier remarked that "the nucleus [wa]s not a star," and this had been confirmed by Barnard using the big Yerkes refractor.[48]

Starting in 1915, Edwin took plates of NGC 2261 with the 24-inch reflector, then used a "blinker" to compare them with others taken by the same telescope eight and 16 years apart. The blink comparator was a type of microscope used to compare two separate images of a stellar object for any differences in brightness, shape, and other details. The observer looked through the eyepiece and switched back and forth between images. As Allan Sandage put it, any variations in a given object would "wink at you."[49] The blinking machine would play a major role in the development of stellar and galactic evolution as well as some folderol at the Mount Wilson Observatory in the years to come.

The results of his investigation of NGC 2261 inspired the young observer. The trailing edge of the nebula appeared to bulge and become more convex and the brighter area near the nucleus had shifted eastward by some five seconds of arc. Interestingly "the center" of the nebula appeared unchanged from the previous images with "a sharply defined brighter wedge-shaped portion pointing to the east." His "first impression" was that the change in the nebula's form might be the result of its "turning about its own axis." In a report to the National Academy of Sciences on March 9th, 1916, the 27-year-old was deferential to the authority of Frost and Barnard in claiming that "the changes [in shape] [wer]e in the nebula itself."[50]

Many decades later, a sage and experienced Allan Sandage would argue that one reason for Edwin Hubble's rise to supremacy in the field was his ability to write and to succinctly apply his logic in ways that captivated and inspired. In his words, Hubble "was of a poetic nature…an intellectual of a most profound type…He was a very good writer."[51]

Edwin was already displaying signs of his prowess with the pen after only a few months of research on NGC 2261. His first paper in March 1916 bore the analytical style common among many astronomers. It was succinct and to the point, opening with "A comparison of photographs has established changes in structural detail in the nebula…"[52]

His very next paper, on the same subject, was published in September of the same year with the title *The Variable Nebula N.G.C. 2261*. The title itself signaled a shift in tone, and Edwin opened with a philosophical flourish:

Recent astronomical research has been especially fruitful in the study of nebulae – a study which has now extended into the realms of dynamics. The spectroscope, with its disregard for the vast distances involved, has reaped, first – radial velocities of planetary, irregular, gaseous, and spiral nebulae; it has also shown internal motion in the great nebula of Orion, and rotation in both spiral and planetary nebulae.

The "realms of dynamics" and the personification of the spectroscope are prime early examples of what would grow to become Edwin Hubble's greatest tool for conquering the astronomy world and dispatching his opponents.[53]

With two papers on NGC 2261 he had made an auspicious beginning and was spurred on by his success. During the next two years, under the watchful eye of Barnard and with philosophical guidance and reassurance from Frost, Edwin methodically accumulated his research on 588 faint nebulae in seven clusters, only 76 of which were previously known.

By the spring of 1917 he was already forming a more educated opinion on the state and characteristics of nebulae. After saying there was "no significant classification" system for the nebulae and that "not even a precise definition" had yet been formulated, he pointed to three "essential features." Specifically, "they [we]re situated outside our solar system, that they present[ed] sensible surfaces, and that they should be unresolved into separate stars." He separated them from the "gaseous nebulae which appear[ed] stellar in the telescope, but whose true nature [wa]s revealed by the spectroscope" and noted that the latter may "differ in kind" and may "not form a unidirectional sequence of evolution." This line of reasoning suggests he might have been envisioning the process by which galaxies like the Milky Way did or did not form.[54]

One of the longest running debates in astronomy was whether the nebulae were clouds of dust within the Milky Way or very remote independent systems of stars dubbed "island universes" by the 18th century German philosopher Immanuel Kant. Recent photographic evidence had revealed evidence that led many astronomers to the notion that at least some nebulae were independent systems.

Photographs taken by James Keeler, a longtime friend and collaborator of George Hale, with the 36-inch Crossley f/6 reflector at the Lick Observatory near San Jose in California in 1898 had clearly revealed the spiral structure of M31 and other galaxies. The Crossley was a descendant of William Lassell's (whose surname may be the origin of the frequent misspelling of Milton Humason's middle name) two-foot f/10 reflector and narrowly predated Yerkes' two-foot f/4 reflector. The largest reflector in use in the United States when it debuted in 1896, the Crossley's mirror, like its smaller cousin at Yerkes, was coated in silver using a new method that helped to usher in the modern era of improved clarity in astrophotography.[55] The Crossley reflector will figure in the story again later.

Keeler, who suffered a stroke at age 42 and died just months after his report on the new telescope was published in the ApJ in June of 1900, was a long-time friend of Edwin Frost. Considering the closeness of their association, there is little doubt that Hubble was keenly aware of Keeler's work and photographs.

By the end of his graduate studies, Edwin was fairly sure on one point, writing "the great nebulosities" and "planetaries, gaseous but well defined, are probably within our sidereal system," by which he meant the Milky Way. On the other hand, "the great spirals, with their enormous radial velocities and insensible proper motions, apparently lie outside our system."[56] These words were published in the same year that Harlow Shapley argued against the existence of external galleries in his debate with Hebert Curtis. In mentioning velocities, Edwin could only have been thinking of Slipher, whose work on galaxy redshifts dominated the field. As no distances had yet been measured to any of the nebulae, he suggested "we must content ourselves with the problem of their distribution."[57]

To this end, he cited the work of Edward Fath who from 1909 to 1912 had used the 60-inch reflector at Mount Wilson to take plates of the Kapteyn fields in "uniform exposures of one hour."[58] The "fields" he was referring to were the 206 areas of the sky that the Dutch astronomer Jacobus Kapteyn suggested would yield a complete picture of the structure of the Milky Way once subjected to an extensive yet generalized photometric and spectroscopic analysis. The Kapteyn Selected Areas was a less ambitious (and more achievable) version of the *Carte du Ciel* (*Chart of the Sky*) that was already in progress.

Fath was a gifted and inventive observer who ended his short association with the Mount Wilson Observatory after 1912, due in part to his inability to get along with then assistant director Walter Adams. Although Adams was seen as being fair-minded, irritating him had a tendency to limit a person's time at the observatory. According to Allan Sandage, some at Mount Wilson reckoned that if he had been able to get along with his boss, "Fath would have become the Edwin Hubble of the observatory."[59] In his future dealings with Adams, Hubble would come to identify with Fath in this regard.

In the course of his research, Fath had discovered in excess of eight hundred previously unknown nebulae and confirmed that the smaller nebulae avoided the galactic plane. This, Edwin suggested, required a Kapteyn-like survey dedicated to mapping the region at the center of the Milky Way. But the 24-inch reflector at Yerkes only caught a fraction of the nebulae in Fath's study and was simply incapable of showing many of their distinguishing characteristics. Further study using more powerful telescopes would be necessary to reach a greater understanding of nebular structure and diversity. In his dissertation

Photographic Investigations of Faint Nebulae, he laid out a comparative analysis of previously published work and a treatment of that which he would like to pursue in the coming years.[60]

Edwin's focused approach and steady work in nebular research was enough to convince Frost that he was going to make a fine observer. The two men posed for a photo with the observatory staff in August of 1916 (Fig. 4.5). Later Frost wrote to Adams to nominate his young protégé for a staff position at Mount Wilson. Adams passed the venerable observatory director's letter on to Hale, who in turn contacted several of his friends at the U of C for their assessment of Hubble's aptitude and overall comportment. With plenty of positive reviews Hale wrote to Hubble offering him a job at the rate of $1,200 per year, a considerable boost from his $320 fellowship in Chicago, effective as soon as he completed his dissertation and made his way west.[61]

Still feeling obligated to his family, Edwin was torn between accepting Hale's invitation and staying put in Wisconsin. In 1916 he had aided his brother Henry in moving his mother and sisters to Madison, about 60 miles southeast of the observatory to get them nearer to him. But these were difficult times for the Hubbles. Henry had a year-long battle with anxiety that caused him to step away from the insurance business, leaving the family to subsist on the meager earnings from Lucy's piano lessons and the money Bill

Fig. 4.5 Edwin Hubble (Standing sixth from the right) poses with the Yerkes Observatory staff for a group photograph, August 1916. (Image Courtesy of Hanna Holborn Gray Special Collections Research Center, University of Chicago Library)

was sending home from Fort Leavenworth. Meanwhile, Helen had entered college at the University of Wisconsin.

The proximity to Edwin gave the Hubble sisters a chance to visit him at the observatory where they toured the grounds and looked through the eyepiece of the great refractor. The Lake Geneva shoreline was only a half mile walk south along a path through the forest, where they enjoyed swimming and getting paddled around in a canoe by their powerfully built brother.[62]

His decision to move the family north to Wisconsin and his desire to stay at Yerkes as Frost's assistant following his graduate work suggests that Edwin Hubble was determined to both live up to his family responsibility and to work at his desired occupation. Once he secured a salary and could do his part for his mother and sisters, he would be free to work as a visiting researcher at any observatory with a reflector sufficiently powerful to advance his nebular research. In addition to the 36-inch at Lick and the 60-inch at Mount Wilson there would soon be the 72-inch at the Dominion Astrophysical Observatory in Canada and the 100-inch at Mount Wilson.

Although it had been his wish also to keep Edwin around, Frost knew there was no way to make that work financially. Besides, the opportunity awaiting him in California was too great to pass up. The seeing was as good as the country had to offer, Mount Wilson had the largest telescope in the world and the Carnegie Institution of Washington (CIW) provided financial security not readily available in Wisconsin. He told Edwin he should wrap up his work and head west.

This would have presented a difficult decision for a young man whose very spirit had been nearly squelched by his late father. Whether or not he and his brother Bill discussed the family's fate, and their individual ambitions is unclear. In contemplating his next course of action and working on his dissertation the almost two-year negotiated peace between Germany and the United States came to a screeching halt.

In January of 1917, with Russia in the midst of its own revolution, Kaiser Wilhelm II moved the bulk of his forces to the Western Front in France and renewed Germany's policy of unrestricted submarine warfare in the Atlantic. At the same time, Germany encouraged Mexico to join the war effort as its ally against the United States and vowed to assist her to regain lost territory in Texas and the southwestern part of the US. This was supposed to keep America occupied while Germany went about trying to capture France, one of its two main objectives in the war.

Word of Germany's menacing behavior regarding the recruitment of Mexico and the sinking of seven American merchant ships in the months that followed prompted Woodrow Wilson (who had recently re-elected on a platform of passivity) to declare war on Germany on April 2nd, 1917. Four days later Congress made this declaration binding.

Edwin Hubble, a resolute anglophile who had long been in a fighting mood on the issue of Germany and her allies, was compelled to go to war on behalf of his country and to save his adopted England. The friendships that he had made among the German men at Oxford would be set aside for the duration. He wrote to Frost seeking assistance in getting through his dissertation as quickly as possible and to Hale asking for a spot to be held open for him at Mount Wilson until the conclusion of the war.

The dissertation paper he submitted did not impress, but it was the best Edwin could do under the circumstances. At his oral examination his mind was on the battlefield rather than the stars. Three days later, on May 15th, 1917, he reported for duty at Fort Sheridan on the shore of Lake Michigan.[63]

–oOo –

The first six pages of the inaugural logbook for *The 100-inch Reflecting Telescope* describe the details of the development of George Hale's latest monument to human ingenuity and engineering. Scrawled largely in the hand of Walter Adams, the heading on the first page reads, "The Mt. Wilson Solar Observatory of the Carnegie Institution of Washington" and lists "Dr. George Ellery Hale, Director, and Walter Sydney Adams, Asst. Director."[64]

The heading is a reminder of the observatory's beginnings as a solar laboratory funded by the CIW with an initial grant of $10,000, which Hale and his associates had received in 1904. Back then the only large telescope on the mountain was the horizontal Snow solar telescope.

In just 13 short years, employing extraordinary guile, carefully detailed planning, precision craftsmanship and sheer brute force labor from man, mule and machine, Hale and company had transformed the mountain wilderness into one of the world's premier solar and stellar research facilities.

By 1908 the 60-foot solar tower telescope and the 60-inch reflector were both in regular rotation and the first recruits for the stellar department began to ascend the old trail to the summit.

First to arrive was Edward Fath, whose tenure was curtailed abruptly at the end of 1912 with Arnold Kohlschütter the beneficiary of his departure. Adrian van Maanen was brought in that year to conduct photometric research. Kohlschütter, a German citizen, left for home at the onset of the war in 1914 and was captured on his way by British troops in Gibraltar and spent the war as a prisoner.[65] Harlow Shapley arrived in 1914 and began what would become an historic run of research.

The observatory was now home to three solar telescopes and three stellar telescopes. In addition to these, dorms had been added to the new monastery. There was a chemistry lab, a workshop, a powerhouse with a five-kilowatt generator, a large emergency water holding tank for fire safety and numerous ancillary buildings and cottages for visiting observers. Although the facility had been intentionally designed for part time use, it was supplied with enough provisions to support the staff and observers on the mountain for up to a month if necessary. Inclement weather and the frequent landslides could make the road up from or down to the valley impassable.

Despite all of this development, as recorded at the beginning of the log-book on Saturday, January 1st, 1916, the observatory was still officially going by its original name as a facility for solar research. Nevertheless, Adams then briefly described the history of the making of the 100-inch and dome, beginning in the summer of 1913 when the ground was graded and more than a hundred tons of cement for the enormous pier were poured.

A retaining wall had been added to keep the soil from washing away in the rainy season. The pier was designed to be isolated from the dome housing to prevent inducing vibrations in the telescope during operation. The first floor of the interior of the three-story pier was used for storage while the second floor housed the water-cooling equipment for the mirror. They would be turned into dark rooms and washroom facilities. The silvering machine was on the third floor, just beneath the observing floor. The next summer the building was taken up to the level of the observing platform. Dozens of heavy steel beams were to support the floor and dome. Concrete had been poured for the ground level of the building and the shell braced, riveted, and painted.

From this point a single steelworker was employed for four months during the winter of 1914-1915 to grind the rails for the dome rotation system. To achieve this a large-radiused black steel beam was placed on the perimeter of the dome. Next a pair of motors were set on pivots anchored at the center of the pier, one carrying a motor and emery grinder and the other a sweep. Using this apparatus, the rails were slowly ground to the desired degree of precision.[66]

Next the double-walled sheet metal dome, which was built, preassembled, and tested in a spare lot by the Morava Construction Company in Chicago prior to being transported to California, was installed along with the slit doors and trucks for the dome's rotation. These steps were completed during the summer of 1915 and the inside of the dome was then given a coat of black paint to keep the interior as dark as possible during observing.[67]

With the platform finally shielded from the elements, the telescope mounting could be installed starting in the fall of 1915, an arduous undertaking that required hauling the heavy steel pieces of the mounting, some of them weighing over twenty thousand pounds, up the newly widened mountain toll road on the flatbed of an "auto-truck," an early version of the Mack semi-tractor.[68]

Building an advanced facility on the peak of a mountain, at the end of a narrow, winding mountain trail resulted in the loss of many a vehicle and the rescue of both man and beast (Fig. 4.6). The dangers involved in the scope of this operation were revealed in scary detail in July of 1913 when Tom Nelson went over the side of the mountain in a brand-new Mack truck that was hauling three tons of cement to the summit. The incident occurred at around 11 o'clock in the morning at a tight turn in the unpaved roadway near Buzzard's Roost. The turn was so tight that Nelson decided to back the truck up to get a better angle. As he guided the truck backwards, the brakes failed, and the truck shot over the edge and down the steep cliff wall. Walter Adams and an assistant had stepped off the truck to try to chock the wheels using heavy stones to stop it rolling and so were unharmed. Clinging to the wheel, Nelson went along for the ride and was thrown from the vehicle when it began to roll some 50 feet down the slope. He was badly beaten up but counted himself

Fig. 4.6 Upside-down 3-ton truck, Mount Wilson. (Image courtesy of the Observatories of the Carnegie Institution for Science Collection at the Huntington Library, San Marino, California)

lucky as the truck was destroyed on the rocks below.[69] It was almost the end for one of the newest members of the observatory who would become a legendary night assistant on the 100-inch and work many long nights alongside his friend and colleague Milton Humason.

In another accident a high steel worker died in a fall during the dome's construction on September 30th, 1915. The incident was recorded in the logbook of the 60-inch telescope:

> Mr. Moore ironworker fell from top of 100-in. dome to cement pier, 70 feet, and instantly killed.[70]

The telescope tube was too large to transport cross-country by rail and had to be shipped around Cape Horn on the S.S. Alaska, on its way to the port at San Pedro Harbor where it could be loaded onto a flatbed trailer for its final trip to the mountain.[71]

In all, 650 tons of material were brought to the summit in the fall of 1915 at the busiest part of the construction of the dome and telescope. One of the final pieces to be built and installed was the clock drive with its enormous 17-foot-diameter drive gear weighing more than two tons. It was installed at the south pier of the massive, closed yoke of the equatorial mount. When completed, the telescope mirror and tube would rotate north and south within the yoke as that rotated east and west to counter the Earth's rotation.[72]

While maintaining an optimistic tone publicly, behind the scenes Hale and his closest advisors had suffered years of frustration concerning the unsuccessful pouring of the four-and-a-half-ton glass blank intended for the primary mirror. The first attempt had been made in September of 1906 at the St. Gobain Glassworks in Paris. To prepare for its arrival, a new fire and earthquake proof laboratory, affectionately known as the "Hooker building," was built at the Pasadena offices for grinding and figuring its surface. On its arrival in 1908 the massive glass disk was deemed useless by George Ritchey, however. Feeling pressure to get something across the Atlantic to their waiting client in California, the St. Gobain team had poured the blank in three separate stages, giving it the appearance of a three-layer cake, and leaving air bubbles in its core.

After several unsuccessful attempts to pour and anneal a more suitable glass disk, it was decided to use the original layered blank as a last resort. For more than five years Ritchey had toiled away, slowly grinding, shaping, polishing, and testing the mirror until its surface was true to a millionth of an inch. The inch-and-a-quarter deep curve of the mirror required 35 gallons of silver for its reflective surface, which could focus light within six thousandths of an inch of the photographic plate placed 42 feet away with a maximum error of one part in 92,000. All but about twenty hours of shaping had been undertaken by Ritchey using a carefully administered machine that he had designed and built specifically for the purpose. Once the exacting optics genius was satisfied and the final tests were completed, the giant mirror was transported to the mountain on July 1ˢᵗ, 1917, accompanied by 200 men to help guard the mirror against bomb threats and aid the truck in the 8-hour ascent to the summit.

As with its 60-inch predecessor the 100-inch telescope used mercury floats to smoothly rotate the 100-ton instrument and the dome had a double wall to keep the interior cool during warm summer days. A new feature was installing cooling coils in the mirror support system to keep the primary mirror at a constant temperature.[73]

With the telescope and facilities nearing completion, Walter Adams started looking for additional talent to advance the observatory's mission.

The first to join the staff in June 1916 was Gustaf Strömberg from Stockholm University as a volunteer assistant in stellar spectroscopy. John Anderson was hired in July, initiating his remarkable career in the design and creation of state-of-the-art instrumentation at Mount Wilson. Seth Nicholson joined the staff on September 1ˢᵗ and Alfred Joy followed a month later.[74] These two, like Anderson, would play key roles in the development of the MWO into a world class research facility in the decades to come. Roscoe Sanford was recruited in 1918 to work on stellar spectroscopy,[75] as was Paul Merrill in January 1919.[76]

–oOo–

At this point in their lives no one could have suspected Edwin Hubble and Milton Humason would establish a history-making partnership at Mount Wilson.

The problem here was clearly not Hubble, who had both the credentials and the job offer and needed only to survive the war to at least make his appearance at the observatory. The only remaining issue in his mind at that point might have been what to do about his family responsibilities.

Humason, on the other hand, was a rancher and former mule team driver who had little or no formal education and no ambition to acquire one, let alone any designs on entering the field of astronomy. Having just successfully reset his life he would have been happy to live out his days on the ranch near the mountain, teaching his boy to fish and camping and hiking with his family and friends. But once again in his strange storybook life the metaphorical stars realigned and, in the fall of 1917, presented an auspicious new opportunity to the 26-year-old rancher.

No record exists that places Milton on the mountain on the day the mirror arrived on the summit, but he was well aware, perhaps more than most people, of the struggle members of the MWO administration went through to finish it. Jerry Dowd was intimately associated with the work on the mountain in his role as chief electrical engineer. In his role he was responsible for wiring the new buildings and additions to the dorms, connecting to the AC power grid that had been run to the top of the mountain and completing the complex DC power controls that would eventually guide the telescope at both the night assistant's table and the pendant close to the eyepiece or camera. Helen Dowd Humason's ties to her parents meant the families were often together and Milton was surely updated frequently on the events surrounding the observatory in detail.

With money they had saved while operating the Storrow ranch for several years, Milton and Helen had bought their own citrus ranch at 225 West Hillcrest Avenue in Monrovia in late 1915, just as the main pieces of the Hooker telescope were being installed. It was near the trailhead to the old toll road, not far from the ranch that had belonged to Ashbel Strain in the 1880s and 1890s.[77] Within a year they were growing bumper crops of oranges and other varieties of fruits and nuts using the system he had learned from his days on his cousin's ranch. Their modest two-story home had a large front porch and could accommodate friends and family during visits. Milton had long since traded in his cowboy chaps and spurs for a pair of bib overalls and a plow horse.

Living in such proximity to the mountain meant that Milton and Helen could visit as often as work and family life allowed. They knew virtually everyone associated with the observatory and the hotel, catching up with news from friends and exchanging stories from the early days while introducing their young son to the wilderness. For the Humasons and Dowds, perhaps more than any other family associated with it, life on Mount Wilson was very much a family affair. They may have left the daily life on the mountain, but they were still a part of its growth. A family picture from the time shows a two-year-old Billy sitting atop one of the steel girders bracing the 100-inch dome during construction.[78]

The addition of the largest reflecting telescope in the world and the scientific discoveries made in the observatory's first decade of operation lent to its stature as a tourist destination. Guests were invited to tour the grounds, or camp and hike the mountain trails by day and attend the frequent viewing sessions hosted by members of the observatory staff by night. The rebuilt hotel was a modern structure with a swimming pool, plenty of rooms and a restaurant. Thousands of tourists visited the mountain and hotel each year, standing in awe of the telescopes in their domes and in wonder at the beauty of their surroundings.[79]

Notwithstanding the commotion over the events at the observatory, there was still much work to do to get the 100-inch dome and telescope into regular use. On the day the primary mirror reached the summit Adams noted in the logbook that the original observing platform had proven to be "too heavy" and this had "necessitated taking it down and redesigning a much lighter one" – a job that had taken half a year to complete.

Condensation from rain had gathered inside the dome walls, requiring the inside of the dome to be covered with a special compressed wood to soak up moisture and prevent water from dripping into the telescope and mirror.

There were complications in billing, cost overruns, and delays in the delivery of various pieces of the instruments as the scope and pace of the operation grew, a source of perpetual angst for Hale and Adams.

As seems inevitable in a small organization, news traveled fast, and as a member of the extended Mount Wilson family Milton Humason was well aware of these and other issues, some of which had impaired relationships within the organization. (The disintegration of relations between Hale and Ritchey will be discussed in the next chapter.)

The mounting of the massive primary mirror in its support system at the bottom of the telescope was finished on September 22nd, prompting Adams to record that the telescope's completion "now seems a matter of some weeks." Polishing the teeth on the giant worm gear and the installation of the clock drive was finished in the final week of October.

Finally, after more than a decade of frustration, planning, design, redesign, construction, assembly, and testing, the 100-inch telescope saw its first light on November 2nd, 1917 (Fig. 4.7). For this maiden voyage into the sidereal realm Hale and Adams were joined by chief optician George Ritchey, Ritchey's former assistant who had been promoted to the telescope's chief designer Francis Pease, astronomers Ferdinand Ellerman and Alfred Joy, Jerry Dowd, the 60-inch night assistant Wendell Hoge and the acclaimed poet Alfred Noyes plus a handful of instrument makers, opticians, machinists, and carpenters who had worked closely on the project.[80]

With work ongoing to improve the new instrument's operation and create spectrographs and other ancillary instruments and equipment to be used with it, a need arose for additional staff to assist around the observatory grounds.

It was this subject that Jerry Dowd wanted to discuss on a visit he made to the Humason ranch around the time of his grandson Billy's 4th birthday in October 1917. It so happened that the janitor at the observatory was leaving and a new man was needed to fill the position. Along with the usual tasks of

Fig. 4.7 The Hooker 100-inch reflecting telescope at Mount Wilson Observatory with its high north pier (left), Newtonian platform (right) mounted to the dome, and Cassegrain platform (lower left with chair) ca 1925. (Image courtesy of the Observatories of the Carnegie Institution for Science Collection at the Huntington Library, San Marino, California)

custodial work the job included training as a part-time night assistant on the large reflectors. It didn't pay well but free housing would be made available on the mountain. There were no guarantees, but Wendell Hoge had become a vital part of the observatory as the night assistant in charge of the 60-inch telescope. If Milton were to display similar aptitude in the maintenance and operation of large telescopes, he might well find himself in line to fill a position as night assistant.

It was an intriguing opportunity, but acceptance would require some sacrifices. They would be leaving behind the comfortable confines of their beautiful home in the valley for a small house nestled in the tall pines on the southern slope of Mount Wilson, just up the road from the monastery. The small measure of autonomy they had fought hard to achieve would have to be discarded, plus the convenience of living in an established community, school, activities, and access to medical care in case of emergency. Living on the mountain meant they would also be exposed to harsher weather conditions, flooding rains, mudslides and heavy snowstorms and wild animals.

All of this, however, was superseded by the romantic notion of returning to the mountain and the joy of sharing it together as a family. One benefit of the move was that it brought Helen closer to her mother, who was more than happy to spend time with young Billy while his parents attended to their needs on the mountain or in the valley. The hotel could provide additional food and shelter, should they get into real trouble, as indeed could the monastery because on the mountain people took care of each other.

To play it safe, Milton and Helen decided to hold onto their ranch and see how things turned out. If nothing else, the time on the mountain would give Billy the same experience living there they had enjoyed years before.

In November 1917 Milton started work cleaning up after some of the best astronomers in the world, knowing that in a matter of weeks he would begin training in the components, modification, operation, and maintenance of the large reflectors on Mount Wilson.

References

1. Gale E. Christianson, *Edwin Hubble: Mariner of the Nebulae*, (University of Chicago Press 1995), pgs. 81-83
2. *Ibid.* pg. 62
3. A Brief History, (The Queen's College University of Oxford), Retrieved April 21st, 2020, https://www.queens.ox.ac.uk/history
4. *Sir Christopher Wren*, BBC History Page, Retrieved April 21st, 2020, https://www.bbc.co.uk/history/historic_figures/wren_christopher.shtml
5. Gale E. Christianson, *Edwin Hubble: Mariner of the Nebulae*, (University of Chicago Press 1995), pgs. 65-77
6. Denis Judd, *The Life and Times of George V,* (Weidenfeld and Nicolson 1973), pg. 92

7. *The World's worst cruise ship disasters,* (www.ship-technology.com), Retrieved Feb. 11th, 2021, https://www.ship-technology.com/features/featurethe-worlds-deadliest-cruise-shipdisasters-4181089/

8. *Ellis Island,* (Ellis Island Foundation), Retrieved Feb. 11th, 2021, https://www.statueofliberty.org/ellis-island/overview-history/

9. Annie J. Cannon, *Herbert Hall Turner,* (Popular Astronomy Feb. 1931), vol. 39, no. 2 pgs. 59-66

10. Gale E. Christianson, *Edwin Hubble: Mariner of the Nebulae,* (University of Chicago Press 1995), pg. 73

11. *Ibid.* pg. 65

12. *Ibid.* pg. 72

13. *Ibid.* pg. 75-81

14. Thirteenth Census of the United State: 1910, State of California, County of Los Angeles, Pasadena Township, Sierra Madre City, Enumerated April 28th and 19th, 1910, (ancestry.com), Retrieved January 13th, 2011

15. Walter S. Adams conversation about the early days at Mt. Wilson [sound recording], circa 1918. by Adams, Walter S. (Walter Sydney) 1876-1956, Call number: AV 5-70-3; 2013-1223

16. Walter S. Adams, *The History of the International Astronomical Union,* (PASP February 1949), Vol. 61, No. 358, pgs. 5-12

17. Walter S. Adams conversation about the early days at Mt. Wilson [sound recording], circa 1918. by Adams, Walter S. (Walter Sydney) 1876-1956, Call number: AV 5-70-3; 2013-1223

18. *Ibid.*

19. Milton L. Humason, 534 Palmetto Dr., Spouse: Helen, Pasadena City Directory 1914-1916, Occupation: Gardener Samuel Storrow

20. Nathan Masters, *Southern California's Lost Resort: The Raymond Hotel of South Pasadena,* (kcet.org 2012), Retrieved April 23rd, 2020, https://www.kcet.org/shows/lostla/southerncalifornias-lost-resort-the-raymond-hotel-ofsouth-pasadena

21. Virginia Linden Comer, *In Victorian Los Angeles, The Witmers of Crown Hill,* (Talbot Press 1988), pgs. 51-59

22. Interview with Ann Humason Bernt, family archives, Retrieved 2006

23. Newspaper clipping of Milton and Helen's wedding announcement, October 14th, 1911

24. Virginia Linden Comer, *In Victorian Los Angeles, The Witmers of Crown Hill,* (Talbot Press 1988), pgs. 77-86

25. Interview with Ann Humason Bernt, family archives, Retrieved 2006

26. Joseph Petterson Witmer, Emergency Passport Application, April 29th, 1915, (ancestry.com), Retrieved March 1st, 2015

27. Virginia Humason, Emergency Passport Application, April 29th, 1915, (ancestry.com), Retrieved March 1st, 2015

28. Letter from Truus Suermondt to her grandchildren, August 4th, 2014

29. *1910 to 1919 Important News, Significant Events, Key Technology,* (thepeoplehistory.com), Retrieved May 28th, 2019

30. Martin Turnbull, *Hollywood Timeline,* (martinturnbull.com), Retrieved Feb. 12th, 2021, https://martinturnbull.com/timeline-3/

31. Nall Ferguson, *The War of the World: Twentieth-Century Conflict and the Descent of the West,* (Penguin Press 2006), pgs. 72-79

32. Charles Seymour, Ph.D., *The Diplomatic Background of the War,* (Yale University Press 1916)

33. Laura G. Humason shipping manifest, S.S. Cameronia, US Department of Commerce and Labor, Port of New York, August 1914, (ancestry.com), Retrieved October 15th, 2007

34. Virginia Humason and Joseph Petterson Witmer, Emergency Passport Applications, April 29th, 1915, (ancestry.com), Retrieved March 1st, 2015

35. Richard Striner, *Woodrow Wilson and World War I: A Burden too Great to Bear,* (Rowman & Littlefield 2014), pg. 30

36. *Ibid.* pgs. 37-38

37. Alice and Joseph Witmer, and Laura Humason shipping manifest, S.S. St. Louis, US Department of Commerce and Labor, Port of New York, September 4th, 1915, (ancestry.com), Retrieved October 15th, 2007

38. Gale E. Christianson, *Edwin Hubble: Mariner of the Nebulae,* (University of Chicago Press 1995), pgs. 84-86

39. *Ibid.* pgs. 87-90

40. *Ibid.* pgs. 90-91

41. *Ibid.* pgs. 91-93

42. V. M. Slipher, *Spectrographic Observations of Nebulae,* Popular Astronomy, Vol. 23, pgs. 21 24, January 1915

43. William Graves Hoyt, *Vesto Melvin Slipher (1875-1969): A Biographical Memoir,* (NAS 1980), pgs. 409-449

44. George E. Hale, *Yerkes Observatory, University of Chicago,* (U of C Bulletin No. 1 Chicago, February 10th, 1896), pgs. 215-219

45. George E. Hale, *The Yerkes Observatory of the University of Chicago. II. The Building and Minor Instruments,* (U of C March 1897), pgs. 254-270

46. G.W. Ritchey, *The Two-Foot Reflecting Telescope of the Yerkes Observatory,* (ApJ November 1901), Vol. 14, No. 4, pgs. 217-235

47. Edwin P. Hubble, *Photographic Investigations of Faint Nebulae,* (University of Chicago Press 1920), pg. 1

48. Edwin P. Hubble, *Changes in the Form of the Nebula N.G.C. 2261,* Yerkes Observatory, University of Chicago, (PNAS March 9th, 1916), pg. 230

49. Interview of Allan Sandage by Bert Shapiro on 1977 February 8, Niels Bohr Library & Archives, American Institute of Physics, College Park, MD USA, www.aip.org/history-programs/niels-bohr-library/oral-histories/32867

50. Edwin P. Hubble, *Changes in the Form of the Nebula N.G.C. 2261,* Yerkes Observatory, University of Chicago, (PNAS March 9th, 1916), pgs. 230-231

51. Interview of Allan Sandage by Bert Shapiro on 1977 February 8, Niels Bohr Library & Archives, American Institute of Physics, College Park, MD USA, www.aip.org/history-programs/niels-bohr-library/oral-histories/32867

52. Edwin P. Hubble, *Changes in the Form of the Nebula N.G.C. 2261,* Yerkes Observatory, University of Chicago, (PNAS March 9th, 1916), pg. 230

53. Edwin P. Hubble, *The Variable Nebula N.G.C. 2261,* Astrophysical Journal, 44, October 1916, pg. 190

54. Edwin P. Hubble, *Photographic Investigations of Faint Nebulae,* (University of Chicago Press 1920), pgs. 1-2

55. James E. Keeler, *The Crossley Reflector of the Lick Observatory,* (ApJ June, 1900), Vol. 11, No. 5, pgs. 325-350

56. Edwin P. Hubble, *Photographic Investigations of Faint Nebulae,* (University of Chicago Press 1920), pg. 1

57. *Ibid.*

58. *Ibid.*

59. Allan Sandage, *Centennial History of the Carnegie Institution of Washington: Volume 1 The Mount Wilson Observatory,* (Cambridge University Press 2004), pg. 88

60. Edwin P. Hubble, *Photographic Investigations of Faint Nebulae,* (University of Chicago Press 1920)

61. Letter from George Ellery Hale to Walter Sidney Adams, November 1st, 1916, MWO director's files, Henry Huntington Library, San Marino, California

62. Gale E. Christianson, *Edwin Hubble: Mariner of the Nebulae,* (University of Chicago Press 1995), pgs. 98-100

63. *Ibid.* pg. 103

64. 100-inch logbook for 1916-1919, Mount Wilson Observatory, Henry Huntington Library, San Marino, California, pg. 1

65. Allan Sandage, *Centennial History of the Carnegie Institution of Washington: Volume 1 The Mount Wilson Observatory,* (Cambridge University Press 2004), pg. 244

66. 100-inch logbook for 1916-1919, Mount Wilson Observatory, Henry Huntington Library, San Marino, California, pgs. 1

67. Mike Simmons, *Building the 100-inch Telescope,* Mount Wilson Observatory 2018, (mtwilson.edu), Retrieved April 29th, 2020

68. 100-inch logbook for 1916-1919, Mount Wilson Observatory, Henry Huntington Library, San Marino, California, pgs. 2

69. *Auto Accident Insurance,* The Adjuster, Vol. 47, July 1913, pg. 153

70. 60-inch logbook for September 30th, 1915, Mount Wilson Observatory, Henry Huntington Library, San Marino, California

71. 100-inch logbook for 1916-1919, Mount Wilson Observatory, Henry Huntington Library, San Marino, California, pgs. 2

72. Mike Simmons, *Building the 100-inch Telescope,* Mount Wilson Observatory 2018, (mtwilson.edu), Retrieved April 29th, 2020

73. *Ibid.*
74. George E. Hale (Director), *Carnegie Institution of Washington Yearbook, n,15, 1916: Mount Wilson Solar Observatory,* (CIWY 1917), pg. 233
75. George E. Hale (Director), *Carnegie Institution of Washington Yearbook, n.17, 1918: Mount Wilson Observatory,* (CIWY 1919), pg. 187
76. George E. Hale (Director), *Carnegie Institution of Washington Yearbook, n. 18 1919: Mount Wilson Observatory,* (CIWY 1920), pg. 227
77. Milton L. Humason, US WWI Draft Registration Card, Monrovia, California, June 5th, 1917, (ancestry.com), Retrieved August 21st, 2006
78. Interview with Ann Humason Bernt, 2006, from family archive
79. Daniel Medina, *Hotels in the Sky: Bygone Mountaintop Resorts of L.A.,* (kcet.org 2014), Retrieved Feb. 14th, 2021, https://www.kcet.org/shows/departures/hotels-in-the-skybygone-mountaintop-resorts-of-l-a
80. 100-inch logbook for 1916-1919, Mount Wilson Observatory, Henry Huntington Library, San Marino, California, pgs. 3-6

5

Merging Orbits (1918-1922)

The war presses Hale and many on the staff at the observatory into service in administering to the United States of America's war effort. Edwin Hubble eagerly eyes his chance to move to the European theater and the front lines. He arrives just as the war ends, leaving him to return home without a story to tell. Instead, he decides to invent a wartime experience that will portray him as the honorable hero he so desperately wanted to become. As the world is gripped by an influenza pandemic, Hubble rides out the storm of sickness in Europe aiding allied efforts to restore peace. Having survived the experience, Hubble makes his way to Pasadena in the fall of 1919 to take his place on the staff at Mount Wilson. Humason, meanwhile, is in the midst of a personal war against his own inner fears of illegitimacy and doubt. Having found his niche as a photographer in the summer of 1918 he begins to cut his teeth in the field with assistance from friends on the staff. As Hubble arrives on the mountain, Humason has already published his first report on astronomical phenomena and is dreaming of securing a permanent position on the staff but pressing family needs force him to consider a different course. Unless he can convince George Hale of his abilities, he will be forced to move his wife and son back down to the valley and resume life as a citrus rancher. Hale resists promoting Humason owing to his lack of education, but a momentary stroke of brilliance convinces Walter Adams, now serving as acting director of the observatory, to promote him.

As 1918 began, Milton Humason was one of the most recognizable faces of all of those that appeared at Mount Wilson Solar Observatory (MWSO) and yet he had zero skill or training in the research being practiced there. He had arrived at Strain's Camp as a 10-year-old boy in the summer of 1902 on the

© Springer Nature Switzerland AG 2021
R. Voller, *Hubble, Humason and the Big Bang*, Springer Praxis Books,
https://doi.org/10.1007/978-3-030-82181-4_5

back of a mule when the observatory was still only a figment of George Hale's imagination. In the early days of the observatory the original five members – Hale, Adams, Ritchey, Ellerman and Pease – might only have had passing glimpses of the teenage Humason unloading luggage from the mules transporting them and their loved ones to the lobby of the old rustic hotel. More than likely the superintendent of construction, George Jones, would have been in closer contact with the young Humason as a cowboy and mule driver than either Hale or his associates. As the years went by his face had become ever more familiar, almost as ineluctable as the surrounding forest, his nature finely attuned to the mountain wilderness.

The name of Humason had taken on new and important meaning after his marriage to Helen Dowd, the daughter of Jerry Dowd. A frequent and welcome visitor to the grounds in the years prior to his return to the mountain, Milton's innate humanity endeared him to many of the men and women he met at Mount Wilson.

It would by no means be a stretch to say that as he arrived in November 1917 in his new role as janitor, groundskeeper and night assistant trainee, Milton Humason was well known and liked by almost everyone on the mountain. Such is the delightful backstory of this once-in-an-era character. His reputation as a man of amiable spirit and earthen charm, with a penchant for mischief and a soulful manner of speech earned him the admiration of the MWSO staff. The talent he was about to show and his ability to quickly grasp and master the field, in theory and especially in practice, would earn him their respect.

For now, though, he was just cleaning up. The increasing presence of researchers on the mountain and the approaching dry season made the clearing of fire breaks and discarding of flammable materials near the observatory buildings of key importance in the spring of 1918. The two-story addition to the monastery that visiting researchers find on their arrival to Mount Wilson today was completed in 1917, just as Humason was arriving back on the mountain.[1]

In addition to his work keeping the facilities clean, there was work ongoing to remodel the "Hooker cottage," to build a bridge way to the 100-inch telescope from the main road, to repair the steel roof of the powerhouse and the roof of the monastery, and to paint various building interiors and exteriors.[2]

A morning on Mount Wilson was usually between 30 and 40 degrees Fahrenheit in those days and it would warm to between 60 and 80 degrees by the afternoon, depending on the time of year. Despite a heavy rain or snowstorm from one year to the next, the weather was generally pleasant.[3]

Fig. 5.1 Six-year-old William Dowd Humason stands in front of the Humason's "little grey house" on Mount Wilson after a heavy snowfall in 1919. (Courtesy, Ann Humason Bernt family)

Starting his day at a necessarily early hour, Milton would have worn a couple of layers most mornings as he closed the door to the "little grey house" he shared with Helen and Billy on the hill. The house was located on a steep slope between the 60-inch dome and the 10-inch Cooke telescope dome, had white trim, a small bedroom and a potbelly stove to keep the interior warm. The front porch was nicely shaded by a tall California black oak tree[4] (Fig. 5.1).

The first order of business was to raise the large American flag to the top of its new 75-foot-tall flagpole. Nationalism was on the rise, and it was decided that the Stars and Stripes, which had stood on the 100-inch dome for the Fourth of July arrival of the telescope mirror the previous summer, ought to have a permanent home. At the request of a patriotic staff, George Jones and his team installed the pole in the fall near the museum at the top of the observatory grounds.[5]

George Hale had been called upon in April 1916 by the National Academy of Sciences (NAS) "to assist in the organization of the National Research Council (NRC)," leaving assistant director Walter Adams to oversee daily operations. The director had summoned Francis Pease to the Council in March 1918 to take the position of chief draftsman, assisted for part of the year by John Anderson, also borrowed from Mount Wilson. Anderson then joined Arthur King and Harold Babcock doing investigative work for the

Navy. George Ritchey lent his considerable abilities in the field of optics to the Ordnance Department.[6]

For most of the men on the staff, though, there was a general feeling of helplessness as their services were never sought out by their country. Coupled with the general paranoia (however misplaced this may seem from today's perspective) and national sentiment, this led them to establish a makeshift militia for the defense of the observatory against a would-be foreign invasion. With Germany and the Central Powers sandwiched between the Allied Powers of Great Britain, France and Russia, an attack on American soil was anything but certain but Hale conceded that the exercises were harmless and allowed them to continue.

Citizens of the United States spent the duration of the war reading of the various battles and the heroes and anti-heroes who waged them. The Medal of Honor was awarded to the former conscientious objector Sergeant Alvin York, who single handedly silenced an entire German machine gun position behind enemy lines, in the process killing 25 and taking 132 prisoners.[7] The German pilot Baron Manfred von Richthofen, also known as the Red Baron, had over 80 confirmed air combat victories prior to being shot down and killed at the age of 25 in April 1918.[8]

The administration of the war brought tough new and controversial legislation, as in the June 1917 signing of the Espionage Act which made illegal the willful support of foreign enemy powers and insubordination or mutiny within military ranks. It was punishable by 20 years in prison and up to $10,000 in penalties.[9]

The signing of the Sedition Act the following year extended the previous law's reach, making antiwar speech or opinion that cast a negative light on the war or the government's effort to execute it illegal, but this law would be repealed by Congress in December 1920.[10]

Also, in 1917 Congress passed the Immigration Act which imposed a literacy test upon prospective immigrants to the United States and largely denied access to people from China and other Asian countries.[11]

Reporting on the war was earning awards and accolades for reporters and publishers. In 1917 the New York Tribune received a Pulitzer Prize for an article on the first anniversary of the sinking of the Lusitania. The estate of millionaire "yellow journalist" Joseph Pulitzer had established the prize for journalism, fiction, and non-fiction writers through Columbia University.[12]

* * *

In the momentous year of 1917, as the Hooker 100-inch telescope was seeing "first light," George Hale and Walter Adams decided it was finally time to drop the word "Solar" from the name of the observatory. For its entry in the 1918 edition of the Carnegie Institution of Washington Yearbook (CIWY) the official name of the facility became the Mount Wilson Observatory (MWO).[13]

It was a fitting change for a growing facility that now boasted the two big reflectors and a 10-inch Cooke astrographic camera. The addition of the nighttime research of extrasolar stars was a welcome addition for Hale, whose overall mission had long been to unlock the secrets of stellar evolution. Recent investigations of stellar spectra, star clusters, novae and other phenomena were revealing deeper questions concerning the birth, life, and death of a star. This research would dovetail nicely with the ongoing work on our native star by Seth Nicholson and others using the 60-foot and 150-foot solar towers.

To power the telescopes, their domes and related facilities, the laboratory's furnace and other instruments, and the various other buildings and utilities at both the observatory and the hotel exceeded the capacity of the original powerhouse with its large dynamo generator. Fortunately, the new AC power supply recently installed by the Southern California Edison Company could provide much of the service needed for the operation. With the addition of 24-hour power, Jerry Dowd designed and installed a generator to convert AC to DC power, as most of the instruments used DC to that point and began the process of converting the older instruments to the new power source. The original plant was maintained as a backup in case of an interruption to the AC supply.[14]

The eagerly anticipated completion of the Hooker 100-inch telescope and its subsequent commissioning was some time off (it would not be available for regular use for almost two years) and in the interim the 60-inch was the principal instrument for nighttime research.

Every precaution was taken in the education of those who would come into contact with the telescopes to ensure their safety as well as that of the $500,000 instrument. To assume his dual role as janitor and night assistant, Milton Humason first had to learn the complete set up and operation of the telescope and then the basics of astronomical navigation to the satisfaction of his mentor, Wendell Hoge.

A former engineer with the Santa Fe Railroad, Hoge was the chief assistant in charge of the 60-inch telescope and a stickler for detail. A hopeless romantic, he would occasionally embrace his muse in written form in the pages of various publications.[15] In charge of recording the daily meteorological readings at the observatory, he once submitted a brief report on the glorious sunrises and sunsets he witnessed during a four-day period in August 1916 for the Monthly Weather Report of the US Department of Agriculture. He described how "a band of crimson…extending some four to five degrees up from the horizon, appears rather suddenly" and then "widens a little and changes rapidly to a deep orange." To his surprise, it seemed, on "several occasions green and orange streamers were sent up, much resembling the aurora," as he recalled the latter in photographs. "The changes are all very rapid," he continued, "and frequently it is a very beautiful sight." This installment captures the sentiment felt by most who came to Mount Wilson and found themselves entranced by the beauty of their surroundings. These magnificent displays of solar light have long since disappeared from the Southern California.[16]

Hoge was, like many, an ardent Humason supporter and in teaching the young man to operate the 60-inch telescope and navigate the sky he would become, in effect, his first mentor in astronomy.

Given the wind and dust, rain and snow, extreme temperature variations between night and day atop the mountain, the care and maintenance of the telescope was of the utmost importance in ensuring its readiness for observing. When not in use, cages were stored with their mirrors edge up to prevent harmful dust particles settling on their surfaces. The main instrument was cleaned on a daily basis by the maintenance crew, who also configured the telescope for an evening's observing run.

To prevent the mirror from overheating, causing distortions in the images of the objects being photographed, the goal was to keep the air temperature within the dome as close as possible to equilibrium with the outside air. Consequently, the dome doors were often kept shut until twilight to allow the outside temperature to fall to that of the dome before opening the slit doors.

Modern telescopes have thin, state-of-the-art, flexible mirrors and computerized micro-adjusting mirror supports, adaptive laser-guided optical image correction and multi-mirror interferometry. A hundred years ago the state-of-the-art Mount Wilson reflectors featured eight to 10-inch thick mirrors and required full cage changes to go from Newtonian to Cassegrain to Coudé focus and separate mirror assemblies attached to the end of the tube for interferometric measurements. These regular fixture changes were routine if done correctly, but extremely detrimental to the telescope or crew (or both) if done incorrectly. This well-choreographed sequence was explained in colorful detail by Allan Sandage in his *Centennial History* of Mount Wilson. Changing of the focus cage started right after lunch with a meeting between the incoming astronomer, the mountain superintendent, and the night assistants. This varied slightly depending on the lunar cycle. The cage change operations were usually done for both telescopes in the same afternoon.

In the case of the 60-inch, where the cages were lighter and smaller in diameter than its larger cousin, the telescope was simply set to a declination of zero degrees, then driven to a position precisely six hours east. Any further and the primary mirror would start to tip against its constraints. If these failed the mirror would likely be damaged or destroyed – a potentially career-ending error for the astronomer and crew. To secure against balance changes during cage replacement, the horizontally staged telescope tube was bolted to the observing floor using a yoke mounted on the frame and a steel rod that ran between the frame and the floor. Then the cage could be unbolted and lifted off the frame with two heavy straps attached to a crane that was mounted to the rotating dome. The dome was rotated, and the cage lowered to its resting place on the dome floor. Then the required cage was lifted into position and bolted to the telescope. Finally, the yoke and rod were removed, and the telescope was ready for its observing run.

In Newtonian configurations, both telescopes required the astronomer to tell the crew which orientation (north, east, south, or west) the cage was to be oriented. Because neither primary mirror had a hole at its center, flat tertiary mirrors had to be bolted to the tube at a position near the primary for either the Cassegrain or coudé focus to direct starlight out of the tube to the astrographic or spectrographic camera.

When the 60-inch cage change was complete the more complex and dangerous change for the 100-inch could begin. Its massive cages had to be stored at the mezzanine level of the dome, one floor below the observing platform, and were accessed through a hole in the floor.

Again, a crane on the dome was used to secure, lift, and move the cages from one place to another. The telescope was placed in the vertical position and the cage secured by four steel cables bolted to the cage frame. One of the night assistants climbed onto the tube from the Newtonian platform 40 feet above the observing floor to bolt the cables to the frame and disconnect the electrical wiring for remotely controlling the telescope. The crane was reeled to place tension on the cables and then a series of lever locks or "dogs" were turned to release the cage from the tube. It was lowered through the hole in the floor. The new cage was hoisted into position and secured in the reverse of the previous operation. Finally, the assistant climbed back onto the telescope to disconnect the crane and connect the electrical wiring that controlled the instrument's movements from the various platforms. In the case of spectroscopic observations, the spectrograph was mounted to the telescope as a last step in the process. This death-defying aerial ballet was performed at regular intervals on Mount Wilson for 50 years without incident before modern safety regulations required safety harnesses, additional equipment, and procedures[17] (Fig 5.2).

Fig. 5.2 Attaching the Cassegrain cage to the Hooker 100-inch telescope – Tom Nelson stands on the telescope frame to lock the cage in place during a changeover. To his left is the Newtonian platform. (Image courtesy of the Observatories of the Carnegie Institution for Science Collection at the Huntington Library, San Marino, California)

There were no modern computers in those days, no automated assistance and no digital readouts, image correction or capture technology. The quality of an image depended entirely on the equipment available and the ability of the man or woman using it to maintain the telescope steady throughout an exposure. For this reason, as well as the safety of the astronomer and instrument, the assistants had to have a good working knowledge of the sky and understand how to navigate quickly and accurately to acquire a given target. Failure to perform his job competently could result in ridicule or worse, dismissal. An angry astronomer could make an assistant's life very difficult if he complained about the assistant's lack of knowledge or attention to detail.

If Milton succeeded in learning the proper use and maintenance of the instruments, he would slowly be introduced to the staff during observing runs, initially overseen by Hoge. Once Hoge was satisfied that he could be left on his own, Milton would take charge of the telescope and in time could become a full-time assistant on one or perhaps both of the large reflectors. If he did not prove competent, Milton would probably have just returned to citrus ranching. The sometimes-harsh weather, remote location and janitorial wages were not an ideal fit when raising a son, especially when there was a nice life to be made in the valley. Fortunately, he worked hard to learn the ropes under Hoge's patient, watchful eye and was able to put in his first solo stretch as the relief night assistant in January 1918.

It is likely that the astronomer he was paired with that week was none other than Harlow Shapley, who was on the mountain for a five-night stretch to take plates of star clusters for a photometric study at the Newtonian focus. Ritchey had used the same set up the previous night, so no cage change was necessary. According to the logbook for the 60-inch telescope none of the five nights had clear skies but progress was made in partial cloudiness on two nights. As a result, Shapley's run yielded just six plates from a total exposure time of an hour and ten minutes.[18]

If the 27-year-old neophyte needed any more inspiration to spur him toward a life in astronomy he could not have found a better man for the job, at a better time in his career, than Shapley. At that time Shapley was awaiting the arrival in print of his groundbreaking report *Globular Clusters and the Structure of the Galactic System* in the February issue of the Publications of the Astronomical Society of the Pacific (PASP). Through his research on star clusters, he had reshaped and resized the Milky Way galaxy and, surprisingly for some, placed the solar system far from its center.[19]

The inclement weather during those five nights would have given Shapley and Humason ample time to get acquainted. Both natural storytellers, Shapley related his childhood on a Missouri farm and a one-room schoolhouse and Humason his upbringing on the banks of the Mississippi River. It is likely that these two Renaissance men established their lifelong friendship during these early discussions.

In the years since his arrival at the observatory, Harlow Shapley had used the pulsation periods of variable stars known as Cepheids (so named for the discovery of the first of the type in the constellation of Cepheus) to formulate a measure of distance for the clusters he used for his analysis of the Milky Way system. In addition to increasing its scale by a factor of 10 and discovering the offset location of the Sun, he would publish a 16-page treatise entitled *The Age of the Earth* in the PASP in September 1918. Here Shapley displayed the withering intellect, broad speculative powers and literary fireworks that were to make him a favorite speaker and author of many consumers of scientific works in the years to come. Comparing analytical evidence available at that time from the field of geology, including the radioactive measurement of rocks and minerals and heat loss from Earth's molten core to his own analysis of stellar radiation and solar radioactivity he led the reader through an inquisition of our planet's age and the age of the Sun relative to the age of the universe. In his theoretical chronology he put the "Origin of Earth" around 3 billion years ago.[20]

His ability to fuse practical evidence across disciplines and his willingness to speculate dramatically in writing made Shapley a media darling and the brightest star in the stellar department at Mount Wilson in the years prior to his departure in 1921. That was a good friend to have in your corner if you were an apprentice observer who aspired to the ranks of the men at MWO. As the weeks turned into months, Shapley would engage Humason as his assistant on laboratory work and introduce him to investigative apparatuses such as the blink comparator.

After just a few months of working with the telescopes and rubbing elbows with the astronomers on the mountain, Milton had become rapt by the romance of the night sky. He would later remark on more than one occasion that his decades long deep-sky explorations reminded him of the poetry of George Gordon (Lord) Byron. Although he never recalled any of his favorite verses, one can imagine the smitten novice observer enjoying this from Byron's narrative poem, *Childe Harold's Pilgrimage*:

Ye stars! which are the poetry of heaven!

If in your bright leaves we would read the fate

Of men and empires, – 'tis to be forgiven,

That in our aspirations to be great,

Our destinies o'erleap their mortal state,

And claim a kindred with you; for ye are

A beauty and a mystery, and create

In us such love and reverence from afar,

That fortune, fame, power, life, have named themselves a star.[21]

It was this life that Humason now hoped to embrace, one of service to the breadth of scientific knowledge. Following the example set by his mentor, Wendell Hoge, he would act as co-pilot and confidant, assistant and adviser to the scores of astronomers who would make their way to Mount Wilson in the years to come.

His goal was to become a fully-fledged night assistant and not just an assistant in relief, perhaps even someday taking charge of one of the mighty reflectors opposite Hoge. It was all new territory for the former cowboy and muleskinner, but Milton had gained a sense destiny he felt compelled to fulfill. This would become a little closer in the summer of 1918 upon meeting Hugo Benioff, a visiting scholar on his summer break from Pomona College.

Victor Hugo Benioff was born to a Russian immigrant father and a Swedish mother in Los Angeles, California on September 14th, 1899. A boy genius, he took an early interest in astronomy and may well have become an astronomer if not for his distaste for the late-night hours and cold weather involved in long observing runs at high altitude facilities. His distinguished career started in 1920 with his selection to Phi Beta Kappa. At the conclusion of his undergraduate studies the next year he would work at both Mount Wilson and Lick observatories before turning to seismology in 1924, a field that he would come to dominate alongside Beno Guttenberg, Frank Press, and Charles Richter at Caltech (Fig. 5.3). He would receive his doctorate from that institution in 1935.

His deep understanding and intuition enabled Benioff to develop sophisticated seismic equipment capable of reading geological events with unprecedented accuracy. These would form the center line for international efforts to understand the causes and effects of seismic activity. These same capabilities would be employed during World War II to improve radar and anti-submarine ranging devices. Throughout his career Benioff showed a diversity of interests that led him periodically into the realms of lightweight bicycle design, exotic kite construction, optics, jet engine mechanics and treatments of various cancers. His interest in musical instruments led him to develop electric violins, cellos, and pianos.

Perhaps Benioff's greatest contribution to science would come between the years 1951 and 1958, when he introduced ideas that would form the basis for the field of plate tectonics in the 1960s and beyond, in particular strain accumulation, release, rebound and relaxation in the evolution of earthquakes. The seismic disturbances that occur in downward moving plates during the process of subduction were named in his honor as Benioff zones.

His myriad and diverse contributions would win him numerous awards and selections to elite organizations in his lifetime, from the NAS to the Geological Society of America (GSA) of which he was president in 1958. He was a fellow of many learned societies, including the American Academy of

Fig. 5.3 The stars of the Seismological Laboratory (From left) Frank Press, Beno Gutenberg, Hugo Benioff and Charles Richter in 1956. (Image courtesy of Caltech Archives)

Arts and Sciences (AAAS), the Royal Astronomical Society (RAS), the Acoustical Society of America (ASA) and the American Physical Society (APS).

It was the fact that Benioff's personal qualities were so aligned with those of Humason that brought these two great friends together. An avid outdoorsman, Benioff loved to spend his leisure time communing with nature and involving himself in the purchase and sale of various properties. He spent time citrus ranching and he loved fishing, camping, and hiking. He was proud but humble, had a keen intuition for human personality and interaction and didn't suffer fools or fakes lightly. Virtually all of these characteristics were a perfect match for his best friend Milton Humason who had lived the life of the old-world mountain man years earlier. Benioff was also highly sensitive to the plight of less fortunate people and his sympathy for the underdog immediately drew him to Humason's aid.[22]

Possessed of an almost indefatigable curiosity, Benioff would have been captivated by Humason's whimsical yarns of the early life on the mountain, and Benioff's optimism and positivity regarding his new friend's potential might have inspired the latter to press on.

In the summer of 1918, the two friends worked together during Benioff's observing runs using the 10-inch Cooke and he began to teach Humason the basics of astrophotography, walking him through how to prepare, load and expose images on a glass plate inserted into the camera. After each run, the two would walk to the darkroom to process the images on the plates using light sensitive silver bromide paper and hypo as a fixer.[23]

Humason proved to be a quick study and a steady hand at direct photography, and it was not long before he was producing high quality images of star fields and nebulae with the Cooke telescope. The clarity of the images he was able to capture may have been due to a particular type of color blindness that enabled him to see contrasts between black and white extremely clearly.[24]

By summer's end, Benioff was convinced Humason had a gift that could be extremely useful in the research at the observatory. He made this known to the acting director Walter Adams, as well as Harlow Shapley and Seth Nicholson, his own mentors in stellar and solar physics respectively. The quality of the relief night assistant's images was undeniable, and soon other friends and admirers began to lend their considerable talents to his education.

Alongside instruction in the techniques of astrophotography from Benioff and the basics of photometry and blinking of plates from Shapley, Humason soon began learning the math involved in the field from Seth Nicholson.

Shapley and Nicholson had published a report on the "main belt" asteroid 878 Mildred (named after Shapley's daughter) in 1917 soon after Nicholson's arrival in Pasadena, and they knew and liked each other.[25]

It may be that Shapley suggested to Nicholson that he instruct Humason in the physics involved in figuring angular diameters, stellar parallax, magnitude, orbital calculations, and other aspects of the field. Nicholson and Humason would form a close friendship and were neighbors along with Adams and his family for a number of years.

Bolstered by the praise that he was receiving, Humason took as much telescope time as he could and spent all his spare time studying. Soon, both Adams and Paul Merrill would figure prominently in his development as well. The transmogrification of Milton Humason, cowboy astronomer, was well under way.

* * *

The USS Imperator was a monster of a ship. Her 909-foot-long hull was more than twenty feet longer than the RMS Titanic, which had sunk in these waters six years before Edwin Hubble's return from the European theater. The pride of the Hamburg-America Line was the first of three sister ships constructed by the Germany-based company as part of a global competition between cruise ship builders to capture the title of the world's largest and most luxurious liner. With a displacement of 51,700 tons the giant steamer sported 11 decks and a Pompeiian-style swimming pool in first class and carried over five thousand passengers and crew when fully loaded.[26]

But on this voyage, it was like a ghost ship. She had been laid up at harbor in Hamburg at the start of the war and remained there for its duration before being pressed into service by Allied forces as a personnel transport to ferry troops to New York from Brest, France. Hubble was one of a scant few aboard the mighty ship on this final transport in August of 1919 before it was to be refurbished and reintroduced as part of the Cunard Lines. The war, like the ship that now carried him home, had left Edwin feeling empty. A year and a half of training and preparation had passed without incident, leaving him and his fellow officers and infantrymen with no stories to tell of the glory and perils of war in the trenches.

Having first secured his future place at MWO with director George Hale, then quickly cobbled together his dissertation, he had gone with all haste to what he was sure would be the aid of the friends of his adopted English motherland whose bodies he had imagined in nightmarish detail lining the battlefields on the Western Front.

He had joined Company 10 of the Officers Reserve Corps as a private, stationed first at Fort Sheridan on Lake Michigan in Chicago for a month of intensive training in which he excelled at all matters of military discipline and command and rose to the rank of Captain. He and his fellow officers were assigned to Camp Grant on the Kishwaukee River south of Rockford, Illinois, a newly established 3,000-acre encampment and training grounds with 1,400 buildings designed to serve as a temporary city with most of its common necessities on-site.

The first train loads of men arrived in early September and shortly thereafter Captain Hubble was assigned command of the 25 officers and 600 men of the 2nd Battalion, 343rd Infantry of the newly formed 86th Army Division. As his ship plowed through the Atlantic, he remembered the whirl of emotions that swept the enormous training facility in the early months at Camp Grant. Under his supervision his officers trained the men in close combat, riflery, camouflage, mine detection, shelling, mustard gas and trench warfare. He recalled with pride his "run-in" with the regimental commander, Colonel Charles R. Howland, who strode onto the shooting range and fired off a score of 46 with a rifle before Edwin casually fired a perfect 50, earning his commander's respect from that moment on. But that turned out to be the high point of his service in the war. After months of training for combat, the morale within the ranks and the officers' corps began to sink. The harshest winter on record made the early months of 1918 almost unbearable as temperatures dropped as low as minus 27 degrees Fahrenheit.

As summer approached, the battalion's departure for the shores of England was finally imminent. The other regiments of the 86th had already moved out for the front. By this time Edwin had been promoted to the rank of Major. The 343rd boarded a ship at Hoboken, New Jersey and sailed through the Hudson and out to sea bound for Europe with a naval escort that included a battleship, a naval blimp and anti-submarine vessels.[27]

It had taken almost seventeen months for his unit to arrive in England for transport to the front. By that time in early October, the Meuse-Argonne Offensive had pushed German forces back behind the all-important Hindenburg Line and the war was nearing its end. On November 9th Kaiser Wilhelm abdicated his throne and two days later Germany signed an armistice that ended the war before Edwin or his men could see any action.

Of the more than one hundred thousand of his fellow countrymen who fell during the war, less than half had died on the battlefield, the greater majority having died of accidental causes or disease, especially the Spanish Flu which was beginning to sweep the world. The epidemic hadn't begun in Spain, as some would have thought, but because Spain remained neutral during the war its press could report on the outbreak of the disease as it spread. The flow of information about the spread of the disease disseminating from the country led the world to refer to the epidemic which overwhelmed other nations as the Spanish Flu. In the end, the global flu epidemic would more than double the total number of deaths caused by the Great War, although an accurate number of the dead caused by the disease was far more difficult to ascertain than those killed in the war, where belligerents tended to fall in isolated areas. Mount Wilson was no exception. A local outbreak obliged staff members "to put on influenza masks for about two weeks," as Adams related to Hale in a note to the director in Washington D.C.[28]

Despite the conditions in Europe, Edwin decided to stay on with occupation forces after the close of hostilities to aid in the effort to reconcile the damage caused and transition the region to peacetime. His love of England led him to Cambridge in the spring and summer of 1919, where he slowly returned to the field of astronomy. He spent many evenings as a guest of the astronomer H.F. Newall, who later recommended him for a membership to the RAS. This led to a serendipitous encounter at a meeting of the RAS in London with three members of the MWO staff: assistant director Walter Adams, the solar physicist Charles St. John and the stellar photometrist Frederick Seares.

Finally, answering a plea from George Hale to return to the US and begin life and work anew as an astronomer, Edwin had boarded the giant liner upon which he now sailed for home. There was much to look forward to in his new life in California. In his letter, Hale assured him the 100-inch telescope was nearing completion and would soon be available for research. In any case his salary of $1,500 a year would be a welcome increase from his time at Yerkes. This entry level salary was sufficient to get him going and Hale promised increases commensurate with results, which he was more than a little determined to deliver.[29]

His brief stint in the officers' corps of the army had proven Hubble a worthy leader. His size, natural athleticism and manner had won the respect of his men as well as his superiors, but his talents were ultimately left unexploited in the field of battle. As was no doubt the case with many of those who yearned to see action in the war, Edwin started to invent his wartime story, imagining himself commanding his men in the trenches during a great battle, bravely leading the living while resisting his natural instinct to rescue the fallen. He would later claim that he was wounded and knocked unconscious by a German artillery shell that left him with a concussion and other minor wounds. In another incident he was trapped in an observation balloon as artillery shells fell in the French city beneath him. In time these fictions would take root, forming a brief but heroic account of his time at war, one he could only reflect upon in a dream.[30]

***

The end of the Great War did little to stop the unrest that had helped create the conflict. In July 1918, as Hubble was receiving his new commission as Major of the 2nd Battalion, the Russian Czar Nicholas II and his family were executed by the Bolsheviks, ending 300 years of Romanov dynastic reign.[31]

As communism took root in Russia, violent anti-government and anti-business groups arose in the United States in protest against labor practices and the curtailing of the freedom of speech under the Sedition Act. The drumbeat of communist sympathy boiled over, and white supremacists began leading attacks on black citizens in race riots that rattled tensions in the country, killing hundreds of Americans by the fall of that year. Union organizers led strikes in Seattle, Boston, Chicago, Pittsburgh, and New York to demand higher wages, cut during wartime to assist in the war effort. Mail bombs from anarchist groups were sent to prominent politicians, businessmen and newspaper executives they believed to be enemies of self-government. The bombings and other uprisings gave rise to the Red Scare of 1919-1920 that caused alarm among citizens and government of the possibility of a communist revolution.[32]

The strain of economic reparations and the loss of a large share of their young working-age men weighed heavily on the German, Austrian and Italian governments. Likewise, the heartbreaking loss of men and infrastructure in England, France, Russia, and smaller allied nations would linger for decades to come.

While US casualties in the war were comparatively low, the public, little more than two generations removed from the carnage of the Civil War, nevertheless recoiled at the loss of life. The second wave of the flu pandemic that took hold in the country before the final bombs fell would ultimately claim the lives of five times as many Americans as the conflict itself. With the scourge of influenza claiming lives with increasing swiftness, much of the nation was yearning for peace. Pacifist magazines sprang up. President Woodrow Wilson delivered his *Fourteen Points* speech on January 8th, 1918, outlining a way forward through increased trade, open markets, and the spread of democracy.[33]

The 18th Amendment to the US Constitution calling for making the manufacture, sale, and distribution of alcohol illegal in every state passed through Congress over a presidential veto and was ratified in January 1919. Its signing into law further fanned the flames of anti-government sentiment. Had he still been alive, John Powell Hubble would have been elated by the passage of the new bill.

In contrast to the chaos caused by disease, labor unions and radical elements within the country, there were signs of social and civil evolution. The US Airmail Service became operational on May 15th, 1918, and the Grand Canyon was designated a National Park by Congress the same year. The 19th Amendment to the Constitution, awarding women in the majority of states the right to vote, was passed on June 4th and would be officially adopted the following year after ratification. Women's suffrage had been legal in Wyoming, Idaho, Colorado, and Utah prior to the turn of the twentieth century but the rest of the country had been slow to adapt to the changing political climate. Puerto Rico, one of the islands won by the naval victory in the Spanish-American War in 1898 became a US territory in 1917, giving its people American citizenship. While President Wilson was helping to guide the establishment of the League of Nations to promote enduring peace between the nations of the world, in 1919 the US signed an agreement with its southern neighbor Cuba, thereby acquiring a small stretch of land called Guantanamo Bay for the creation of a naval station.[34]

***

On arriving in New York City, Edwin began his long journey cross-country to Los Angeles, eager to access the most powerful telescope ever constructed to resume his romance of the stars. He made a brief stop in Chicago to visit his mother and siblings who were pleased to see him back safely from the war. The next day he boarded a train for San Francisco where, a week later, he was honorably discharged at the Presidio.

In his biography *Edwin Hubble: Mariner of the Nebulae,* Gale Christianson gives only a single sentence to Hubble's stay in Chicago with his family before going on to California. Why would a man who had just come back from war not spend at least a few days with his family?

The easy answer would be that Hubble was reluctant to push back his arrival at Mount Wilson any further for fear of jeopardizing his position there. Although Hale was eager to have him, it is doubtful that a few more days would have made a difference.

Another answer might be that the while serving in the army during the war Edwin began to distance himself from the responsibility that had played such a pivotal role in his decision making in the past. He desperately wanted to begin his career in astronomy and there was a healthy salary waiting for him in California. The time he spent in England prior to sailing home likely rekindled his love of that country and culture and sewed the root of Anglophilia that he would embrace for the rest of his life. His adopted persona of an English gentlemen was incongruous with the life he had lived prior to his days as a Rhodes Scholar at Oxford, and alien to his family and friends back home. But there was, perhaps, an even darker issue rooted in the relationship he had with his father.

In an interview with Bert Shapiro in 1977 astronomer Nicholas Mayall mentioned that Hubble "dropped his middle initial early in his career." When pressed, Mayall said that he had inquired about this fact with Hubble's wife, Grace, who had replied curtly that it "was something [she didn't] wish to speak about."[35]

Hubble's middle name was Powell, the same name as his father John Powell Hubble. A glance at his bibliography (Appendix 2) reveals that Edwin used his middle initial through his years at Yerkes and in the first article he published at Mount Wilson. That first paper was a joint report published with Frederick Seares, who was then mentoring the new recruit and was the editor of the MWO contributions for many years. It is likely that Seares drafted this report and used Hubble's middle initial based on the recorded history of his published papers. The first paper that Hubble published solo was *A General Study of Diffuse Galactic Nebulae,* which appeared in the ApJ in 1922 without his middle initial. Thereafter he used it only twice prior to 1927, and once afterward, and that was on the transcript for the Rhodes Memorial Lectures that he delivered in 1936.

While not absolutely definitive, these facts suggest that Edwin Hubble at the very least attempted to distance himself from his past and may have decided to omit his middle initial in a final act of rebellion against the man who had nearly snuffed out his dream of a life in astronomy.

On reaching California, Edwin spent a day at the Lick Observatory on Mount Hamilton, which he would visit often in future years, then continued to Los Angeles (Fig. 5.4).

He reported for his first day of work on Wednesday, September 3rd, 1919, the same day Woodrow Wilson set off by train on a 22-day, 8,000-mile tour of the nation to promote the League of Nations. On arriving in Los Angeles on the 20th Wilson and his wife Edith were greeted by throngs of tens of thousands of people when their motorcai paraded through the streets. The following week he fell ill after giving a speech in Pueblo, Colorado, 250 miles southeast of the Rio Blanco Ranch where Hubble would suffer a massive heart attack 30 years later. On October 2nd, back in Washington D.C., Wilson suffered a nearly fatal stroke which effectively ended his presidency as well as stifling any hope of realizing the creation of the League of Nations.[36]

Finding the housing situation lacking, Hubble bounced around from one room to another until settling in a room at 55 Euclid Avenue, home to the head of the humanities department at Caltech, Clinton Judy, and seismologist Harry Wood of the Carnegie Institution. Judy had spent time at Oxford and Hubble felt quite at home in the small, pleasant confines and book-filled shelves. It was only a mile walk northeast to the observatory offices, no struggle for the sporting 30-year-old astronomer, initiating the daily routine he would follow for the rest of his life of walking to and from the observatory offices.[37]

After waiting six weeks for his first turn on the mountain, Hubble embarked on his first evening of viewing on Mount Wilson using the 10-inch Cooke telescope in October and followed that up with a three-night run at the 60-inch. His name appeared near the bottom of page two of logbook number four of the 60-inch for the years 1918 and 1919. Among those listed with him were Paul Merrill, one of the men he would first befriend and later distance himself from, and Roscoe Sanford, who had joined the staff a year earlier. Also, present was Hugo Benioff, who spent the summer of 1919 working in both the solar and stellar departments, as he had the previous year. The entry five lines above Hubble's name and date of arrival read: "Milton L. Humason – Relief night assistant and janitor."[38]

Fig. 5.4 Edwin Hubble shortly after his arrival to California ca 1919. (Image courtesy of Caltech Archives)

George Hale returned from the east coast with his health renewed by the war work he had been doing for the past couple of years. For reasons neither he nor his physician could understand his mental state, which had betrayed him for so many years, had been stabilized during administration of research and development on behalf of the country. It seemed he was fine administering to the management of facilities or other official duties but whenever he turned to solar work, he suffered a break down.[39]

Rejuvenated, Hale inserted himself into the activities at Mount Wilson with a vigor and enthusiasm he had not shown for nearly a decade. In volume 18 of the CIWY for 1919, he laid out an extensive history of the mission of the observatory and its subsequent evolution. In retrospect it reads like the account of a man firmly aware of his past infirmities and the prospects for their imminent return, who, while still firmly in control, was determined to define afresh his intentions for the future of the organization that he had nurtured from its inception. It is a remarkable piece of scientific history which warrants review in relation to the developments of this story.

"The purpose of the Observatory," as stated in his introduction, "was to undertake a general investigation of stellar evolution, laying special emphasis upon the study of the Sun, considered as a typical star; physical research on stars and nebulae; and the interpretation of solar and stellar phenomena by laboratory experiments."

He cautioned against the temptation of straying too wildly from the overall mission, an impulse that on at least one occasion had unfolded with damaging effect on science, if not the observatory, slowing progress in telescope technology.

In Hale's determination a difficult course needed to be steered "between the dangers of atrophy that may result from fixed procedure and endless routine, and the losses inevitable in an unstable and shifting policy."[40]

The former "danger" could have been a reference to the Harvard Observatory director Edward C. Pickering, who had died in February. With the help of Williamina Fleming he had published the first *Henry Draper Catalogue* in 1890, a comprehensive and exhaustive guide to the spectral makeup of a quarter million stars photographed using the 8-inch Bache telescope.[41]

The latter "danger" likely referenced the still burning resentment Hale felt toward his former friend, colleague, and chief instrument maker and optician George Willis Ritchey whose insubordination had nearly sent Hale over the mental cliff during the creation of the 100-inch telescope (about which there will be more in the next section).

However, given recent developments in physics, instrumentation, and discovery he had concluded that some expansion of the overall mission would be necessary. As examples he noted the discovery of the relationship between stellar motion and spectral type by Jacobus Kapteyn and William Campbell, the work on dwarf stars by Ejnar Hertzsprung and Henry Norris Russell, and the work on globular clusters and galactic structure by Harlow Shapley.

Recalling the history of the solar investigations that preoccupied him as far back as 1890 when he was inventing his spectroheliograph at the Massachusetts Institute of Technology (MIT) and his subsequent use of that instrument with his 12-inch refractor at the Kenwood Observatory at his family home in the Hyde Park section of Chicago, he briefly stated the evolution of the solar facilities at Mount Wilson, culminating with the creation of the 150-foot solar tower telescope whose 75-foot vertical spectrograph plunged deep underground to project an image of the Sun 16.5 inches in diameter. By contrast, his original instrument could deliver a solar image only two inches across.

The 60-foot solar tower that was commissioned in 1908 had successfully alleviated the shimmering effect of heat radiation which impaired the Snow horizontal telescope but had deficiencies of its own. In the first place it was buffeted by sometimes high winds streaming over the treetops where the dome was situated. In the second place the solar image, which was coalesced on a large observation table in a small lab at the base of the telescope, was still bedeviled by heat radiation. The frame-within-a-frame design and deep underground spectrograph of the newer tower, along with its higher position on the tree line, eliminated these characteristic flaws with the added advantage of providing a larger and more detailed image. By 1919 the 150-foot solar tower was the state of the art for solar spectroscopy and direct photography.[42] In 2019, if you are fortunate enough to know one of the handful of wonderful amateur solar photographers who regularly view the roiling surface of our native star in awesome filtered detail in the privacy of their back yards, you'll find it quite a sight to visit during the few hours of the day that their telescopes are in operation.

While the solar towers may seem quaint by modern standards, it bears remembering that the MWO telescopes as a whole and the discoveries they enabled played an important role in the development of telescope technology and cosmology into the new millennium.

For instance, the discovery of the method of spectroscopic parallax by Walter Adams in the second decade of the 20th century was a direct result of the discovery in 1906 using the 60-foot solar tower of the cooler temperature variations in sunspots. This led Adams and Kohlschütter to relate these temperature variations with the spectra of stars, then correlate these measurements with their absolute magnitudes. Combining the results with parallactic measurements to nearby stars they devised the means of computing the absolute magnitude and distance of stars by the measurement of their spectra. This new highly efficient means of determining magnitude and distance spectroscopically would later be borrowed to create a mean absolute magnitude measurement for nebular distances by Hubble in his attack on the expansion problem with Humason.[43]

Hale devoted a great deal of time in his opening discussion to the history of successes in the physical laboratory in corroborating experimental data from direct photography and spectroscopy, and the planned expansion of its import on the observatory. Underpinning this development was the General Electric 500-kilowatt generator that had been installed in the lab in Pasadena, yielding 4,000 amps of 220-volt DC power to the expanding plant. It was being used to operate a high energy electromagnet for research on the Zeeman effect, the splitting of a spectral line into several components by a magnetic field. On noticing this effect in the spectra of sunspots in 1908 Hale had inferred that the Sun possesses a magnetic field.

Finally, Hale circled around to the 100-inch reflector that was by then in regular use on the north side of the summit of Mount Wilson. Its unprecedented light gathering capability would "bring into view hundreds of millions of stars beyond the reach of the 60-inch." The new reflector's greater aperture and focal length (134 feet at the Cassegrain focus) would facilitate spectroscopy of stars down to the 11th magnitude, well beyond the 8th magnitude limit of the 60-inch, and thereby allow Paul Merrill to investigate long period variable stars in a program to which Milton Humason would contribute in the years to come.[44]

Of primary importance to Hale was the extension of the research on the "star streams" noted in the Kapteyn Selected Areas already underway at Mount Wilson. A photographic and photovisual magnitude scale system had been created by the titan of photometry on the mountain, Frederick Seares to about the 18th magnitude and the new telescope would allow him to extend his research to around the 21st in the next 10 years. This work would also be performed by Humason after his mentorship with Seares on stellar color indices of the stars of the North Polar Sequence in the early 1920s.

Perhaps in anticipation of the arrival of Hubble, whose interest in nebular research was apparent in his as yet unfinished dissertation, Hale pointed out the 100-inch's potential for studying nebulae, with particular regard to their composition and motion.[45]

Strictly speaking, Hubble's desire to create a classification system for nebulae wouldn't quite have suited Hale, whose overall vision for the observatory was based on cosmological pursuits and not merely the classification of stars or systems of stars. In the director's view, these long-term compilation projects were better suited to the programs run at Harvard and other institutions that lacked the light gathering power he was making available on Mount Wilson.

If he was under any apprehension about Hubble's intentions, he would never have mentioned it in his annual report to Carnegie. Hubble's reputation as a gifted and promising observer preceding him and Hale was doubtless willing to allow the man some tether while he took his measure.

As befitted the demeanor of someone whose life's work was steeped in the courtship of wealthy donors, Hale's political acumen and poise was on full display during the testing of the new reflector's operational and technical capabilities. At least some of the "unavoidable delays," as he called them, were basically unforced errors in the heat of the prolonged and anxious problem solving during the telescope's creation. Bruised and battered egos fueled by passionate individual beliefs, and the general anxiety of the telescope's late benefactor John D. Hooker, eventually eroded trust and caused not just delays but otherwise avoidable deficiencies in the telescope's design.[46]

The telescope had several tedious mechanical flaws, some of which could be overcome and others that would have to be worked around. The discovery of a periodic error in the drive-clock was to challenge observers for years to come until it was fully understood and carefully timed. Whereupon the problem of losing a target could be solved once you knew how. Other issues concerning the edge support for the four-and-a-half-ton primary mirror, the overheating of mercury in the floatation system, and comatic aberration (distortion) of point sources away from the axis of the parabolic primary exacerbated by astigmatism caused by atmospheric disturbances around the telescope were also workable but required a great deal of careful fine-tuning and preparation. Nevertheless, it was clearly capable of yielding unprecedented results.

"Since the armistice" and the "completion of the optical [system]" for the "Ordnance Department," the job of finishing and fine-tuning the 100-inch had been in full swing. With "work on the Cassegrain spectrograph, double-slide plate holder, tube balancing system, coudé mechanism, cage clamps, mirror temperature control, mirror silvering equipment, Newtonian and Cassegrain cages and mirror mountings, driving clock, sidereal indicator, burnishing apparatus, observing platform, dome ladders and stairs, and dark room" all now approaching completion, the world's most powerful telescope was poised to start its reign.[47]

Edwin Hubble was among those who would be expected to exploit the telescope, but no one would have predicted Milton Humason's rise, nor his future collaboration with Hubble as the principal driver of the evidence for the Big Bang.

***

At this point it is appropriate to step aside to consider the story of George Willis Ritchey whose long career as chief optician and instrument designer set several industry standards, some of which remain to present day.

Ritchey's formidable skill and pioneering approach to developing large-scale telescope optics helped to create both the world's finest astronomical instruments and the renown of George Hale during their 24 years of collaboration. But his greatest vision for the future of telescope design and optics, which would result in the modern Ritchey-Chrétien Telescope (RCT), became the victim of financial circumstances, professional recklessness, and outright stubborn arrogance. His clash with Hale and Adams over the 100-inch Hooker telescope led to his dismissal from the MWO in the fall of 1919 for misconduct and his declining stature in America. It was an event that set modern telescope technology back half a century and directly impacted the work of Hubble and Humason.

Born in Tuppers Plains, Ohio on the last day of 1864 to a family of moderate means, Ritchey was educated as a furniture maker. He got his first taste of the field of astronomy while working at the Cincinnati Observatory as a student assistant in the late 1880s.

In 1888 he moved his family to the south side of Chicago and took a job at the Chicago Manual Training School where he had access to much of the tooling he needed to continue his experiments in optics and machining small telescopes. He was encouraged by the dean of the school to join the Chicago chapter of the Astronomical Society of the Pacific (ASP) whose secretary was none other than George Ellery Hale, then a young MIT graduate.

Hale and Ritchey immediately recognized they could be of service to one another, and a friendship formed out of mutual respect. Ritchey began working as Hale's Superintendent of Instrument Construction at the Kenwood Observatory in 1895, going on to design and build the 24-inch reflector at Yerkes that Edwin Hubble later used for his graduate research.[48]

As Hale began his move to California in 1903 Ritchey was among the first to join him, others being Adams, Francis Pease, and Ferdinand Ellerman. Ritchey's 1904 paper, *On the Modern Reflecting Telescope and the Making and Testing of Optical Mirrors*, written for the Smithsonian Institution, had swept through the field of optics. His photographs of the Moon were used by Hale in his argument to Carnegie seeking funds to create an observatory on Mount Wilson.[49]

In 1908 Hale and company presented the world with the 60-inch reflector designed almost entirely by Ritchey, who also shaped and polished its parabolic primary and a host of flat, parabolic, and hyperbolic secondary and tertiary mirrors. By this time Francis Pease was working as Ritchey's assistant and probably shaped and polished many of the smaller mirrors in addition to aiding in the design of the mechanical elements of the telescope.

Differences of opinion between Hale and Ritchey regarding astronomy's usefulness and the course of research had surfaced while at Yerkes. From his earliest days tinkering with small telescopes in his family home in Chicago, Hale's central mission had always been to unravel the mysteries of stellar evolution. Ritchey, on the other hand, essentially believed astronomy's greatest purpose was the exposition of the beauty and majesty of the stars as evidence of the glory of God. This ideological divide was already starting to fray nerves in the first decade on Mount Wilson.[50]

As usual for Hale and his team, planning for the 100-inch telescope was already underway before the 60-inch reflector was even finished. In 1907 the New York Times ran the headline *The Greatest Telescope in the World: Monster Instrument Ordered by Carnegie Institution Will Far Exceed in Power All Other Watchers of the Skies.*[51]

Around 1906 Hale had met and befriended John Daggett Hooker, a wealthy hardware magnate who was intrigued by Hale's stories of developments in astronomy in general and at the new observatory. The 70-year-old philanthropist had a burning passion for science and, keying on Hale's knowledge and enthusiasm, decided to commit $45,000 toward an even larger mirror to be placed in a new technologically superior telescope that would bear his name.

After some deliberation, Hale and team decided on an aperture of 100 inches (Hooker liked the round number) for their latest marvel, and Hale soon won tacit support from the Carnegie Institution for the money to complete it if and when a suitable glass disk became available.

It was at about this time that the Hale-Ritchey partnership started to veer off course. In addition to his prowess as an optician and designer, Ritchey was a superior photographer, producing pictures that bore the same marks of precision and attention to detail evident in his lab work. While reviewing photographs he had taken with the 60-inch, he had become disenchanted with the parabolic configuration of the primary mirror of the telescope. The paraboloidal shape in combination with the secondary mirrors yielded comatic aberration or coma, a problem that makes stars away from the optical axis (center) appear to have comet-like tails. Astigmatism and spherical aberration that causes objects to appear off axis or in pairs was also common, as it is with smaller telescopes even today. In the case of large telescopes, it is amplified greatly and can render images useless for detailed or precise measurements.

Ritchey sought to remedy these problems in developing the shape of the 100-inch mirror when it was set to be curved and polished and started to consider what type of shape might best be used for the purpose. He was still pondering the complexities of the issue when the Parisian optical genius Henri Chrétien arrived at Mount Wilson in the winter of 1909 and 1910.

Serving his apprenticeship on the Journal de Mathématiques Elémentaires (Journal of Elementary Mathematics), an educational manual published by Imprimerie Chaix, the firm well known for its series of Les Maitre de L' Affiches lithographs of the late 19th century, Chrétien was introduced to the mathematics that would lead him into the field of optics to which he would lend his own groundbreaking results. Impassioned by his experiences in scientific publishing, Chrétien went back to school, graduating in 1902 with a Bachelor of Science in physics and mathematics. In 1905, ever curious, he took a job at the observatory in Nice while earning an engineering degree at L'ecole Supérieure d'Électricité, a graduate school in engineering also known as Supélec. The following year the young astronomer was recruited by members of the observatory to create an astrophysics department at the facility.

Overwhelmed by the opportunity, Chrétien decided upon a tour of some of the world's great astrophysics facilities to gain a sense of the state of the craft. He first spent time at the Imperial Russian Observatory at Pulkovo before heading to the Potsdam Astrophysical Observatory in Germany and finally Mount Wilson.

It was during this visit that Ritchey and Chrétien became friends, each marveling at the other's abilities – Ritchey, the superb, exacting, and inventive craftsman, and Chrétien, the theoretical creative genius. The son of an upholsterer, Chrétien found a kindred spirit in the older optical pioneer and expert astronomical photographer. The two shared ideas about recent developments in physics in general and optics in particular. Although the visit lasted only a couple of months Chrétien made a lasting impression on Ritchey and they vowed to keep in touch.

After Chrétien's departure, Ritchey returned to the problem of determining the correct mirror set up for the Hooker telescope. Specifically, he wondered whether a more refined hyperboloid secondary mirror might reduce or perhaps even eliminate coma altogether. He wrote to his new friend, now back in France, to see if he could come up with a workable solution. After some complicated calculations, Chrétien realized that a highly specialized set up that utilized both a hyperboloid primary and secondary mirror, one concave and the other convex would be capable of reducing coma to zero.

This "new curves" design, as Ritchey and Chrétien called it, required both primary and secondary mirrors to be figured as hyperboloids rather than as combinations of paraboloids and flat correctors. Configuring the new mirrors as complex hyperboloids had never been tried and a lot of careful planning and testing would be needed before it could be attempted on the blanks for the 100-inch. Still, Ritchey was convinced the new mirror design would revolutionize telescope optics and, probably believing that his boss would instantly see the merit of the idea, presented it to Hale and Adams as a preferred approach for the 100-inch. To his surprise and consternation Hale would have none of it.

In theory, Ritchey ought to have been able to understand Hale's apprehension. He was fully aware that by this time problems concerning the immense 100-inch glass disk were threatening to derail the whole project. Nobody had ever attempted to form a single piece of glass that big, and the St. Gobain Glassworks in Paris was struggling valiantly to create a viable one.

The original blank had arrived in Pasadena a couple years earlier, on the day the 60-inch telescope was to see its first light. Upon opening the crate and seeing the disk, Ritchey had immediately rejected it as unusable. Complications, initially in the forming and then in the annealing process had obliged the glassworks to try forming it in three separate pours. This strategy, although successful in producing a full disk, had left bubbles between the layers of the five-ton chunk of glass.

This set off a chain of events that began to pit ally against ally and friend against friend until it devolved into an irreparable impasse. Needing to ensure a good working disk could be produced, Hale first sent Ritchey on one of several trips to France to see if he could help the team at St. Gobain overcome their production problems. But these visits by the reputed Mount Wilson optician proved fruitless and nerves began to fray back home in Pasadena.

The problems with the 100-inch mirror were frustrating its benefactor, John D. Hooker, and to make matters worse the aging millionaire was growing hostile toward Hale owing, in part, to Hale's developing friendship with Hooker's wife Katherine, a travel writer and socialite. Hale knew he had to keep Hooker happy and convinced that both the mirror and the telescope would soon be completed. The experienced and artful director also knew that the funds to build the dome and mounting for the instrument would only be given after the disk was approved for shaping and polishing; Andrew Carnegie's way of hedging his bets.

As frustration mounted, Hale began to lose control of his emotional faculties. After years of dealing with one eccentric millionaire after another, seeking money for his increasingly large telescopes and the state-of-the-art research facilities to support them, his nerves were starting to unravel. As mentioned earlier, the monastery that housed the astronomers during their stays on the mountain had burned to the ground in December 1909, only a few months before Carnegie was to visit the mountain. When the retired steel baron and philanthropist arrived in Pasadena in March, Hale's mental state was completely exhausted. Summoning the will that had so often propelled him in the past, he managed to crawl out of his sick bed to join Carnegie at the summit and participate in the celebration of the observatory's many successes.

Hence, when Ritchey approached Hale and Adams with his idea to design the 100-inch as the world's first RCT, Hale had neither the nerve nor the patience to listen to the longtime optical engineer. Ritchey, who had known and enjoyed Hale's trust for more than a decade, would have understood the difficulties facing the stalwart director and might have shown some deference toward him but he was sure his new concept would revolutionize large telescope construction.

In theory, there were clear benefits to his design. The optical configuration solely as a Cassegrain focus and its relatively short focal range meant that the telescope tube would be shorter, in turn allowing the dome to be smaller and the associated instruments to be fewer in number. All this would save money. Better still, Ritchey believed, the telescope would provide far superior clarity.

The key factors that squelched Ritchey's fiery hope for a new model were financial and experiential. The "new curves" model may have presented a superior mirror type (and Hale remained certain of the brilliance and ingenuity of his longtime chief optician even during the final days of their alliance) but the new approach had never been tried and extensively tested, and no one, not even Ritchey or Chrétien, knew for sure if they were correct in their assertion that the design was superior. Acquiring the confidence in the new approach would require years of careful development and testing to determine whether it worked. Devoting additional funds to an unproven design when the project was already running behind schedule and over budget would be irresponsible and might give Hooker and Carnegie the impression his team were not sure they could meet the challenge they had undertaken.

In the light of these mounting pressures the idea was out of the question to Adams. Hale, who had never had reason to doubt Ritchey, was less adamant and supported letting Ritchey try his idea on several smaller disks that were already in the observatory's possession. But, like Adams he was against moving forward with the 100-inch as a "new curves" telescope.

It is entirely possible that Hale, an engineer by training who was always seeking out ways to improve on optical and mechanical design, thought that if Ritchey's idea were to prove out then it could be used in the next, even larger telescope. Never a man of small dreams, Hale was doubtless already envisioning a grander design amid the tumult and confusion of the 100-inch. This may have contributed to the psychological congestion his mind underwent in the ensuing years.

Around this time, Ritchey made his first and perhaps most egregious error in judgment, one that sent his career and the dream of his new telescope optical design careening off course.

As the handwringing about the casting of a suitable disk for the mirror persisted, Hale and Adams took another look at the blank they had received in 1908. After considering the matter with Ritchey and other members of the shop team, including Francis Pease, Adams became convinced that the bubble-filled blunder might be useable after all, and suggested that Ritchey begin shaping it in the Pasadena shop while the team at St. Gobain worked on an alternative. If it appeared the disk would hold up under the grinding process, then they could thank St. Gobain for their time and effort and push on with the project. Desperate for solutions to appease Hooker, Hale agreed and began to shine and polish his explanation for the move for the telescope's simmering sponsor.

Believing (as he apparently always did) that he was correct about the catastrophic flaws in the first disk, Ritchey became increasingly vociferous in his contempt for the policy, the plan, the process, and the leadership in charge of the project.

If he had been able to summon the wherewithal to step back and allow time for his ideas to gain favor while working in cooperation with his director and assistant director, Ritchey would surely have prevailed, eventually. But he was insistent that his plan was better. One can only imagine the burden of obsession involved in seeing so clearly a path to a brighter future, but the fact remained that he wasn't in charge and Hale was facing extreme pressure to produce results. Instead of backing down, Ritchey began speaking to Hooker behind the scenes, seeking to convince the magnate of the merits of his ideas over those of Hale and Adams. As he started to win Hooker's favor, Ritchey even went so far as to suggest that he pull his financial resources from Mount Wilson in favor of a new observatory built under Ritchey's supervision with the intent of producing the world's finest telescope and using it to create beautiful images of the heavens. This, he argued, was a far more culturally popular plan than the mole work that Hale had his staff were engaged in at Mount Wilson. But he had gone too far.

Hooker thought of Ritchey as a sympathetic genius of the highest order and trusted his ideas but knew there was no real avenue to success in leaving Hale and the backing of the Carnegie Institution. Whatever his misgivings about Hale or the process at the time, he had to respect the Mount Wilson director's enormous standing within the science community and his refined pedigree. Neither Ritchey's bearing nor his standing was a match for Hale's formidable accomplishments. Hale's charisma, enthusiasm and knowledge was a far more appealing combination than Ritchey's imperiousness and self-aggrandizement. Although Hooker began to distance himself from Ritchey he wrote to Hale asking if Hale supported any of his ideas, in effect giving away Ritchey's insubordination. When Hale realized the extent of Ritchey's betrayal he went through a cycle of surprise, grief, anger and eventually mistrust that triggered the start of Ritchey's long downward slide. To make matters worse Hooker died in 1911, leaving Ritchey virtually without a sympathetic ear at Mount Wilson. He was relegated mainly to the optical shop in Pasadena for the remainder of his career at the observatory and worked solely on the 100-inch's mirror. The instrument's mechanical design would be handled by his former assistant, Francis Pease.

When Hale was called to Washington D.C. to assist with the war effort, Ritchey slowly and meticulously worked to fine tune the enormous disk until, finally, in the spring of 1917, he finished figuring and testing the mirror. From there it was brought up the mountain and into history. Ritchey spent most of the remainder of his time at the observatory honoring his agreement with Hale to complete the ancillary mirrors.[52]

Letters between Hale and Adams while Hale was in Washington D.C. confirm Ritchey's insubordination. Adams, touching on "a few of the high spots" of the administration of the observatory in a letter he wrote on December 8th, 1918, remarked on Ritchey's "assumption of a military commission," referring to himself in writing as "Commanding Officer, Mount Wilson Observatory."[53]

By February, Adams was boldly suggesting "the question of letting Ritchey go," adding that with the completion of the telescopes on the mountain Ritchey's services would be of "no real use," and that researchers with a "willingness to handle things in a scientific way," were needed in the future.[54]

George Willis Ritchey's ties to his beloved observatory were severed forever on October 31st, 1919, but that was only the beginning of his trouble. Seeking to diminish his claim to a better telescope design, Hale and Adams began a campaign to annihilate his reputation within the American scientific community, blacklisting their former colleague and friend for his insubordination. An optician of extraordinary skill and creativity was thus relegated to life on his family farm for several years while he waited for the seas to calm. Fortunately for the future of astronomy, that is what happened.

Five years after being first fired from Mount Wilson and then ostracized within his own country, Ritchey was summoned to France at the behest of his old friend Henri Chrétien, who was eager to see their "new curves" idea developed in practice, for better or for worse. Things got off to an auspicious start on April 8th, 1924, about a week after his arrival in Paris, with the French

Academy of Sciences (FAS) awarding him the Jules Janssen gold medal by for his work on astronomical instruments.

By now Chrétien was professor at the Institute of Optics and beginning his development of the hypergonar lens system (originally developed to help tank drivers see a wider field) which in 1955 would win him an Academy Award (the only astronomer to win an Oscar) after it was used to create the CinemaScope widescreen film process.

Ritchey set up an office at the institute and a laboratory at the nearby Paris Observatory and continued to develop his ideas for more refined modern telescopes. These included a fixed vertical telescope employing six segmented six-meter mirrors and a smaller if equally ambitious six-meter segmented RCT. This absurd increase in aperture over the Mount Wilson telescopes suggests that Ritchey had it in mind to develop a telescope of such devastating scale and superiority as to banish Hale and Adams to the sidelines of telescope technology forever.

The trouble for Ritchey was that the French, who had suffered great financial, structural, and human losses during the war, simply lacked the funds, the will, and the confidence in the success of the project Ritchey was insistent on. He lasted only ten years in his adopted home before his pride and ego once again got the better of him. Despite his never deigning to learn even a modicum of French and carrying himself with the self-righteous dogmatism that had destroyed his reputation in the United States, the French never abandoned Ritchey entirely.

In the end, it was Chrétien who settled him into a good working environment at the Paris Observatory where the two built not a four-meter or a six-meter RC mirror, as Ritchey had favored, but one with a diameter of only half a meter as a proof of concept. Years after his propitious arrival, Ritchey was forced to curb his ambition and accept incremental growth in seeking a larger gain.

After a year, Ritchey and Chrétien had completed the first ever hyperboloidal telescope system, which they mounted on a base borrowed from the observatory in Nice. It debuted in July 1930 to a less than rousing response, due in part to the inadequacy of its mount. But the two men believed the design held promise and continued to speak about the creation of their telescope inside the astronomy community.

At the less than successful conclusion of his work in France, Ritchey climbed back onto a boat for home with his wife and children, uncertain of his future. Not long after his arrival home, however, he was contacted by the United States Naval Observatory, which had learned of his return and progress regarding new optics. The USNO wanted to commission him to design a new 40-inch reflector for its facility in Washington D.C. It was there that the first fully realized "new curves" telescope was mounted in 1934 and used for astronomical work, although its performance was undercut by the poor seeing conditions in the city. Lacking both funding and support for its transfer to a more suitable site, the instrument languished until well after Ritchey's death in November of 1945.

Although Ritchey did not live to see the glory of his visionary telescope design his ideas have been fully vindicated. The first large RCT was the 84-inch (2.1 meter) reflector at Kitt Peak National Observatory (KPNO) in Arizona. The project was begun by the first director, Aden Meinel, and completed by his successor, Nicholas Mayall, with first light occurring in 1964. Five decades after Ritchey's original conception, the RCT was ready to take over the world, which it has done in stunning and emphatic fashion. Today these systems can be found in every large reflecting telescope, both ground based and in space. The last large telescope to use a parabolic primary turned out to be the 200-inch designed by Pease under the supervision of Hale and Adams, whose disparagement of Ritchey never waned until their deaths.

It is unlikely that Humason and Hubble would have had a 100-inch RCT available, given the circumstances facing Hale at the time it was being built, regardless of what the state of relations between the "big three" at Mount Wilson might have been. Sadly, that connection was lost over time as the vagaries of daily life and the operation of the observatory soured their relationship.

Had cooler heads prevailed however, and Hale and Ritchey had continued with the kind of joy and spirit of invention they shared during their time at Yerkes, and Ritchey had been able to pursue his "new curves" idea through the 1920s and 1930s, then the 200-inch Hale telescope might well have been designed as the first large RCT reflector. Imagine it as an f/8 RCT Cassegrain (the Hale is f/16). Even considering the advances in technology that lay ahead, it is likely that the Hubble-Humason duo would have established the visible horizon out to distances far in excess of the already impressive measures they achieved during their long and successful collaboration. One is left to wonder what the state of technology would be today if things had turned out differently for Ritchey and his visionary idea.[55]

For a detailed account of the story of Ritchey and Hale, read Donald E. Osterbrock's book, *Pauper and Prince: Hale, Ritchey and Big American Telescopes*, from which much of the preceding was taken.

* * *

From the start of his apprenticeship, Milton Humason struggled to balance his daily routine of cleaning up after astronomers and managing the grounds with the excitement of working beside them at the controls of the instruments. He also had to juggle this off and on nightly schedule with the routine responsibilities of family life.

These contrasting environments, with their conflicting levels of expectation, interest, and obsessive enthusiasm, must have been hard on the young Humason family in the four years they spent on the mountain. The presence of her father and mother on the Dowd family's nearby ranch would provide some support for Helen Humason, and she certainly loved the life on the mountain as much as her husband. It isn't hard to imagine the limits of such a solitary existence on the psyche of a wife and mother who had grown accustomed to living among close family and friends in the valley, especially as their son Billy approached school age. Long nights of fighting off the cold mountain air while her husband was off on his exciting new adventure at the observatory wore on Helen, but she did her best to hide it. The longer they stayed the harder the conversations became over when they should leave.

Out of the mist of these sometimes-tumultuous years for his mother and father came the happy memory of a long-ago Christmas morning by Bill Humason 93 years later.

Snug in his bed, he was awakened by the sound of sleigh bells and footsteps on the roof of the family cottage. Arriving home from a night assisting on the 60-inch telescope Milton had climbed onto the roof of the little house, which was easy to do from the north side as the sloping ground made a natural platform to stand on. Shaking the bells and stomping his feet, he gave his young son the impression it was jolly old Saint Nick himself and the boy was beside himself wondering what he had left him. Once he was sure Santa was gone, he sprang from his bed to find a brand-new train set under the giant Christmas tree, which was so tall that Milton had to cut it off four feet up the trunk to get it in the cabin. Later that day the family went sledding on Jones' Hill, behind the 100-inch dome. Real conversations about the family's future were no doubt offset by many moments of pure joy.[56]

The intensity of work on the grounds varied from day to day, month to month and season to season. Fire breaks had to be cut on the sides of the surrounding hills in preparation for the forest fires of the dry season, rainstorms brought mudslides and flooded buildings, and snowstorms could dump as much as five feet of snow on the mountain in a single weekend.

Heavy snow was a particularly treacherous event for the grounds crew at the observatory as they worked throughout the storm, tunneling from building to building to enable people and supplies to be moved from one place to the next. And snow on domes had to be cleared from the slit doors so that when they were opened snow would not fall onto the telescope. The slit of the 60-inch was opened vertically toward the top of the dome, and it would jam under a snow load, while the horizontal movement of the slit doors on the 100-inch simply dropped the snow into the dome itself. It was precarious work, standing atop the icy dome and sweeping snow from the top of the doors six to 10 stories above the ground. Although the grounds crew would sometimes be asked to sweep the domes the job was usually given to the night assistant for the telescope. As a member of both the grounds crew and the night assistants, Humason often drew the short straw.

Humason's friend and supporter, Seth Nicholson was one of the stars of the observatory staff in both solar and stellar physics. Like Benioff, he was possessed of an expansive and open mind. In addition to his decades of research on the solar cycle, sunspots, solar flares and magnetism, Nicholson made significant contributions in orbital, planetary and stellar dynamics and at various times discovered a total of four Jovian moons.[57]

Nicholson was Humason's age and was raising a family with his wife Alma who he had met as an undergraduate at Drake University in Des Moines, Iowa. They had fallen in love while photographing and computing the orbit of a minor planet Joel Metcalf had discovered in the winter of 1909. They later convinced Metcalf, an ordained minister and stargazer, to name the object Ekard, the name of the school spelled backwards. The adventure brought them together and they entered graduate school together at the University of California at Berkeley. They married in 1913 and had their first child, Margaret, in 1915, as Nicholson was working for his doctorate at Berkeley.[58]

So here was this triumvirate of Shapley, Benioff and Nicholson coming to Humason's aid in the hopes of helping him achieve a goal that a year earlier none of them would have expected. Shapley, the visionary photometrist, orchestrated most of these connections and Humason made the most of the opportunities presented by making friends and excelling at every stage of his development.

It may have been Shapley's enthusiasm and colorfully descriptive tales of his adventures mapping the galaxy that spurred Humason to investigate the whereabouts of Pluto a little later on. In any event, Humason learned much from Shapley regarding how to interpret the information he gleaned from spectroscopic plates and direct photographs, as well as some of the tricks of operating the large telescopes.

In 1918 the total number of nights he would typically work as an assistant on the 60-inch in a given month was five, and with Wendell Hoge in control of the instrument there was not much room to improve on that number. The pending introduction of the 100-inch for regular use was an opportunity for advancement for Humason, if he showed the overall competence required for the job.

At the turn of the new year, Milton was summoned by his father-in-law Jerry Dowd to hunt down and kill a mountain lion that had been snatching goats. Besides the frustration of losing his livestock, Dowd complained, the presence of a large predator in such proximity to the house where six-year-old Billy often played was dangerous. Milton was a resolute preserver of the balance of nature but when his son's life was invoked in the cause he felt he needed to act. Tracking the lion's steps and scat, he fell upon a point where the animal crossed frequently and set a trap using the dead carcass of one of Dowd's goats as bait. The big cat outsmarted him at first, dragging the carcass off and leaving Humason no choice but to reset the trap and try a second time. Again, the trap failed to snare his wily prey. This time, as he knelt to inspect the empty trap, he looked up to find the lion staring him in the face only a few yards away. Before he could think, he raised his 22-caliber rifle and shot the lurching beast between the eyes. For Humason this was an episode he would willingly have forgotten, and he may well have done were it not for his father-in-law, who saw fit to photograph it. This appeared in an article in the Los Angeles Times Magazine on December 3rd, 1933, in an article that detailed the lives and exploits of some of the stars of astronomy of the day.[59]

In the meantime, the story served to raise Humason's folk hero status at the observatory, even prompting Wendell Hoge to write a poem about his former mentee's exploits in the logbook for the 60-inch telescope on January 19th, 1919:

A mountain lion got Dowd's goat

So Milton went a-hunting

And now he has a lion's skin
To wrap up Baby Bunting.

The lion was a monster beast

More than a hundred weight

From tip of nose to end of tail
She measured five feet eight.

Astronomers are very brave

To work up here all night

With lions roaming all around
Most folks would faint with fright!

They're brave all right but then we're sure

They'll all be glad to note

No more will roam around the dome
The beast that got Dowd's goat.[60]

While the staff was focused on the events of his hunting expedition, Humason sought to elevate his status at the observatory. One way to do that was to get his name in print. At Benioff's suggestion, he compiled the light curve (variation in brightness over time) for a nova in the constellation Aquila and, with Nicholson's help, prepared it for publication in the PASP, where it appeared in February of 1919. The report detailed a comparison of its magnitude over a four-month period relative to stars listed in a recently published Harvard Bulletin.[61]

Later in the year Nicholson suggested a learning project to Humason in which he would photograph a comet spotted by Nicholson's old college professor, Joel Metcalf in August. Once they had a few good plates of the object, he and Humason would go over the details of the comet's appearance, determine its angular diameter and estimate its size. Humason made his run on the comet in September and the two submitted their joint report to the PASP in October.[62]

In this manner, Humason was getting a crash course in advanced astronomy on the job without ever having completed so much as a year of high school and was performing well, so well in fact that Shapley began to champion his cause to Adams.

Paul Merrill was the first person to benefit from Humason's growing aptitude, starting in the summer of 1919. Merrill had been having some success using the rarely used hydrogen alpha lines in spectra from the 10-inch Cooke and was anxious to continue the project. Although he never let it dampen his spirits, Merrill suffered from arthritis for much of his career and may have been looking for someone to take on the laborious work of gathering spectra of hundreds of stars using this new method. Shapley recommended Humason. Although the latter wasn't yet on the staff, Shapley assured Merrill that Humason could do the work and added that "it would be a favor to Humason."

On Shapley's recommendation, Merrill asked Humason, who jumped at the opportunity, volunteering his time in the evening after completing his janitorial duties. As Merrill would later recount, "he would take the spectroscopic photographs and soon was able to interpret them, even better than I was. He had gained experience in the interpretation of the plates."[63]

The success of this survey led Merrill to pursue, with Humason's help, the long period variables and B-type stars that would so occupy his time in years to come. Merrill became the fifth star in Humason's constellation, and soon Walter Adams would become the sixth.

In the fall of 1919, as the orbits of Edwin Hubble and Milton Humason were beginning to slowly merge at MWO, their independent levels of self-esteem could not be more starkly contrasted.

Humason was already suffering from what is nowadays known as impostor syndrome. His general lack of education and rapid rise left him feeling as if he didn't belong, even as others assured him that he did. Climbing a ladder to wipe hypo from the ceiling of the darkroom had helped to mitigate these feelings when he was still primarily a janitor, but he no longer had that to lean on.

On page two of logbook number five of the 60-inch for 1920 and 1921 his name is given as "Relief Night Assistant," having been relieved of his janitorial duties by a "W. H. Carroll." His relief status for that telescope would later be in conjunction with his full-time work on the 100-inch, which was only then going into regular operation.

A self-described war hero and the youngest member of the RAS, Edwin Hubble bore the confidence of a man firmly in control of his destiny. His name appeared two lines above that of Humason, with "Direct Photography Nebulea + C." Given Hubble's desired path of research the "C" probably meant Cluster. As regards the misspelling of the word nebulae, it is not clear if this was Hubble's mistake – he was a notoriously poor speller – or some other member of the staff.[64]

Starting where he left off at Yerkes, Hubble looked to carry on his work in pursuit of a new classification system for galaxies, which at that time were still referred to as nebulae. The inability of even a telescope as powerful as the 60-inch to resolve stars in the fainter nebulae and distant spirals kept the wind in the sails of Shapley and his fellow supporters of the single universe theory.

Shapley was listed in the same logbook as working on "Stellar photometry, variable stars and 'spl' (spectral?) work on clusters," related to his ground-breaking work on galactic structure. This research dealt extensively with the study of variable stars for determining distance. His report *Nineteen New Variable Stars*, was published in the July 1919 issue of the PASP and detailed some of the findings that led him to his conclusions on the size and disk-like shape of the Milky Way system.[65]

Shapley was the big man on campus at Mount Wilson and everyone, including Hubble, knew it. As status influenced the allocation of observing time on the telescopes, if Shapley had stayed, he, not Hubble, may well have made the key discovery of Cepheid variables in M31 in Andromeda indicating it to be a system of stars comparable to the Milky Way. But his departure for Harvard in March 1921 denied him prolonged use of the 100-inch, whose aperture was large enough to reveal the true nature of the spiral nebulae.

Like the Nicholsons, Harlow Shapley and his wife Martha (Betz) had been brought together under the stars. They were married after their postgraduate work was finished, Martha at Bryn Mawr College and Harlow at Princeton University supervised by Henry Norris Russell, who was the leading American astrophysicist of the time. An expert mathematician, Martha Shapley was continuing her research on eclipsing binary stars even as she assisted her husband with calculations for his study of Cepheid variables in globular clusters. Their daughter Mildred would have a career as an editor and writer of astronomy books and their son Lloyd would win the Nobel Prize in Economics in 2012.[66]

As Humason's star continued to rise a growing chorus began to sing his praises, and as proof of the affable opsimath's talents reached Adams the latter forwarded it to Hale with his assertion that Humason might be a uniquely talented observer. Merrill would later recall that by 1919 Hale already "thought highly" of his abilities.[67]

Humason was first mentioned in the observatory's contributions to the CIWY using the 10-inch Cooke for Merrill's research. In a letter dated October 23rd, 1920, Hale wrote to Adams asking his assistant director to "kindly take up" the program of "the spectra of faint stars" in the Kapteyn Selected Areas using the 10-inch Cooke and asked if he would "start Humason on this work."[68]

In his response Adams wrote that he "had already discussed this work with Humason," and that it was "under way." Conferring with the apprentice observer, Adams "advised him to begin classifying the spectra to see what accuracy [could] be obtained." The innovative spectroscopist suggested "six hours on Seed 30 plates should give suitable spectra of 12th magnitude stars," and added that "if 14 x 17-inch plates [were] used…they [would] give…a complete map of the sky."[69]

When he wasn't working, Humason was reading constantly, soaking up as much as he could on the subject of astronomy. Among the works of interest was Shapley's recently published article for the PASP, *On the Existence of External Galaxies*, in which he laid out the argument against the evidence "growing in general acceptance" that the nebulae were so-called "island universes." His own results, he argued, based largely on the "work on star clusters," appeared to indicate that "the hypothesis that spiral nebulae should be interpreted as separate stellar systems" must be rejected.[70]

By now, the argument over whether the nebulae were interstellar dust clouds or distant galaxies was centuries old but, in 1920 conventional wisdom still held that the Milky Way system was essentially the Universe. Like so many great philosophers and physicists before and after him, Shapley simply could not fathom a universe full of universes, as the "island" crowd suggested. He decided what the truth was and then made his research fit his preconceived conclusion, which depended largely on the parallactic measurements of his colleague and good friend Adrian van Maanen. In the years prior to Shapley's departure, Humason got to know van Maanen well and he and Helen were frequent guests at his home.

As he read up on the subject, Humason began to question what he was seeing when he gazed at the night sky. As happens with all astronomers, the mysteries of the heavens were overwhelming him. And just as happened to Hubble 20 years earlier, Humason was developing an itch he had to scratch. Fortunately for him he happened to be at the world's preeminent scratching post for itchy stargazers.

Hubble was scheduled for his first run on the 60-inch in February of 1920 and Humason wanted to take the measure of the newest member of the staff. As his good friend Nicholas Mayall would later point out, Humason was "shrewd…in evaluating character…He had an uncanny knack of locating…the pseudo-types, the elite-types that looked down on others." When he did, Mayall added, he "unerringly – it didn't take him long – found that some big-name men had feet of clay."[71] Being one who didn't suffer charlatans politely, he could not have been impressed by the stories circulating the observatory about Hubble. The affected English accent, the military dress and insistence on being referred to as "Major Hubble," and the haughty, brooding pretentiousness all betrayed the selfish intentions of a man of outsized ambition and ego.

Hubble had the telescope set up for direct nebular photography at the Newtonian cage. This being his first time with the large reflector, he was assisted by Hoge. The logbook for February 24th records the dome being opened at 6:40 pm and Hubble starting his observing run at 11 pm.

Humason slipped into the dome for a peek at Hubble once the telltale steady movement of the telescope tracking across the starry firmament began. The Newtonian focus required the observer to stand on a platform well above the concrete floor of the dome. Humason watched as Hubble guided the telescope along the sky, an event he immortalized in moving and poetic terms in a sympathetic eulogy for his fallen friend 34 years later:

> His tall, vigorous figure, pipe in mouth, was clearly outlined against the sky. A brisk wind whipped his military trench coat around his body and occasionally blew sparks from his pipe into the darkness of the dome.[72]

This memory, looking back fondly on the life of a colleague and friend, was the result of years of close partnership and the slow building of mutual trust, respect, and admiration. As Humason entered the dome on that cold winter night in February of 1920, Hubble would surely have been an object of suspicion.

Battling the internal demons of his early life, Hubble had determined to reinvent himself in his own image before he arrived at Mount Wilson. None of this was clear to the men and women already working at the observatory. How much of this new self-image was born of his early environment and how much was raw egotism can never be known and certainly was not understood by many who knew him. It would take years for Humason to discover Hubble's deeper truths and develop the sense of empathy that he most certainly had gained by the middle of the 1930s. Nevertheless, some of Hubble's darker secrets likely remained hidden entirely in life and in death.

Hubble was certainly impressed by the quality of the plates he produced in his first run at the 60-inch. If these were made under "extremely poor" seeing conditions he was going to get incredible direct images of even some of the fainter nebulae that had been well out of reach of the 24-inch reflector at Yerkes.[73]

Idiosyncratic though Ritchey's second great reflector may have been, it was a marvelous astronomical instrument. The telescope's sheer size was intimidating to everyone at first. The operation of the big reflectors offered a unique set of complications. The observer had to simultaneously rotate the telescope, dome and observing platform, all while keeping the target centered in the guide scope. Failure to properly manage these simultaneous acts could have grave consequences in damage to the instrument and/or injury to the observer. And of course, these complications would be amplified in the case of the 100-inch Hooker telescope.

Francis Pease and Harlow Shapley had done the lion's share of the viewing at the 100-inch during Hubble's first fall at Mount Wilson while the final adjustments were being made to the mirrors, Newtonian platform, and spectrograph. One by one, the staff of the stellar department were invited up to work with the new telescope and familiarize themselves with its unique set of mechanical peculiarities.

For an astronomer, observing with the 100-inch was the ultimate experience because its light gathering ability dwarfed even that of the newly completed 72-inch reflector of the Dominion Astrophysical Observatory in British Columbia, Canada.[74] It was massive, 100 tons of rotating steel and glass slipping effortlessly across the night sky. Working at the Newtonian focus platform high above the deck gave one the feeling of sitting on top of the world.

It was the kind of experience one could never forgot. The kind of exhilaration that could extinguish the feelings of futility from inactivity during the war years. A path to glory had been illuminated for Edwin Hubble through the skies above Los Angeles.

His first run with the giant reflector came on Christmas Eve of 1919. He spent 12 hours observing through lousy weather conditions trying to get useable images. Three days later he was back, opening the slit at 6:30 pm, but gave up four hours later in frustration as the weather still wouldn't cooperate. It was a time of great experimentation with equipment and observing methods and in the possibilities for research these experiments revealed. A notation in the logbook for December 30th shows both Pease and John Anderson working on the "Michelson experiment," almost certainly a reference to Albert Michelson's 20-foot interferometer that he and Pease were to use in the coming months to measure the diameter of the star Betelgeuse with incredible precision.[75]

Despite the weather, Hubble's early nights on the great Mount Wilson reflector yielded striking images of some old friends, in particular NGC 2261, the primary component of his dissertation work at Yerkes.

Unlike the cold dampness of the rolling hills of Wisconsin, the mountainous region of California was magical. Going between the domes and the monastery in the predawn light, breathing the cool mountain air and gazing at the vast San Gabriel Valley stretching out to the sea gave one pause to reflect on his fortune. Despite a wealth of useful discoveries past and present, astronomy was still looked at more as a hobby than a science by the scientific community at large in 1920. These attitudes would begin to change in the next decade and beyond as generations of astrophysicists brought theory to the observational foundation of the field.

The monastery was the place where Hubble felt closest to home. He reveled in his days and nights on the mountain, pacing the hallways between his tiny one-bunk room with its simple arrangement of a desk and chair, and the library study room near the kitchen. He enjoyed many an afternoon sipping tea with others on the back porch looking out over the edge of the mountain at the beautiful skyline.[76]

Early on he tried to make the most of his time on the mountain, humbly working beside the likes of Frederick Seares and Francis Pease, eager to soak up whatever he could from these leaders of the observatory and the field. At the time, Seares, with experience as a professor and as a photometrist, was pioneering in stellar magnitudes and color indices, and would later tackle interstellar absorption and the brightness of the Milky Way in relation to external galaxies, the breadth of which would earn him a Bruce Medal in

1940. Joining Seares one evening in February 1920 Edwin wrote "Hubble" in the box where a night assistant would usually identify himself by the first letter of his surname. After Humason started working as a night assistant his mentor, Hoge, would differentiate himself from his younger colleague by circling the "H" when he was at the helm of the 60-inch.[77]

Of the men at the observatory, Hubble probably preferred the night assistants to the astronomers, if for no other reason than because they didn't represent an existential threat to his outsized ambitions.

Once the 100-inch telescope went into regular operation, three assistants were needed to support the large reflectors working the same schedule as the solar assistants. One was assigned for each of the two instruments and would serve 10 straight nights while a "swing assistant" would work five nights on one and another five on the other. The staggered shifts meant that the telescopes could be manned all 30 nights of the month. Housing was built to accommodate all six assistants, as well as the grounds crew and staff for the monastery. As his observing duties increased, Humason would take on the role of swing man for the crew.[78]

This was no ragtag bunch of ne'er-do-wells like the cowboys who still ran the mule and wagon trains up the south trail. These were men of varied backgrounds but usually with a focus on engineering, mechanics, or technology in general. All in all, they formed a good cohesive unit of strong-minded observers who were skilled at navigating the stars with the giant reflectors and easy to command if respected for their worth, as the men in Hubble's infantry regiment had been.[79] In addition to what appears to have been a difference of opinion as to just how to spell his name, the night assistants didn't quite know how to take Hubble initially. In his case, it seemed, you could take the man out of the army, but you couldn't take the army out of the man.

An ardent planner, Hubble treated his observing runs like he would any other mission, whether in the field of battle or a hiking or camping trip. He prepared extensively for each observing run and let the night assistant in charge of the telescope know as soon as he reached the mountain what his plans were. This gave plenty of time to prepare the telescope and other instruments. He trusted the assistant to work on his own preparing the instrument and he expected his instructions to be carried out properly. He wouldn't tolerate delays due to improper preparation. When he found the assistants work lacking, he let him know of his displeasure, but always respectfully. In time, the men appreciated this straightforward and detailed approach, including Humason who worked with Hubble on a handful of occasions.

Usually when the weather was not cooperating the assistants and supporting staff held card games, sometime with the frustrated astronomers joining in. Even Hubble found his way into the occasional game.[80] He was primarily a bridge player but his natural tendencies proved to be an asset at the poker table. However, Humason was by far the toughest to beat by accounts. An inveterate gambler, Humason's prowess at the card table compelled Robert Robertson to later proclaim him "the best poker player I ever saw."[81]

After the start of Prohibition in January 1920 Humason could be found commanding the table in the break room beneath the 60-inch, taking occasional swigs from a tin flask filled with the most vile tasting moonshine ever unleashed on the tongue. Questions abounded about where he got the rotgut, but Humason never divulged the origin of the devil drink. No doubt he was distilling it himself at any one of the family ranches in the valley. When Prohibition ended in 1933 this "panther [juice]" as Humason referred to it (the actual word more than likely rhymed with "bliss") was discarded for a more refined drink. Through it all, Milton's demeanor betrayed nothing about either the hand he was holding or the origins of the elixir that accompanied him at the table, further substantiating his growing legend.[82]

As noted, Hubble's intention was to create a classification system for the nebulae, one that he hoped might reveal something of their nature and evolution. Like astronomers who compiled large compendiums before him, he was prepared to make it his life's work. The world in 1920 was barely starting to recover from the ravages of war and the flu pandemic. Einstein's relativity was turning physics on its head. As Hubble set about his self-appointed task, the possibility that the universe might be expanding was the farthest thing from his or almost anyone's mind.

Much of his early work centered on the collection of direct photographs of the objects he sought for classification, and then putting them into buckets based on their appearance. Examinations of their spectra were useful for insight into their composition, rate of rotation and line of sight velocity.

Whether nebulae would be resolved into stars (indeed, whether stars were constituents of the nebulae or not) was a popular debate that would necessarily impinge upon Hubble's research.

In 1908 a report appeared in the Annals of the Astronomical Observatory of Harvard College. The details were assembled by a member of the computing team named Henrietta Swan Leavitt and set out a relationship between period and luminosity Cepheid variable stars.[83]

A star represents a balance between gravitational contraction and the internal pressure of its core. In some cases, a star achieves a stable state but in others this balance is dynamic, and the star varies. The outer layers expand outward, propelled by the hot inner core, which upon cooling cause the outer layers to fall in toward, heating up the core anew and causing the star to expand. As the diameter of the star and the temperature of its surface varies, this causes its brightness to fluctuate. Some stars vary in an irregular way, others cyclically. In the case of Cepheid variables, they pulsate with a distinctive light curve with a relatively well-defined period.

This pattern was first discovered in 1784 in the star Delta Cephei by an amateur English astronomer named John Goodricke, a descendant of English baronets dating back to the reign of Charles I. Like Leavitt, Goodricke was taken ill as a boy and lost his hearing. His death at age 21 of pneumonia preceded his appointment to the RAS and a Copley Medal that was awarded posthumously for his incisive work on eclipsing binaries and the Cepheid variable stars Delta Cephei (for which this type of star is named) and Beta Lyrae.[84]

Henrietta Leavitt was born on Independence Day in 1868 to a Massachusetts minister, George Roswell Leavitt and his wife Henrietta Swan Kendrick. After attending Oberlin College for two years, she transferred to Radcliffe College (then called the Society for the Collegiate Instruction of Women). A highly intelligent and curious student, she excelled in classes from analytic geometry and calculus to classical Greek and philosophy. During her senior year she took a semester of astronomy and was so taken with the field that she later applied for a position as a computer at Harvard College Observatory under the direction of Edward Pickering. The work there was largely focused upon the study and standardization of stars according to their brightness, spectral type and various other determining factors.

Leavitt impressed the director and not long after her arrival at Harvard in 1895 he made her the head of the stellar photometry department. In those days, irrespective of how well-educated women might be, they weren't hired as astronomers but as computers to examine, detail and record data for use by their male counterparts.

In 1907, as photographic plate technology continued to improve, Pickering expanded his plan of providing photographic magnitudes for as many stars as could be recorded with the Harvard telescopes. His team included Williamina Fleming and Annie Jump Cannon, among others. He put Leavitt in charge of the program. She began with the 46 stars in the North Polar Sequence that would serve as references for astronomers measuring how bright a particular star was. Her creativity in solving the problem of measuring a star's brightness led her to improve and advance Pickering's original program for stars in the Northern Polar Region. She published her findings in the Annals in 1912 and in 1917.[85] In all, Leavitt established magnitudes for stars in 108 areas of the sky. This work was adopted by the international community of photometrists which was later headed by Frederick Seares at Mount Wilson, who would complete the color sequences down to the 18th magnitude with the aid of Humason in the 1920s and 1930s.[86]

Essential though Leavitt's work on the North Polar Sequence was for astronomy, it was her contribution on variable stars that really set the stage for the explosion in astronomical advances in the coming century. She accounted for the discovery of more than half of all known variable stars prior to 1930. This required long hours of painstaking work blinking plates taken on many nights, repeatedly measuring the brightness of stars relative to scales she had already determined to find those that varied and the manner in which they did so, all the time testing, altering figures and formulas and retesting before finally recording her findings.

It was early in her study of variable stars that Leavitt had her stroke of genius. In 1908 she published a paper *1777 Variables in the Magellanic Clouds* in reference to the Large and Small Magellanic Clouds, a pair of dense clusters of stars in the southern sky. Sixteen of the variables in the Small Magellanic Cloud (SMC) were of the Cepheid type. Studying their pulsation periods lasting between roughly a few days and several months, she realized that the brighter a star appeared the longer was the interval between the times of maximum brightness.[87]

Then in a 1912 paper published with Pickering, Leavitt drew a graph with details of 25 Cepheids in the SMC. The stars were presented in two separate columns corresponding to their minimum and maximum brightness and the length of their period. On the assumption that all of the stars in the SMC were at roughly the same distance from Earth she concluded that the fundamental brightness of a Cepheid is directly related to the period of its pulsation. This meant the relative distances of two stars that had the same period could be measured, the dimmer one being farther away according to the inverse square law in which a star that is four times dimmer is twice as far away. This period-luminosity relationship for Cepheids would later become known as Leavitt's law.[88]

The importance of this discovery for the development of modern cosmology cannot be overstated. It provided the means to create a formula for measuring distances to clusters in which Cepheids were present (in astronomers' parlance, using them as standard candles). But first it was necessary to calibrate the relationship by measuring the actual distances of a handful of Cepheids.

At the time the best way of measuring the distances to stars was direct triangulation with the parallax across the baseline of the Earth's orbit around the Sun. This worked for nearby stars, but when the triangle was so long that the angle was unmeasurable this was impractical. The SMC was well beyond this limit. Calibrating Leavitt's law would therefore require measuring the distances to nearby Cepheids. Unfortunately, none showed a measurable parallax.

In 1783 William Herschel (Chapter 1) had perceived the Sun's motion in space by noting the progressive displacement of stars in certain constellations, year by year. Ejnar Hertzsprung, a Danish astronomer married to Jacobus Kapteyn's daughter, Henrietta (they named their daughter Rigel after the star in the constellation Orion), used the best-known calibration of this motion as a baseline for a parallactic study of nearby Cepheids and then used Leavitt's law to calculate the distance to the SMC as approximately 30,000 light years, a result that he reported in 1913.[89] Although his math was off, he and Leavitt had given astronomers a way of measuring the distances to extremely remote star clusters.

It was Harlow Shapley's (Fig. 5.5) use of the Cepheids in his research on globular clusters that enabled his seminal work on the structure of the Milky Way and made him an overnight sensation in the scientific community. It also fed into the long-fought debate about galactic structure that boiled over in 1920.

The wisdom of the day, which had changed little since William Herschel's work on the subject, centered around the work of Kapteyn, the revered Dutch astronomer and frequent Mount Wilson visiting researcher. Kapteyn had concluded that the Milky Way was a lens-shaped disk about 25,000 light years in diameter and roughly 5,000 light years thick, with the Sun situated very nearly at its center.

From 1915 to 1919, Shapley wrote his classic series of 14 papers (I to III were published only in the CIWY) in which he detailed his plan of attack on the clusters, his justification, his objectives (which included a solution to George Hale's "problem of stellar evolution"), and the steps he took that led him to his ultimate conclusions starting in 1918.

At that time there were about one hundred globular clusters known, most of them lying in the direction of the constellation of Sagittarius, and very little was known about them. Shapley had been steered to this research by the Harvard astronomer Solon Irving Bailey, who had pioneered the study of variables in globulars whose light curves resembled those of Cepheids but with shorter periods, typically less than a day. Bailey's "cluster variables" are now classified as of the RR Lyrae type. The pulsation of RR Lyrae itself was discovered by Williamina Fleming at Harvard around 1900.[90] Having found them to be common in globulars, Bailey urged Shapley to use the greater capabilities of the 60-inch at Mount Wilson to dig deeper into the matter. Allan Sandage described Shapley's pursuit in the *Centennial History* in some detail.

Fig. 5.5 Harlow Shapley after his arrival at Harvard ca 1922 (HUP Shapley, Harlow (20A), Harvard University Archives)

Shapley began this epic journey with a three-year photometric study of thousands of the brightest stars in several clusters, classifying them by color and magnitude using techniques gleaned from his former professor and Mount Wilson mentor Frederick Seares. In 1917 he proposed three methods for determining distances to clusters. The first was a recalibration of Hertzsprung's period-luminosity relation for Cepheids to allow him to calibrate absolute photovisual magnitudes, which he used to calculate a reliable ratio of magnitude to distance for RR Lyrae variables. Next, he cleverly applied his mean magnitude values for RR Lyraes to calibrate the absolute magnitudes of the 25 brightest giant stars in the clusters M2, M3, M5, M13, M15, M22 (the numbers indicating their positions in the Messier list) and Omega Centauri (too far south for Messier to discover), all of which had almost the same apparent magnitude. This gave a simple means of measuring the distances to such clusters. Shapley then used one or both of these methods along with his assumption that the linear diameters of the clusters were statistically the same to further corroborate his distance measurements to them.

Shapley's 1918 paper (number VII in the series) on galactic structure is an astronomy classic and it propelled him to international renown. Using his newly devised methods for measuring distances, he produced a spectacularly proportionally accurate conception of the shape and structure of the Milky Way. At 300,000 light years in diameter, it was ten times greater than any previous estimate. The distribution of globulars in the sky placed its center in the direction of Sagittarius, some 50,000 light years from the Sun. His determination of the overall structure of the galaxy, the direction of its center and our eccentric position were foundational discoveries that reverberated around the world.

The assumptions Shapley made based on his foray in the globulars were breathtaking in approach, breadth and interpretation in the eyes of his fellow astronomers and showed the fearless certainty he exhibited during his seven-year stint at Mount Wilson. The enormous size of his galaxy, and the comparatively small size of the nebulae combined with estimates of their rotation rates by van Maanen, led Shapley to further infer that the entire universe was subsumed by the Milky Way system. The errors in judgement that he made on galactic and universal scale stemmed from various misunderstandings, some of which wouldn't be cleared up for 30 years.

First, in the absence of evidence to the contrary, Shapley assumed that cluster Cepheids had the same absolute magnitudes as their non-cluster cousins. The fact that the two types were in different stages of their evolutions would not become apparent until Walter Baade's discovery in the 1940s that there are two stellar populations. We now know that RR Lyrae variables are low-mass (half a solar mass), low metallicity (Population II) stars which have evolved beyond their red giant phase, burning off most of the hydrogen from their cores to leave helium as their remaining fuel.

Second, the presence of interstellar gas and dust can significantly attenuate the passage of light, making distant stars fainter than they otherwise would be. Although the presence of such material had been predicted, it would not be properly investigated until the 1930s. With no data on attenuation to correct his apparent magnitude measurements, Shapley had seriously overestimated his distances.[91]

In 1920 the fact that Shapley's estimate for the diameter of the Milky Way was so much greater than that made by Kapteyn fed into the centuries-old controversy about whether the nebulae were really galaxies like our own.

George Hale decided to put together a debate as part of a lecture series that was funded by an endowment from his late father William Hale for the NAS. Hale regarded Shapley's intellectual facility and personal qualities with the utmost respect but the young astronomer's soaring bravado and bold self-confidence on the subject, based on grossly overstated assumptions, gave the Mount Wilson director pause. Few in the field were better read than Hale, and he was well aware of Kapteyn's work on proper motions and galactic rotation as well as Slipher's velocity measurements as possible indicators that the nebulae might be literally out of this world.

As we have seen, Edwin Hubble, who had been studying the nebulae since his college days at Yerkes, was a cautious supporter of the island universe theory.

Another Mount Wilson man who was forming a personal opinion on the external galaxy question was Milton Humason, who somewhat famously approached Shapley with a pair of plates of M31 he had been comparing using the blink comparator for Shapley's cluster program. In doing so, he noticed the telltale pulsing of a Cepheid variable star and circled the suspect on the plates. Showing them to his mentor he asked if these might be used to determine the distance to the M31? Wouldn't this be a useful means of checking Shapley's previous measurements? Shapley would have none of it. Wiping the marks from the plates he admonished Humason not to be swayed by the island universe contingent, assuring him that the nebulae were all a part of the Milky Way.[92] Shapley's failure to be open to additional evidence would cost him a second major breakthrough when Hubble discovered Cepheids in M31 and other nebulae several years later, some on the same plates Humason had highlighted. For Humason, his failure to convince Shapley to look into the question would become the big one that got away.

At that moment Shapley's attention was focused on proving his case in front of a panel of his peers. The debate was held in Washington D.C. on

Monday, April 26th, 1920, between Shapley, representing the single universe theory, and Heber Curtis, who was preparing to depart the Lick Observatory to assume directorship of the Allegheny Observatory later that year, arguing the island universe case.

For Shapley the stakes could not have been higher. Beside the threat to his burgeoning career, a loss in the debate meant he might lose his chance at a coveted promotion. As has been mentioned, Pickering had died in February 1919 and Shapley was one of two primary candidates to replace the longtime director of the Harvard College Observatory. In a letter to Hale in February 1919, Kapteyn wrote: "Shapley is a brilliant man and personally I, who know him mainly through his scientific work, would think him best fitted for the position."[93]

Hale sent Kapteyn's letter to Harvard University president Abbot Lawrence Lowell (the younger brother of the Lowell observatory founder) supporting the recommendation. Hale's second choice was Frank Schlesinger at the Allegheny Observatory followed by Henry Norris Russell at Princeton. However, the opinions of Hale and Kapteyn were countered by George Agassiz, a prominent member of the astronomical community who pointed out that they were "both doubtful of [Shapley's] ability to run the practical part of the Observatory," and wondered whether there was "some way of getting some better idea about this before going ahead?"[94] The matter languished through the remainder of 1919.

In the meantime, Schlesinger was chosen to assume the directorship at Yale University Observatory vacated by William Lewis Elkin. Schlesinger was vacating the directorship at Allegheny Observatory earmarked for Curtis.

This left only Shapley and his former mentor and champion Russell as prime candidates. In the interim, Agassiz had read up on Shapley's recent research and decided he was a more capable observer than Russell.

By February of 1920 Lowell had all but made up his mind to invite Shapley to Harvard. On writing to Hale to convey his decision he was surprised and somewhat perplexed at the Mount Wilson director's reply:

> As the directorship of the Harvard Observatory is a post of great importance, I venture to suggest that Professor Frederick H. Seares, Astronomer and Superintendent of the Computing Division of this Observatory, be seriously considered for the position.[95]

Having spent months pouring over letters and articles regarding who would be the best to take the helm at his institution's prestigious observatory, Lowell had been thrown a curveball by the Mount Wilson director. Writing glowingly of Shapley, noting his "versatile, daring and industrious" ways and his "constructive imagination" that made "him decidedly the most promising of all the younger group of American Astronomers," Hale nevertheless said he had "not yet reached complete maturity." Therefore, he was proposing Seares, 12 years Shapley's senior and an established steady hand.[96]

Even in hindsight Hale's note is confusing. He concluded his wavering letter by saying he believed Shapley "would prove a great success at Harvard" should he be elevated to the post.[97] Hale was doubtless hedging his bet but it seems as though he really was of two minds on the issue of Shapley, and in suggesting Seares he was offering a much more conservative option to Lowell (and himself) in the event the Harvard president might be looking for someone with strong credentials who was a little less adventuresome. It is possible Hale was influenced by Edwin Frost, who was visiting during the early exchanges. After Lowell pressed him further on the matter, Hale reiterated his comment on Shapley's overall promise, saying he "would show great originality and ability" in the role as director but his "comparative youth" represented a degree of risk.[98]

Grappling with the disparate information and opinions, Lowell asked Agassiz to attend the NAS debate to assess Shapley's competence for the directorship. The event was largely considered to be a draw, with Shapley and Curtis backing their stances with evidence from their research. In the opinion of Agassiz though, Shapley appeared meek, lacking his usual swagger. He wrote Lowell in unambiguous terms saying that the Mount Wilson man lacked "maturity and force and does not give the impression of being a big enough personality for the position." On the face of it, this assessment portended the end of Shapley's candidacy.[99]

Given the options of gaining the directorship at Harvard College Observatory or staying at Mount Wilson with the myriad opportunities for future discoveries it presented, Shapley wasn't exactly a hard luck case. But the directorship at the HCO carried with it considerable prestige that undoubtedly coveted.

Much has been made of Shapley's missed opportunity in regard to universal structure, but he had shown the ability to reassess his work when he correctly revised his conclusion from papers II and III in 1915 that some of the globular clusters must reside outside the Milky Way. And of course, he could have made a similar reversal on Humason's suggestion of the Cepheid in M31. Having a six-year head start on Hubble, it is not out of the question that he could have revised his conclusions in the face of more conclusive evidence.

As it turned out this was not the end of Shapley's attempt to gain the prized position at Harvard. In the wake of the debate, Russell had wondered whether

Shapley could actually handle the job solo. But after word reached Russell that *he* was now the man being sought for the position his attitude changed. After confessing that he was no fan of administrative work he informed Lowell that not only did he not want the job, but Shapley was also the right man.

Nearly at his wits end, Lowell now took a slightly different approach, offering Shapley in writing the position of "Assistant Professor and Astronomer."

Seeking assurance that he would assume a more authoritative role on his arrival, Shapley suggested the title of "Astronomer and Scientific Director." But Lowell wasn't comfortable sending so strong a message to his staff and student body without knowing the capabilities of the man being placed over them.

Once again, a stalemate ensued and this time it was broken by George Hale who offered to send Shapley to Harvard on a year's leave of absence to try out the new job.[100] Lowell leaped at the chance and Shapley left Mount Wilson on March 15th, 1921 "to become identified with Harvard College Observatory."[101] Within six months it was clear that he was indeed the right man for the job. Among Shapley's students in the coming years were Georges Lemaître and Cecilia Payne-Gaposchkin. Charles Whitney, a graduate student at Harvard in the 1950s, would later opine of Shapley, "I have never seen a quicker mind, a more agile sense of humor, or a more complete absence of what usually passes for humility."[102]

Unfortunately for the heroine of this story, Henrietta Leavitt (Fig. 5.6), the answer to the structure of the universe would not come in her lifetime. She succumbed to stomach cancer and died a few months after Shapley's arrival at Harvard. He is said to have visited her on her death bed and regarded her highly.

Leavitt's feat of intuition was certainly deserving of better treatment than she received. Notwithstanding her enormous contribution to the field, she retained the title of "assistant" to the end.

Although the Cepheid period-luminosity relationship was her discovery, the 1912 report of it bears only Pickering's name. His contribution may well have been the first sentence saying the "following statement…has been prepared by Miss Leavitt." He didn't allow her to develop a formulation to apply the relationship in practice. It is another example of the difficult path to women's equality and the probable negative consequences to our collective knowledge of the world we live in.

In 1925 Leavitt did receive some praise when a Swedish mathematician named Gösta Mittag-Leffler, unaware of her death four years earlier, wrote to her of his plan to nominate her for the Nobel Prize in Physics. This didn't stop Shapley from delivering one final slight to the late astronomer in his letter of

Fig. 5.6 Henrietta Swan Leavitt – Discoverer of the period luminosity relation in Cepheid variable stars

reply to Mittag-Leffler. On the one hand he credited Leavitt's work for having "led to the discovery of the relation between period and apparent magnitude" of Cepheid stars and then he went on to claim, "it was my privilege to interpret the observation by Miss Leavitt, place it on a basis of absolute brightness, and [apply] it to the variables of the globular clusters [for] my measure of the Milky Way." Shapley was angling for a nomination for himself even though his work relied on Leavitt's discovery.[103] In Shapley's defense, he had already credited Leavitt for the period-luminosity discovery in his seminal 1917 work on globular clusters (Paper VI) and would continue to celebrate her work in the years to come.

Although Shapley was still very much a part of the MWO in 1920, Edwin Hubble was not without his own political capital. As Frost's former assistant at Yerkes and already a member of the RAS he had come highly touted both as a student and a researcher, and his status as a war veteran and officer carried considerable weight as well. His respect for those in command, learned as a child under his father's iron fist, had intensified during his time as an officer in the army. The praise and support he received from Hale, Adams and Seares no doubt eased his transition into the staff in Pasadena.

Where leadership on Mount Wilson was concerned, Hubble probably believed Hale had acted smartly in naming Adams as assistant director. Adams was direct, frank and fair and had some of the conservative traits Hubble idealized. He was also an accomplished and well-regarded observer, winner of both the Henry Draper Medal and the Gold Medal of the RAS, of which he was a member, "for his investigations in Stellar Spectroscopy, including the determination of Absolute Magnitude."[104] He would later be awarded a fellowship in the AAAS (1922), a Valz Prize (1923), Prix Jules Janssen (1926, the highest award for astronomy given in France), Bruce Medal (1928) and a Janssen Medal (1934, a separate honor from the Prix).[105]

However, as the years wore on and Hubble began to see himself as the rightful heir to the directorship of the observatory the relationship would suffer an ethical divide, although professional respect remained. For the moment, however, Adams was a respected leader in Hubble's eyes.

Hubble knew little about the remaining old guard at the observatory, Charles St. John and Ferdinand Ellerman, both of whom were solar observers with fine reputations in their own right. In addition to his regular research for Hale's broader stellar evolution program, St. John spent many years applying Einstein's relativity to issues in solar physics. He was among the astronomers Einstein sought out for conversation on his arrival in Pasadena in 1931. Ellerman was as conspicuous a presence for his neatly trimmed beard and handlebar mustache as he was for his theatrics, which always managed to attract an audience. He was famous for pausing midway to the top of the 150-foot solar tower in the "bucket" (Quite a ride!) and lighting his pipe by focusing the Sun's rays through a magnifying glass. Beside his regular duties as part of the solar department and as a skilled mechanical handyman, he served as the official photographer on the mountain for many years. His association with Hale dated all the way back to the Kenwood Observatory.[106]

Hubble probably got to know Francis Pease a little during his first year on the mountain, especially in learning to operate the 100-inch telescope, dome and platform. A wonder of modern engineering, the telescope nevertheless had some idiosyncrasies that needed to be understood for both the efficiency of the observing run and the safety of the observer and the instrument. There was the periodic error in the drive clock that had to be corrected for to prevent going off course and ruining an exposure. A design flaw in the mercury floatation system caused the heavy liquid metal to froth from the heat of the friction in the gear housing. This caused the telescope to vibrate, and the astronomer had no choice but to halt his run until the frothing float was attended to. In those days, mercury's deleterious effects on human health weren't yet understood and the men spent many hours chasing the expensive lubricant on the observing floor, scooping it up in little beads to be reintroduced into the float. To all those who worked at the observatory in the first half of the 20th century it was all part of the romance of the heavens taking place on the western slope of the Sierra Madres.

Pease was another of the highly respected originals at Mount Wilson. The staff knew very well the trouble George Ritchey's rift with Hale and Adams had caused, sending their beloved director into early retirement in Pasadena. At the very height of the turmoil Pease, who had served for years in Ritchey's shadow, was thrust into service in the design of the new instrument and the management of its construction. It had been a mighty undertaking, especially with Ritchey still charged with shaping and polishing the primary and ancillary mirrors, a task with which Hale would only have trusted Ritchey.

It was Pease who aided Albert Michelson in the design and mounting of Michelson's 20-foot interferometer on the top of the tube of the Hooker telescope. Michelson had used an earlier type of the device to investigate the theory that light travels through a mysterious invisible medium referred to as "luminiferous ether" in an experiment with Edward Morley in 1887. Their debunking of this widely held belief shocked the physics world. (Chapter 1)

The idea for this latest contraption formed out of a report by Hippolyte Fizeau (Chapter 1) to the FAS in 1868 where he put forth the notion of measuring the angular diameter of a fixed star by the method of interference of separate light beams from the source. The idea was to use a mirror array at the end of the tube of a reflecting telescope whose primary was covered save for two slits, spaced some distance apart. The light from the star reflected off the primary would take the form of two "pencils" of light that would be reflected to the eyepiece or camera of the telescope and brought together to produce interference fringes of alternating light and dark bands. Once synchronized, the amplitude of this interference would provide otherwise unobtainable information about the size, shape and brightness of a star. It would give photometrists, spectroscopists and astrophysicists a more precise tool for extrapolating other generalized measurements like stellar and (later) nebular distances.

Michelson himself had not offered anything further on the subject since his 1890 paper titled *On the Application of Interference Methods to Astronomical Measures* where he de-emphasized the need for large apertures in favor of a small periscopic arrangement using four small mirrors set in a frame. He proved his concept a year later using only two mirrors fixed above the 12-inch refractor at the Lick Observatory to measure the Jovian moons. In addition, following Fizeau's speech the French astronomer Édouard Stephan had used this method to detect fringes and it had proven useful in the measurement of double stars. While testing the new device on the 100-inch John Anderson even devised a way of determining the distance between the components of the Capella binary system, which was impossible by direct means.

Up to then, the consensus had been that atmospheric conditions would be unfavorable for stellar measurements even using larger apertures. The commissioning of the 100-inch and the exceptional seeing at Mount Wilson eliminated all doubt that the theory could now be implemented in practice.

With help from the design, instrument and optical shop, Pease and Michelson created a new instrument comprised a pair of 21-foot channels, each 10 inches wide and separated by 12-inch cross sections with thin steel plates riveted at either end. The light path of a star was reflected off two six-inch adjustable mirrors at either end of the channel to another pair of six-inch mirrors set at 45-degree angles and at a fixed distance of 45 inches apart. From there the separated light beams passed through the telescope to the primary and secondary mirrors in the usual manner to the camera and spectrograph at the Cassegrain focus.

Comparing data already gleaned from observations of the Sun and parallactic studies of nearby stars, the angular diameter of a star was calculated from wavelength measurements of its spectrum and the proportional distance of the mirrors of the interferometer, which for alpha Orionis (Betelgeuse) was determined to be 121 inches (with a possible error of plus or minus 10 percent). This gave an angular diameter of 0.047 seconds of arc. Using known parallax (distance) measurements for Betelgeuse by Adams and several others, Michelson and Pease used the weighted mean value to calculate the red supergiant's linear diameter as 240,000,000 miles, much greater than the diameter of the Earth's orbit around the Sun (Modern measurements of the distance to Betelgeuse have shown the star to be a whopping 1.14 billion miles in diameter!).

Measurements of other stars in Orion, Perseus, Taurus and other constellations as could be achieved at the time were also made. The real significance of this achievement was that accurate measurements of linear diameters and distances could now be taken for stars well beyond the bounds of the parallactic method. Pease and Michelson published their findings in the Proceedings of the American Philosophical Society in 1921.[107]

Francis Pease was a proud man, a capable observer and a great instrument designer. His masterpiece, the 100-inch, would go on to make history with the (practical) discoveries of external galaxies, the expanding universe, stellar populations and research on supernovae to name only a few.

Of all those on the mountain, Frederick Seares was by far the most influential to Edwin Hubble in his first year at MWO. In need of a mentor to help him break the ice as he took his first tentative steps into the field, a young astronomer could not have asked for a better tutor than Seares. The middle-aged photometrist was no doubt enlisted by Hale and Adams to take Hubble under his wing until he was comfortable. One of Hubble's first tasks was to assist Seares with a comparative test of the resolving power and clarity of the two large reflectors. For the test, photographs were taken of the same target objects at both telescopes using emulsion plates from the same batch, exposing them simultaneously and developing them in the same tray.[108]

Another project which stemmed from Hubble's solo work on nebulae was determining a plausible explanation for the vast color fields in "nebulous stars."

A quick Google search of this phrase today will get you a page full of links to color and activity books for young girls. But in 1920, before the world was presented the undeniable reality of external galaxies, supernovae, diffuse gaseous star nurseries and other deep space phenomena, the phrase "nebulous stars" referred to objects that seemed to be lurking within the inner regions of some nebulae (or in some cases the nebulae themselves).

While examining these "stars" at the telescope Hubble found their color to be yellowish but their spectra showed they were early type stars. Hale asked Seares to investigate further, so the veteran and his protégé carried out a survey of 42 of these objects using comparison photographs and spectra exposed at the 60-inch and 10-inch telescopes.[109]

The result of the study was that the "stars" overall were "appreciably redder than would be inferred from their spectral type" and the range of color apparently varied wildly. In his description of "the phenomenon" as it pertained to stars such as the bright one in the upper region of the M20 nebula in Sagittarius, Hale pointed out that "long-exposure photographs indicate…that none of these objects is nebulous in the sense defined by Herschel." Two possible explanations he suggested, were either "that the light of these stars is scattered by the surrounding nebulosity" or some "resonance excitation" was inducing the glow. As we shall see, it would be some time before the categorization of nebulae was extended beyond diffuse and planetary types.[110]

The "sense" Hale was referring to from Herschel may have come from a theory the latter reported on after his lengthy study of some twenty-four-hundred nebulae. Published in the Philosophical Transactions of the Royal Society in 1791 his *On Nebulous Stars, Properly So Called* suggested the presence of a "shining fluid or luminous matter" which somehow confined the stars within these nebulae and perhaps helped to create them.[111]

This marked a reversal for Herschel, who had earlier suggested the spiral nebulae were external galaxies. The idea of an enormous mass of liquidity in open space was novel to be sure, but he was closer to reality with his idea of luminous matter and stellar development. Some classical nebulae, emission nebulae and dark nebulae, and supernovae remnants are indeed essentially luminous clouds of gas. However, it is unclear which of these assertions Hale was referring to.

It is thrilling to envision 18th century astronomers making the kind of presumptive leaps that occupy Herschel's 1791 paper based on data collected using primitive, relatively small aperture telescopes and without the support of photographic equipment. For comparison, consider Percival Lowell's compelling (though highly inaccurate) depictions of canals and intelligent life on the surface of Mars from 1895 through the first decade of the 20th century. It would be hard to find two better examples of astronomers' romance of the heavens than these.

Even the far superior light gathering power of the 60-inch telescope was insufficient to establish the reflective nature of the bright star in M20 to the Mount Wilson team beyond doubt.[112] Images of M20 today reveal each of the three classic HII diffuse types of nebulae – emission, reflective and dark – all within the one object. And it is two to three times larger than Messier, and Herschel would have seen it. Their telescopes would have shown only the reddish (filtered) reflection nebulosity at the lower end of the nebula, giving it the appearance of having three distinct parts. That was why Messier named it Trifid, from the Latin word trifidus meaning "partly or wholly split into three divisions or lobes."

In addition to its historical interest, this small section in the CIWY for 1920 establishes the degree to which Edwin Hubble was already engaged in reaching to the core of nebular research. He was also credited with the discovery of three new planetary nebulae and two globular clusters on direct photographs taken at the Newtonian focus. His study of possible "rotation" and the source of the magnitude fluctuations in what were referred to as "variable nebulae" was outlined. He collected sixteen additional spectra (bright-line and continuous) for nebulae and "nebulous stars" not previously known. The inclusion of his four methods for determining distance to the nebulae by means of spectroscopic and photometric parallax reinforced the fact that he was already absorbed in an effort to understand the distances to these objects as they related to the oddities he was seeing in their spectra, magnitude, color and overall structure.[113]

Hale and his organization were in complete solidarity with the staff in pushing forward almost any thread of investigation that might help solve the various problems of stellar and galactic evolution. It was an extraordinarily propitious time to be a staff member at MWO and for a first-year man like Hubble it must have felt as though the metaphorical stars were aligned. There were explosive developments underway in every aspect of research and the mountain was literally and figuratively popping.

In the lab, John Anderson built a 1 microfarad condenser capable of operating at 26,000 volts of electricity and he and Harold Babcock were using it to study the "spark spectra" of metals such as iron, titanium and nickel to serve as references for spectroscopic research. Operated using a five-kilowatt transformer the condenser produced "brilliant sparks of extremely violent character" when discharging. Anderson's principal work involved the study of the spectra of exploded wires to try to mimic the conditions a meteor meets when entering the atmosphere of a star such as the Sun.[114]

Our parent star was being studied constantly by Hale, Seares and Nicholson for insights into its magnetic fields, sunspots and cycles, while Nicholson carried on his calculations of the ninth moon of Jupiter, Sinope, which he had found while studying for his doctorate at Drake College years before.[115]

The stellar staff and administration – which by then included Hale, Adams, Alfred Joy (Secretary), Seares (Superintendent of the Computing Division and editor of observatory's publications), Pease, Shapley and Paul Merrill – had seen a shift in its gravitational center with the arrival of Albert Michelson.

The use of Michelson's stellar interferometer in the measurement of stellar diameters had occupied members of the drafting, optical and instrument shops including Pease who produced the drawings.[116] Michelson was also exploring his other great muse, an improved measurement of the velocity of light. A 36-inch mirror with a focal length of 900 feet and a 22-inch plane was figured and polished in the optical shop and preliminary trials made to identify the intensity of the light source required to conduct the test at the necessary distance of 37.5 km.[117]

By the next year word would reach the mountain that Hale and Adams were talking of designing a massive instrument which, according to Pease's vision for the telescope, would have a primary mirror 300 inches in diameter. This concept, somewhat scaled down, would eventually be realized as the 200-inch Hale Telescope on Mount Palomar, 40 miles to the south of Mount Wilson where the seeing was not impaired by the continuing growth of Los Angeles.

So, we see that as Edwin Hubble entered into his new career, he was riding the crest of an almost two-decade-long wave of technological, optical and observational creativity. To make matters even more exciting, in June 1920 he met a young woman through his friend and colleague, the Lick Observatory astronomer William Wright. They had become friends during a visit to Mount Hamilton, and Wright was making his way to Mount Wilson at the invitation of Hale to demonstrate his new ultraviolet spectrograph in conjunction with the 100-inch telescope.

As Hubble later learned to his eternal delight, Wright had brought company in the form of his wife Elna and her sister-in-law Grace Lillian (Burke) Lieb. Instantly taken with her beauty and charm, Hubble was dismayed to find that Grace was already married to a mining engineer from a well-connected California family.

Wright and his lady companions were staying at the Kapteyn Cottage, a small but quaint two-bedroom affair near the summit overlooking the valley built for Jacobus Kapteyn and his wife during his almost annual visits to the mountain between 1909 and 1914. The onset of war had ended the Dutch astronomer's journeys west and Hale had since converted the cottage to a lodging for visitors in Kapteyn's honor.

Edwin and Wright dropped by the cottage to chat over tea with the ladies and it quickly became clear that a chemistry was developing between Edwin and Grace. She was a petite, athletic woman with short cropped wavy brown hair, parted from the right. She had a sharp wit, a love of books, theater and film and she shared some of Edwin's conservative views. She was the elder of two daughters born to John Burke and Luella Kepford Burke. In 1891, when Grace was only two years old, the family had moved to San Jose where John Burke became vice president of the San Jose and Santa Clara Railroad. On being admitted to the bar in 1901 he moved the family south to Los Angeles, which was beginning to eclipse its northern neighbors in both size and stature. There he rose to become vice president of the First National Bank, solidifying the family's fortune and providing a charmed childhood for Grace and her sister Helen. They were enrolled in the Marlborough School where Grace excelled, focusing her studies on English literature.

Coincidentally, her father's position in the Los Angeles banking community, in which Milton Humason's uncle Henry Clayton Witmer was a central figure in the first decade of the century, and her love of riding horses would become a source of conversation between Grace and Humason in the years to come. She proved to be a consummate hostess. Indeed, on Albert Einstein's visit to Pasadena in 1931 she would make the time to chaperone him around town.[118]

In 1920, however, that future was far from certain and for the next year Edwin and Grace courted their feelings for each other in silence, sharing brief acquaintances on her father's visits to the mountain. Having no desire to tell Grace the truth about his childhood, Edwin made little or no reference to his family, choosing instead to beguile her with edited tales of his early exploits at Oxford, his athletic prowess and law practice, and leading his men into battle in the Great War. As had been increasingly the case since his arrival at Yerkes, he spoke little of his family with whom he had increasingly little contact. Better that Grace (and the world) see him as he would prefer to be seen, without the baggage of his youth or the embarrassment of his family's station, which he viewed as mundane and mediocre.

The depth of their relationship prior to her husband's death has not been uncovered but Edwin was nevertheless considering the impression he must make on her father. Whether Wright was partial to her first husband has been lost to history, but Edwin had to consider whether his relatively low status as an entry level astronomer would stack up well against the future prospects of a mining engineer who currently held Grace's attention, regardless of how divided it may have become.

Holding his thoughts and desires about Grace close to his chest he threw himself at his work in search of a discovery or contribution that would bestow real status upon him in the world of science. It seemed probable there was more to be found out about nebulae, but he couldn't be sure yet what their mysteries were. However, in his reports to Hale and Adams for the CIWY he had already begun to use the term "extra-galactic nebulae" for those that were located well away from the plane of the Milky Way.

Bound by his restless ambition, Edwin Hubble worked tirelessly in those first few years at Mount Wilson seeking a career-defining discovery. The year-books were sprinkled with details of his investigations, but he took his time before formally publishing his findings.

His first report appeared in the June 1921 edition of the PASP. The paper called *Twelve New Planetary Nebulae* merely laid out his discovery of these new objects on plates from the 10-inch Cooke astrographic camera and confirmed by observations using the two large reflectors. He pointed to four criteria for the determination of a planetary: (1) it should have a bright line spectrum, (2) it should be symmetrical in form, (3) it should be formed around a central star and (3) it should have a well-defined border.[119]

The year 1922 was not a good one for seeing and the winter of 1921-1922 was the worst on record for precipitation at Mount Wilson (whose fiscal year ran from September through August). Over 60 inches were recorded by Hoge, who collected meteorological data from a small station near the 60-inch dome as part of his routine; 75 percent above normal. From December 17 to 26, 1921 a storm dumped over 29 inches of snow on the mountain and the overall snowfall for the year was 84 inches. Although the mean temperature was a pleasant 56 degrees, the highs and lows reached 95 and 15 respectively, in both cases straining man and machine. Wind velocity was above normal for the year as well, an all-together different problem because high winds buffeting the telescope would cause the instrument to shudder. The average wind for the year was just over ten miles an hour but much higher speeds had been recorded, including 80 mph on the 21st of December 1921.

Still, in what had been a historically bad year for seeing overall the big reflectors were in operation for all or part of 277 evenings and for complete nights on 193 of them. When the weather cooperated, Hubble was making good use of his time on the mountain, and it didn't take long for him to have an impact.[120]

In 1922 he published his findings on "galactic nebulae" from the previous two years' at Mount Wilson using the Lick quartz ultraviolet spectrograph mounted on the 100-inch. His paper, *The Source of Luminosity in Galactic Nebulae*, signaled his intention to take the lead in the field of nebular research.

The principal insight outlined by Hubble that year was that all diffuse nebulae have stars associated with them and that two separate types of the nebulae could be identified by the spectral type of the star associated with their nebulosity. Luminous gassy nebulae showing emission spectra were located near hotter stars of spectral type "B0 or earlier" while those that merely reflected light from nearby somewhat cooler stars of type "B2 or later" showed continuous spectra.

In the first case, Hubble had identified the so-called HII regions that were later found to be dense in hydrogen gas.[121] "The luminosity," as Henry Norris Russell put it in a 1921 paper, was "probably due to excitation of the individual atoms" by nearby stars.[122] This ionization process through absorption of ultraviolet from stars would be confirmed separately by Donald Menzel and Herman Zanstra in 1926.

The second type were not glowing as such, instead they depended on the stars for their luminescence and unlike their luminous cousins their continuous spectra tended to correlate strongly with that of their associated stars. This naturally led to their being called reflection nebulae.

In describing and substantiating the idea of the effect in luminous nebulae, Hubble made the analogy of gases in the Earth's atmosphere being excited by radiation from the Sun and causing auroral glows.[123]

In two papers published that year Hubble made reference to the history of developments that prompted his conclusions. An earlier paper published in October in the ApJ cited both William and John Herschel (Chapter 1) and the mid-19th century observations of Lord Rosse in Ireland as having established the field of nebular types from "liquid" or gassy to planetary with involved stars.[124]

He opened his seminal paper of August with a mention of Slipher's earlier revelation of the agreement between nebular and stellar spectra in certain nebulae which suggested they were illuminated by reflected starlight. Until now, nebulae had been separated in their basic form between diffuse and planetary. Hubble had introduced two subdivisions to the diffuse class: HII or emission nebulae and reflection nebulae. A third type, dark nebulae, had been introduced by Russell as being formed from interstellar dust that obscured the surrounding light, as in the dark patches that delineate the lobes in the Trifid Nebula. These types remain as three of the four classic types of diffuse nebulae. The fourth, supernova remnants such as the Crab Nebula (M1), would not be recognized until the middle of the next decade.[125]

Hence by the summer of 1922 Edwin Hubble was enjoying for the first time the fruits of his many long years of work in becoming an astronomer. He had set his goals, planned for and executed them to achieve the academic milestone in the shape of his doctorate from Yerkes shortly after his arrival in California. Now, with mentorship mainly from Seares, which he had put to effective use in his recent studies, he was starting to make his way into the murky world of scientific research in the company of some of the giants of astrophysics. He was finally living his dream and had announced his arrival on the scene with the papers published in the summer of 1922.

***

In contrast, Milton Humason's life was becoming increasingly uncertain and untenable by the close of the fiscal year 1922 and he had probably begun to rue his decision to step away from school as a teenager in 1905 in favor of a life on the mountain. The buzz created by the building of the observatory was surely a draw, and he doubtless took part in the public observing nights that Hale and his associates offered as part of the observatory's land lease agreement.

Whether these glimpses of the stars captured the imagination of the young Humason is lost to history. Given his family connections he was in the unique position of being able to choose whatever route he wanted to pursue. It is perhaps easiest to infer from his decisions at crucial points in his young life that his natural tendencies led him toward a more general attitude to life with a naturalist's penchant for involving himself to the fullest extent in his surroundings. These and other attributes that would remain dormant until much later would also make great qualities for an astronomer during the first half of the 20th century.

In 1920 Humason had completed two years of what might loosely be described as an abbreviated undergraduate degree focused primarily on advanced mathematics with bits of history and humanities thrown in as they were introduced to him by various friends in and around the observatory. He had accumulated some broader experiential knowledge sets in business and management through his family and farming connections, and he was widely recognized as a serious student and a meticulous observer. He had already published his first semi-professional articles in astronomical journals prior to Hubble's arrival and with the help of Adams, Seares, Nicholson and Merrill he was receiving his advanced education in the field. In sheer volume, Humason's output before the end of August 1922 far exceeded Hubble's with 16 published papers, either solo or in partnership, covering a broad range of topics and discoveries. But despite his improbable rise, the way forward remained unclear, and the window of opportunity was closing.

It is easy to imagine Humason sitting on the exposed roots of "the perch" near the 100-inch dome looking across at Mount Baldy in the spring of that year and wondering how he was ever going to find his way onto the observatory staff.

The excitement that he and Helen had felt when they arrived back on the mountain four years earlier with Billy in tow was fading like the planet Venus in the morning sky. Now, Billy was nine years old and the need to return to the valley so that he could be near to kids of his own age and attend public school was growing as fast as he was.

Thanks to contributions from the likes of William Wirt, Booker T. Washington and John Dewey, the American education system was improving in both scope and outreach. High schools were increasingly offering more rounded education and nearly two million children aged 14-17 were enrolled by 1920, a tenfold increase from 1890. This progressive approach to broad secondary education in America was unprecedented for the time and would raise the percentage of high school educated Americans from nine percent in 1910 to 50 percent by 1940.

More importantly, Milton Humason, the high school dropout, was going to make sure his son got an education. The question that weighed heavily on him at that moment was how, and in what capacity he would provide for it.

To shed more light on just how simultaneously precarious and promising his situation was, one need only examine the various records of the observatory. From his first mention in the CIWY for 1920 his title was given as simply "night assistant." Yet in the 60-inch logbook for that year his title is given as "relief night assistant," as was fitting since Hoge was the chief assistant in charge of the telescope.

But Humason's initial appears very little, if at all, in the logbooks of the 100-inch for the same period and, although he wouldn't have known it at the time, Tom Nelson, the driver who slipped over the cliff in the observatory's truck who had been associated with the facility since 1915 and was almost as well liked as Humason, would soon take over charge of the big telescope. At that time there were several well qualified men waiting to take their turn as night assistants who had backgrounds in engineering or other trades which were well suited to the position. Humason was more than likely acting as the swing man after 1919, with Merrill, Adams and even Hale himself using his prowess at the 10-inch astrograph for various projects.

Hale was clearly aware of Humason's growing skill and the appreciation he was earning on the staff, but the director was not at all inclined to promote him. Pitching a guy with no educational background whatsoever to the CIW would have been a tall order regardless of his aptitude. He was likely unaware of the statement by Andrew Carnegie in his trust deed to the CIW in 1902 as to one of the aims of the fledgling organization:

> To discover the exceptional man in every department of study whenever and wherever found, inside or outside of schools, and enable him to make the work for which he seems specially designed his life work.[126]

On the other hand, Humason was the only one of the night assistants apart from Hoge, and by far the most often mentioned in the CIWY, which the Carnegie board read eagerly, so they must have been aware of his work. There were two central reasons for his inclusion in the official reports.

The first was he was good, very good, and in Adams's mind, if you were good you were going to find work. In the years since his first publication Humason had worked alongside a growing list of present and future luminaries. On his own he had discovered a planetary nebula, the 17th and 21st novae in M31 and made magnitude measurements to chart the light curve of Nova Aquilae no. 3. Along the way he also discovered a very faint eclipsing binary star by analyzing 18 plates that he had taken with one-hour exposures. Soon he was moving between the 10-inch Cooke, the 60-inch and the 100-inch doing direct photography and spectroscopy for various members of the observatory staff.

By 1922 he had captured the spectra of dozens of stars with bright hydrogen alpha lines together with Merrill on his blue giant variable star program. In his first three years he had worked with nearly everyone on the mountain including Hubble, for whom he had obtained spectra for two planetaries. In addition, he worked with visiting researchers like the Swedish astronomer

Knut Lundmark (with whom Hubble would have one of his bigger disputes in the years to come). Lundmark liked Humason and appreciated his stories of the mountain as much as his skill with the telescopes.

Humason's ability to photograph very faint objects was already becoming apparent and this led to his working with Seares to update the photographic scale for 15th magnitude and photovisual scale for 13th magnitude stars of the North Polar Sequence using the 10-inch. It was a kind of master class and Seares was impressed. Exposing hundreds of plates each year, Humason was honing his considerable talent extremely rapidly.

The second equally important reason was that Adams was increasingly running the show and had become convinced that Humason belonged on the team. Whether his role as night assistant was intended more to signify an assistant observer in Adams's mind is not clear, but Humason's progressively more abundant mention in the Carnegie records indicates at least an effort to provide Hale every reason to alter his point of view. Humason, who was only just finding his sense of personal ambition, would have accepted his fate had Hale absolutely denied him the opportunity. His respect for Hale bordered on reverence. This might have made all the more frustrating the assurances of Adams, Shapley, Seares and others that his work was worthy of consideration.

Increasingly desperate for a chance to prove himself worthy of the legendary director's praise, Humason had kept his eyes equally on the skies and on the prize, seeking out every opportunity to shine. One opportunity presented itself in the winter of 1920 with a search for the planet Pluto (at that time referred to as "Planet X") long suspected of perturbing the orbit of Neptune. After studying the matter and discussing possibilities with Nicholson and others, Humason took a number of plates of the region of the sky where it was thought the ninth planet might reside. He later tried again, this time with the 60-inch. He wrote little or nothing in the logbook for the last evening in November. The single "H" in the assistant column indicated he was working alone with "Search for planet" in the work column. He made no attempt to indicate what time the dome was opened or closed, how many hours of viewing were involved or the number of plates that he exposed. That anything was written indicates he was there, but the lack of detail suggests either he had some trouble with the telescope or the weather, or simply got cold feet. Whatever the answer, he returned the next night, seemingly in lighter spirits. When van Maanen walked in the following evening, the comment he read in the work column for Humason the night before was "Planet hunting." With a good night of seeing, Humason had taken eight plates to blink for signs of a possible planet moving across the sky.[127]

He wasn't one to dwell on such matters but the run-in with Shapley over the possibility of variables in M31 had also been deflating for Humason and although Shapley continued to champion his rise in the ranks at MWO Humason was left somewhat untethered by the incident. He later told his friend and longtime Lick Observatory collaborator, Nicholas Mayall, of a recurring nightmare that began around this time. In the dream, Humason found himself leaving the dome building of the 60-inch telescope after a night of observing. As he walked along the path under the slowly lightening morning sky, he heard the slit door screech to a halt as it was being lowered, then breaking free of its track and hurtling down the path after him like a giant inch worm of steel. Running for his life Humason inevitably found himself pinned against the wall of the monastery as the killer drew down on him. He was saved only by abruptly awakening from the nightmare. The humor of the story wasn't lost on Mayall, and neither was its underlying meaning.[128]

Incidents like that with Shapley, along with Hale's continued insistence that Humason was not worthy of promotion, had prompted his predicament that day at "the perch." None other than Frederick Seares, Chair of the International Committee on Magnitudes, thought highly enough of the fledgling astronomer's work to take him under his wing. But not even Seares's recommendation was enough to convince Hale.

Evidence that Humason's frustration was boiling over into his work may be found in a tale about his intolerance for disrespectful visiting astronomers, most notably Henry Norris Russell (Sandage refers to him as "a famous visitor from back east") who treated the night assistants more like indentured servants than skilled assistants. Having spoken to him in a condescending tone during the run, Russell curtly instructed Humason to go to the midnight lunch shack to prepare his boiled egg, tea and biscuit so that they were ready for him when he arrived as scheduled.

Humason did as instructed, but to "save time" he prepared Russell's tea using the same water he had boiled the preening astronomer's farm fresh and slightly soiled egg in. Sitting across from Russell looking on intently with his best poker face Humason feigned surprise when Russell spit out the eggy tea and started grousing about the taste of sulfur in the water. Russell evidently continued to voice his displeasure at the quality of the drinking water on the mountain to Walter Adams on his visits to the observatory over the next several years.[129]

The events that precipitated Humason's final elevation to the ranks of the astronomers at the observatory unfortunately stemmed from the declining health of its director. George Hale coped readily with work that would overwhelm most people of promoting institutions, raising funds for instruments and facilities, and administering the day-to-day aspects of the observatory, but from time-to-time things really got on top of him.

He had fully invested himself in the transformation of Throop College into what became the California Institute of Technology (Caltech) in 1920. In pursuit of this goal, Hale had lured his old friend Arthur Noyes, builder of the famous chemistry lab at his alma mater of MIT, to the university as a professor and an educational and cultural advisor to its board of trustees. The resignation of James Scherer, the president of the college who had overseen its transformation into a highly respectable technical institute, obliged Hale to search for a suitable replacement. He decided to ask another old friend Robert Millikan to take on the role as Caltech's first president, but the future Nobel laureate was firmly entrenched at the University of Chicago and was also being pursued by both the Carnegie Corporation and the Rockefeller Institute. To woo his friend, Hale waged what he later described to Adams as "the most difficult campaign I have ever been through, involving scores of interviews and discussions, with all the cards against [Caltech]."[130] In the end the gambit paid off and Millikan would go on to lead the university to greatness.

But the struggle had been exhausting and when Hale tried to go back to his solar work his mental state deteriorated again. In May 1922 he and his wife Evelina left "to go abroad for a year's rest" to attempt to recover from his terrible headaches and profound depression. They traveled extensively and he found the travel delightful, but the pain in his head would not subside and he decided to resign as director of the observatory. Adams's name appears as "Assistant Director" in the observatory's report to the CIWY in 1922 because Hale was seeking to elevate his friend's standing prior to stepping away.[131]

Despite his frustration with Hale, Humason, like all those around the observatory, felt real remorse on hearing of Hale's resignation. They remembered the joy etched on his face when he and his cohorts first climbed the mountain, his boundless energy as he (most often) rode up the trail on the back of a mule, and his encyclopedic knowledge of science, poetry and music. In the almost two decades Hale and Humason had coexisted on Mount Wilson they'd had very little direct contact. Yet to hear of his deteriorating mental condition and forced departure from the institution and the science he loved most, while still only in his fifth decade of life was difficult for Humason to comprehend. It did, however, open a door.

Adams may well have waited until Hale's resignation was announced officially in July 1923 to bring Humason on, but another incident convinced him otherwise. That spring, a defect was discovered in the dome tracking system of the 100-inch telescope. The defect was exacerbated by the heavy snowfall and extreme weather the previous winter and the dome trolley wheels were in need of repair. The dome would have to be jacked up while the work was being carried out, rendering it immobile and the telescope useless for several months. When at some point Adams spoke of the issue to Humason he suggested to Adams that facing the dome slit due East would allow for observations of any objects rising in the sky within the slit aperture. The solution was limited perhaps, but at least some work could continue. A smile formed on the acting director's face. His belief in his friend and colleague was now fully realized in a simple act of ingenuity. With the telescope being prepared for renovation, the last page of the 100-inch logbook for that year bears the note: "Slit opening: 93.0 – June 25, 1922 – Hum."[132]

With the director having set sail for Europe the day before, Adams would have time to contemplate how and when to promote Humason either to Hale or to the board at Carnegie. How the acting director managed this is not known but Humason's name appears without the title of "night assistant" behind it for the first time in the yearbook for 1922, which was published in January 1923, three months before Hale tendered his resignation to the board and six months before it was made public. It is possible that in view of his ongoing illness, and the arguments that Adams and Humason himself were making, Hale simply relented on the issue.[133]

Just two months shy of his 31st birthday, Milton Humason had finally earned the title of astronomer. Five years earlier his fascination with the stars wasn't much more than that of any other backyard stargazer – a passing fancy, a momentary reflection and a cause for wonder. Unlike earlier notable astronomy enthusiasts such as the 16th century Tycho Brahe, he had no vast fortune to underwrite his growing passion. Blind luck, natural ability and an acute situational awareness were his acolytes, and, in this endeavor, they served him well.

In a lecture at the University of Lille in 1854 the famed French microbiologist Louis Pasteur famously remarked, "In the fields of observation chance favors the prepared mind." Milton Humason, rogue naturalist turned vulgarian astronomer, had taken his chance, and this time it would lead him and the rest of the world deep into the cosmos.[134]

References

1. George E. Hale (Director), *Carnegie Institution of Washington Yearbook, n.16, 1917: Mount Wilson Observatory,* (CIWY 1918), pg. 203
2. George E. Hale (Director), *Carnegie Institution of Washington Yearbook, n.17, 1918: Mount Wilson Observatory,* (CIWY 1919), pg. 217

3. *Ibid.* pg. 198
4. Interview with Ann Humason Bernt, 2006, from family archive
5. George E. Hale (Director), *Carnegie Institution of Washington Yearbook, n.16, 1917: Mount Wilson Observatory,* (CIWY 1918), pg. 203
6. George E. Hale (Director), *Carnegie Institution of Washington Yearbook, n.17, 1918: Mount Wilson Observatory,* (CIWY 1919), pg. 181-187
7. *Sergeant York, War Hero, Dies; Killed 25 Germans and Captured 132 in Argonne Battle,* (New York Times, Sept. 3rd, 1964)
8. Kevin Robbie, *A Gallant and Worthy Foe: The Death of the "Red Baron,"* Thursday Review April 22nd, 2016, Retrieved Feb. 15th, 2021, http://www.thursdayreview.com/RedBaronVonRichthofen.html
9. *June 15th, 1917: U.S. Congress passes Espionage Act,* (history.com 2013), Retrieved August 31st, 2019, https://web.archive.org/web/20130313041013/http://www.history.com/this-day-in-history/us-congress-passes-espionage-act
10. Geoffrey R. Stone, *Perilous Times: Free Speech in Wartime From the Sedition Act of 1798 to the War on Terrorism,* (W.W. Norton & Co. 2004)
11. Steven G. Koven, Frank Götzke, *American Immigration Policy: Confronting the Nation's Challenges,* (Springer 2010), pg. 130
12. Seymour Topping, *History of The Pulitzer,* (Pulitzer.org 2021), Retrieved August 31st, 2019, https://www.pulitzer.org/page/history-pulitzer-prizes
13. George E. Hale (Director), *Carnegie Institution of Washington Yearbook, n.17, 1918: Mount Wilson Observatory,* (CIWY 1919), pg. 186
14. *Ibid.* pgs. 217-218
15. *Wendell P. Hoge, An Astronomer, 72,* Obituary notice, The New York Times, November 15th, 1939
16. Wendell P. Hoge, *Twilight Colors at Mount Wilson, Cal., August-September, 1916,* Monthly Weather Report of the U.S. Department of Agriculture, November 1st, 1916
17. Allan Sandage, *Centennial History of the Carnegie Institution of Washington: Volume 1 The Mount Wilson Observatory,* (Cambridge University Press 2004), pgs. 185-187
18. 60-inch logbook for January 1918, Mount Wilson Observatory, Henry Huntington Library, San Marino, California, pgs. 148-149
19. Harlow Shapley, *Globular Clusters and the Structure of the Galactic System,* (PASP February 1918), Vol. 30, No. 173, pgs. 42 54
20. Harlow Shapley, *The Age of the Earth: A Discussion of Recent Evidence from Geology and Astronomy,* (PASP 1918), pgs. 283-298
21. Lord Byron, *Childe Harold's Pilgrimage,* (Cassell & Company 1894)
22. Frank Press, *Victor Hugo Benioff: 1899-1968,* (NAS 1973), pgs. 25-40
23. Interview with Don Nicholson, 2006
24. Letter from Truus Suermondt to Ann Bernt, September 16th, 2014
25. Seth Nicholson and Harlow Shapley, *The Orbit and Probable Size of a Very Faint Asteroid,* (ApJ May 1917), vol. 30, iss. 710, p. 127-128

26. *S.S. Imperator/R.M.S. Berengaria,* (atlanticliners.com 2015), http://atlanticliners.com/hapag_home/imperator_home/

27. Gale E. Christianson, *Edwin Hubble: Mariner of the Nebulae,* (University of Chicago Press 1995), pgs. 103-111

28. Letter from Walter S. Adams to G.E. Hale, February 1st, 1919, Walter S. Adams Manuscript Collection, Henry Huntington Library, San Marino, California

29. Gale E. Christianson, *Edwin Hubble: Mariner of the Nebulae,* (University of Chicago Press 1995), pgs. 109-110

30. *Ibid.* pg. 108

31. Ėdvard Radzinskii, *The Last Tsar: The Life and Death of Nicholas II,* (G.K. Hall 1992), pgs. 609-616

32. Ferdinando Fasce, *An American Family: The Great War and Corporate Culture in America,* (Ohio State University Press 2002), pgs. 119-127

33. Thomas J. Knock, *To End All Wars: Woodrow Wilson and the Quest for a New World Order,* (Oxford University Press 1992), pgs. 170-173

34. *1910 to 1919 Important News, Significant Events, Key Technology,* (thepeoplehistory.com)

35. Interview of Nicholas Mayall by Bert Shapiro on 1977 February 13, Niels Bohr Library & Archives, American Institute of Physics, College Park, MD USA, www.aip.org/history-programs/niels-bohr-library/oral-histories/4766

36. Richard Striner, *Woodrow Wilson and World War I: A Burden Too Great to Bear,* (Rowman and Littlefield 2014), pgs. 224-230

37. Gale E. Christianson, *Edwin Hubble: Mariner of the Nebulae,* (University of Chicago Press 1995), pgs. 120-121

38. 60-inch logbook No. 4, 1918 and 1919, Mount Wilson Observatory, Henry Huntington Library, San Marino, California, pg. 2

39. Memories of George E. Hale and Mt. Wilson, 1953 by Paul Merrill Call: av 5-70-7; 2013-1226

40. George E. Hale (Director), *Carnegie Institution of Washington Yearbook, n.18, 1919: Mount Wilson Observatory,* (CIWY 1920), pg. 217

41. Edward C. Pickering, *The Draper Catalogue of Stellar Spectra,* (John Wilson and Son University Press 1890)

42. George E. Hale (Director), *Carnegie Institution of Washington Yearbook, n.18, 1919: Mount Wilson Observatory,* (CIWY 1920), pgs. 217-219

43. *Ibid.* pg. 219

44. *Ibid.* pgs. 219-221

45. *Ibid.* pgs. 222-223

46. *Ibid.* pg. 221

47. *Ibid.* pg. 259

48. Donald E. Osterbrock, *Pauper & Prince: Ritchey, Hale & Big American Telescopes,* (University of Arizona Press 1993)

49. George W. Ritchey, *On the Modern Reflecting Telescope and the Making and Testing of Optical Mirrors,* (The Smithsonian Institution 1904), 84 pgs.

50. Donald E. Osterbrock, *Pauper & Prince: Ritchey, Hale & Big American Telescopes,* (University of Arizona Press 1993)

51. Garrett P. Serviss, *The Greatest Telescope in the World: Monster Instrument Ordered by Carnegie Institution Will Far Exceed in Power All Other Watchers of the Skies,* (NY Times January 27th, 1907), Retrieved September 6th, 2019

52. Donald E. Osterbrock, *Pauper & Prince: Ritchey, Hale & Big American Telescopes,* (University of Arizona Press 1993)

53. Walter Adams to George Hale at National Research Council, Washington D.C., December 8th, 1918, Henry Huntington Library, San Marino, California

54. Walter Adams to George Hale at National Research Council, Washington D.C., February 2nd, 1919, Henry Huntington Library, San Marino, California

55. Donald E. Osterbrock, *Pauper & Prince: Ritchey, Hale & Big American Telescopes,* (University of Arizona Press 1993)

56. Bill Humason, *Santa Visits Mount Wilson,* (MWOA 2013), Reflections Winter Quarter, pg. 4-5

57. Interview with Don Nicholson, 2006

58. Paul Herget, *Seth Barnes Nicholson,* (NAS 1971), pg. 205

59. Elias Ransome Sutton ("Ransome"), *Astronomy Stars That Are Human,* (LA Times 1933)

60. Wendell Hoge, 60-inch telescope Logbook No. 4, January 1st, 1919, Henry Huntington Library, San Marino, California

61. Milton Humason, *The Light Curve of Nova Aquilae No. 3,* (PASP February 1919), vol. 31, no. 179, pgs. 43-44

62. Seth B. Nicholson and Milton Humason, *Metcalf's First Comet,* (PASP October 1919), vol. 31, no. 183, pg. 280

63. Memories of George E. Hale and Mt. Wilson, 1953 by Paul Merrill, Call: AV 5-70-7-2013-1226

64. 60-inch logbook No. 4, 1918 and 1919, Mount Wilson Observatory, Henry Huntington Library, San Marino, California, pg. 2

65. Harlow Shapley, *Nineteen New Variable Stars,* (PASP 1919), vol. 31, no. 182, July 1, 1919, pg. 226

66. I.J. Falconer, J.G. Mena, J.J. O'Connor, T.S. Peres, E.F. Robertson, *Harlow Shapley,* (University of St. Andrews 2018), https://mathshistory.standrews.ac.uk/Biographies/Shapley_Harlow

67. Memories of George E. Hale and Mt. Wilson, 1953 by Paul Merrill, Call: AV 5-70-7-2013-1226

68. Letter from George E. Hale to Walter S. Adams, Oct. 23rd, 1920, Henry Huntington Library, San Marino, California

69. Letter from Walter S. Adams to George E. Hale, 1920, Henry Huntington Library, San Marino, California

70. Harlow Shapley, *On the Existence of External Galaxies,* (PASP 1919), vol. 31, no. 183, pg. 261

71. Interview of Nicholas Mayall by Bert Shapiro on 1977 February 13, Niels Bohr Library & Archives, American Institute of Physics, College Park, MD USA, www.aip.org/history-programs/niels-bohr-library/oral-histories/4766

72. M. L. Humason, *Obituary Notices: Edwin Hubble,* (MNRAS 1954), Vol. 114, pgs. 291-295

73. *Ibid.* pg. 291

74. J.S. Plaskett, *The 72-inch Reflecting Telescope of the Dominion Astrophysical Observatory, Victoria, B.C.,* (Popular Astronomy 1919), Vo. 27, pgs. 210-216

75. 100-inch logbook for 1916-1919, Mount Wilson Observatory, Henry Huntington Library, San Marino, California, pgs. 20

76. Gale E. Christianson, *Edwin Hubble: Mariner of the Nebulae,* (University of Chicago Press 1995), pg. 123

77. 60-inch logbook No. 5, February 1920, Mount Wilson Observatory, Henry Huntington Library, San Marino, California

78. Allan Sandage, *Centennial History of the Carnegie Institution of Washington: Volume 1 The Mount Wilson Observatory,* (Cambridge University Press 2004), pg. 183

79. *Ibid.* pgs. 188-190

80. Gale E. Christianson, *Edwin Hubble: Mariner of the Nebulae,* (University of Chicago Press 1995), pgs. 123-125

81. Robert Robertson, Milton L. Humason, (Griffith Observer 1972), pgs. 9-14

82. Interview of Nicholas Mayall by Bert Shapiro on 1977 February 13, Niels Bohr Library & Archives, American Institute of Physics, College Park, MD USA, www.aip.org/history-programs/niels-bohr-library/oral-histories/4766

83. Henrietta S. Leavitt, *1777 Variables in the Magellanic Clouds,* (AHCO 1908), pgs. 87-110

84. Michael Hoskin, *Goodricke, Pigott and the Quest for Variable Stars,* (JHA 1979), Vol. 10, pgs. 23-41

85. George Johnson, *Miss Leavitt's Stars: The Untold Story of the Woman Who Discovered How to Measure the Universe,* (W.W. Norton & Co. 2005), pgs. 56-58

86. Frederick H. Seares and Milton L. Humason, *Further Evidence on the Brightness of the Stars of the North Polar Sequence,* (ApJ 1922), pgs. 84-96

87. Henrietta S. Leavitt, *1777 Variables in the Magellanic Clouds,* (AHOC 1908), Vol. 60, No. 4, pgs. 87-111

88. Henrietta S. Leavitt and Edward C. Pickering, *Periods of 25 Variable Stars in the Small Magellanic Cloud,* (ACOH 1912), Vo. 173, pgs. 1-3

89. George Johnson, *Miss Leavitt's Stars: The Untold Story of the Woman Who Discovered How to Measure the Universe,* (W.W. Norton & Co. 2005), pg. 55

90. E.C. Pickering, H.R. Colson, W.P. Fleming, L.D. Wells, *Sixty-Four New Variable Stars,* (ApJ April 1901), pgs. 228-230

91. Allan Sandage, *Centennial History of the Carnegie Institution of Washington: Volume 1 The Mount Wilson Observatory,* (Cambridge University Press 2004), pgs. 287-307

92. Interview with Don Nicholson, 2006
93. Owen Gingerich, *How Shapley Came to Harvard of, Snatching the Prize from the Jaws of Debate,* (JHA 1988), Vol. 19, issue 3, pg. 201
94. *Ibid.* pgs. 201-202
95. *Ibid.* pgs. 202-203
96. *Ibid.* pg. 203
97. Ibid.
98. Ibid.
99. *Ibid.* pg. 204
100. *Ibid.* pgs. 205-206
101. 60-inch logbook No. 5, 1920 and 1921, Mount Wilson Observatory, Henry Huntington Library, San Marino, California, pg. 3
102. Allan Sandage, *Centennial History of the Carnegie Institution of Washington: Volume 1 The Mount Wilson Observatory,* (Cambridge University Press 2004), pg. 309
103. George Johnson, *Miss Leavitt's Stars: The Untold Story of the Woman Who Discovered How to Measure the Universe,* (W.W. Norton & Co. 2005), pgs. 118-119
104. R.A. Sampson, *The President's Address on presenting the Gold Medal of the Society to Mr. Walter Sydney Adams,* (RAS Feb. 1917), pg. 395
105. Alfred H. Joy, *Walter Sydney Adams: A Biographical Memoir,* (NAS 1958), pg. 14
106. Interview with Don Nicholson, 2006
107. Francis Pease, *Measurement of Star Diameters by the Interferometer Method,* (American Philosophical Society 1921), Vol. 60, No. 4, pgs. 524-534
108. F.H. Seares, W.S. Adams, *Comparative Tests of the 100-inch and 60-inch Reflectors,* (PASP Feb. 1921), Vol. 33, No. 191, pgs. 31-34
109. F.H. Seares, E.P. Hubble, *The Color of the Nebulous Stars,* (ApJ July 1920), pgs. 8-22
110. G.E. Hale, *Contributions of the Mount Wilson Observatory,* (CIWY 1920), Vol. 19, pgs. 236-237
111. William Herschel, *On Nebulous Stars, Properly So Called,* (Philosophical Transactions of the Royal Society of London 1791), Vol. 81, pgs. 71-88
112. G.E. Hale, *Contributions of the Mount Wilson Observatory,* (CIWY 1920), Vol. 19, pg. 236
113. *Ibid.* pgs. 232-237
114. *Ibid.* pgs. 262-263
115. *Ibid.* pg. 220
116. *Ibid.* pgs. 262-265
117. *Ibid.* pgs. 253-254
118. Gale E. Christianson, *Edwin Hubble: Mariner of the Nebulae,* (University of Chicago Press 1995), pgs. 163-164
119. Edwin Hubble, *Twelve New Planetary Nebulae,* (PASP 1921), Vol. 33, No. 193, pgs. 174-175

120. G.E. Hale, *Contributions of the Mount Wilson Observatory,* (CIWY 1922), no. 21, pg. 220
121. E.P. Hubble, *The Source of Luminosity in Galactic Nebulae,* (ApJ 1922), pgs. 400-438
122. Henry Norris Russell, *Dark Nebulae,* (PNAS 1922), Vol. 8, pg. 117
123. E.P. Hubble, *The Source of Luminosity in Galactic Nebulae,* (ApJ 1922), pgs. 400-438
124. E.P. Hubble, *A General Study of Diffuse Galactic Nebulae,* (ApJ October 1922), Vol. 56, pg. 163
125. E. P. Hubble, *The Source of Luminosity in Galactic Nebulae,* (ApJ 1922), pgs. 400-438
126. Andrew Carnegie, *Carnegie Institution of Washington Yearbook,* (CIWY 1903), no. 1, pg. xiii
127. 60-inch logbook No. 5, 1920 and 1921, Mount Wilson Observatory, Henry Huntington Library, San Marino, California
128. Letter from Nicholas Mayall to Helen Dowd Humason, Nov. 9, 1972
129. Allan Sandage, *Centennial History of the Carnegie Institution of Washington: Volume 1 The Mount Wilson Observatory,* (Cambridge University Press 2004), pgs. 192-193
130. Helen Wright, *Explorer of the Universe: A Biography of George Ellery Hale,* (E.P. Dutton & Co, Inc. 1966), pg. 337
131. G.E. Hale, *Contributions of the Mount Wilson Observatory,* (CIWY 1922), no. 21, pg. 198
132. 100-inch logbook for 1922, Mount Wilson Observatory, Henry Huntington Library, San Marino, California, pgs. 153
133. G.E. Hale, *Contributions of the Mount Wilson Observatory,* (CIWY 1922), no. 21, pg. 209
134. M. Pearce, *Chance and the Prepared Mind,* (Science 1912), Vol. 35, No. 912, pg. 941

6

Rising Stars (1923-1927)

The mid-1920s mark the period of ascendancy for both Edwin Hubble and Milton Humason. Hubble's discovery of Cepheid variable stars in M31 and other nearby nebulae leads him to conclude that they are external galaxies, catapulting him to international fame. Humason's virtuosity at the helm of the 100-inch telescope increases his reputation, particularly in the exposure of the spectra of faint stars. In 1927 he formally announces himself to the astronomy world with a clear spectrum of the faintest star ever recorded. But their public and professional reputations are diametrically opposed. Known to the public as the man who unlocked the door to the greater universe, Hubble adopts a quarrelsome, pretentious and privileged attitude that serves to undermine his reputation in the field. In contrast, entirely unknown to the public, Humason's reputation as capable, conscientious and collegial grows within the astronomy community.

When the Great War finally reached its conclusion with much of Europe in ruins, the next phase of American economic and political dominance on the world stage got underway.

Meanwhile, within the wreckage of the old German Reich, a young radical named Adolf Hitler organized a riot that was sniffed out and put down by authorities. For his role in this Beer Hall Putsch Hitler received five years in prison, where he wrote his fascist manifesto *Mein Kampf.*

In Italy, Benito Mussolini marched on Rome in 1922 leading a coup which installed him as the new fascist ruler of the Italian state.

© Springer Nature Switzerland AG 2021
R. Voller, *Hubble, Humason and the Big Bang*, Springer Praxis Books,
https://doi.org/10.1007/978-3-030-82181-4_6

The leader of the recently formed United Socialist Soviet Republic, Vladimir Lenin died in 1924 in Moscow and Joseph Stalin took over as the Soviet dictator.

As the decade wore on, war reparations would drive the German economy into collapse and by 1927 there were growing concerns that the entire world economy could be in danger as well.

In the US, President Woodrow Wilson, the recent recipient of the Nobel Peace Prize for his efforts prior to, during and after the war was incapacitated by a stroke. His wife, Edith Bolling took over as de facto president, screening issues and bringing to his attention only those she deemed most pressing, a role that she performed for 17 months until his departure from the presidency in March of 1921; just as Harlow Shapley was leaving Mount Wilson for Harvard.

Wilson was succeeded by Warren G. Harding, whose mismanagement of the office led to the Tea Pot Dome Scandal that embroiled a number of his cabinet members. With public sentiment waning in the wake of this bribery scandal, Harding suffered a heart attack and died on a tour of San Francisco attempting to try to quell public dissent. He was succeeded by Calvin Coolidge, who subsequently won the office outright in 1925. He resided over the remainder of the decade as America and the rest of the world slid into economic depression.

In the meantime, the US led a cultural boom, ushering in the big band musical style, bob hairdos, flapper dresses and boas of the Roaring Twenties. For the first time in its short history the United States began to influence lifestyle and musical trends that swept through the western world. The American century was picking up steam. In September of 1921 the first Miss American Pageant was held in Atlantic City, New Jersey. The population of the country exceeded 100 million. Fear of rampant immigration (despite Coolidge introducing legislation in 1924 to constrain it) and lingering Southern antipathy to African Americans caused membership in the Ku Klux Klan to rise to its highest historical levels, topping four million in 1920.

With Conservatism on the rise, the Scopes Trial convicted schoolteacher John Scopes for teaching Charles Darwin's theory of evolution to his class at a high school in Tennessee, but the conviction was eventually tabled, and Scopes was fined $100 for his trouble.

The onset of Prohibition was viewed by many as a violation of basic human rights and many millions took part in the willful protest of the amendment. Public demand for illicit alcohol fueled the development of the mob and created the notorious gangsters Al Capone, Bugs Moran, Machine Gun Kelly and others.

Progressive policies like the 19th Amendment, which gave women the right to vote, were protested by many in southern states who also resented the idea of black suffrage, be it for men or women. The decade would also see the election of the first woman governor, Nellie Tayloe Ross, sworn in on January 5th, 1925, in Wyoming. Miriam Amanda "Ma" Ferguson became the second two weeks later, in Texas.

American literature was starting to make its mark on the world stage as well. Playwright Eugene O'Neill's *Beyond the Horizon*, the first of his four Pulitzer Prize winning works, opened in February 1920. Reader's Digest and Time Magazine were in regular publication by 1923.

The inaugural season of the American Professional Football League started in the fall of 1920, led by the American sports legend Jim Thorpe. Two years later it changed its name to the National Football League. In the Bronx the New York Yankees were building a new stadium nicknamed "the house that Ruth built" in honor of Babe Ruth, who had helped the team win three championships since joining its ranks a decade earlier. Ruth would lead the team to three more World Series titles during the 1920s and seven in all during his career.

One sign of America's burgeoning preeminence was the growing list of national parks, natural resources and historical landmarks awarded protection under law. The Appalachian National Scenic Trail, Lincoln Memorial and Aztec Ruins National Monument (a World Heritage site since 1987) were among more than a dozen landmarks to be dedicated during the decade.

Motion pictures, long the domain of Charlie Chaplin, Buster Keaton and the silent film generation debuted the first sound on film motion picture *Phonofilm* in the Rivoli Theater in New York City on April 15th, 1923; the same month in which Warner Brothers Pictures was formed in the Hollywood neighborhood of Los Angeles. In October of 1927, Al Jolson would star in *The Jazz Singer* in New York, the first full length movie to be presented in sound.

Not to be outdone, tens of thousands of "radio boxes" were finding their way into American homes. The Grand Ole Opry transmitted its first radio broadcast in November of 1925 and NBC Radio Network began regular broadcasts on 24 stations the following year.

The United States continued to flex its technological muscle as well. On March 16th, 1926, Robert Goddard tested the first liquid fueled rocket in Auburn, Massachusetts, and in May of that year Floyd Bennett and Richard Byrd made the first flight to the North Pole in a three-engine monoplane. The pair were awarded the National Medal of Honor for their achievement. One year later, almost to the day, Charles Lindbergh made the first solo flight across the Atlantic, steering his single-engine aircraft Spirit of St. Louis from Long Island, New York to a landing in Paris a little over thirty hours later.[1]

*   *   *

At the start of 1923 Edwin Hubble and Milton Humason had both established themselves in the work they would be recognized for in decades to come, Hubble in the field of nebular research and Humason in direct and spectroscopic photography.

Hubble's conclusive analysis showing a link between stellar and nebular luminosity was early evidence of his potential for Shapleyesque interpolation of apparent facts to revise an established theory and to make new discoveries.

Humason was already starting to take on the bulk of the observing effort for numerous studies of stellar type, color, absolute and apparent magnitude and radial velocity for both classification and interpretation.

On February 23rd, 1922, Hubble mailed an outline of a general classification of nebulae with a letter of explanation to Vesto Slipher, president-elect of the IAU Commission on Nebulae and Star Clusters. Hubble had been given a seat on the commission and hoped to present his system at the Rome meeting in July with the intention of making it the standard system for nebular research. Hubble's proposal, which he admittedly started "flamboyantly," was based on two general categories: Galactic and Non-galactic nebulae. Galactic nebulae were subdivided into Planetary and Diffuse, with the latter distinguished between Luminous and Dark nebulae. Non-galactic nebulae were subdivided into Spiral, Elongated, Globular and Irregular, with the Elongated nebulae further divided into Spindle and Ovate. Laying out first the problems concerning the nebulae and the investigations to be undertaken, Hubble went into considerable detail on the courses to be pursued in research, including on variable stars. However, he made no mention of Cepheids in his letter to Slipher.[2] It was generally a good system and Slipher is thought to have liked it, but with five of the 14 members of the commission working on classification schemes of their own the incoming president was reluctant to choose one over the others without further review and discussion.

Although he had not had much real contact with anyone at the observatory for years, the resignation of George Ellery Hale on July 1st, 1923, dealt a blow to the morale on the staff. Hale was revered by everyone at Mount Wilson and was one of the most highly regarded men of science anywhere in the world, a sentiment that was further fueled by the reverberating stories of his decade-long battle against mental exhaustion. His resignation was formally announced in the MWO contribution to the CIWY for that year with this note by Adams:

> The entire conception of the Observatory, its development, the scope of its scientific research, its equipment, and the methods of investigation are to so large an extent a direct product of Dr. Hale's foresight and his ability in the organization and conduct of research that any change in his relationship to the Observatory is necessarily a matter for profound concern.[3]

In reality, Hale hadn't been involved directly in the day-to-day operations of the facility for some years and, as Adams went on to say, he was to stay connected in the capacity of "Honorary Director" lending his support and guidance on "general policy, the problems of research, and the development of new methods and new instruments." In other words, their leader was going to remain as close to the observatory as his health allowed.[4]

In the desperate hope of being able to continue observing his favorite subject, Hale was designing a home-based solar laboratory at his house in Pasadena that would be fitted with a special spectroheliograph of his own design. The Hale Solar Laboratory, first mentioned in the MWO contributions the following year, would be used to study the Sun in conjunction with the solar telescopes on Mount Wilson.[5]

Milton Humason (Fig. 6.1) was working with both Hubble and van Maanen on nebular research that year. Hubble's focus was primarily on galactic nebulae and Humason's work centered on taking precise photographs of the spirals for van Maanen's investigation of their internal motions. As Adams explained about the latter, "it is essential that the images [of stars] be very small and round, a combination frequently difficult to secure in the case of exposures extending over several hours."[6]

The requirement for precision carried over into other areas. Humason had photographed directly a number of planetary nebulae and other objects for classification in a variety of programs for Hubble and Seares while continuing his investigations of stars in the Kapteyn Selected Areas for Adams and Joy using the slitless spectrograph on the 60-inch telescope. Some of these stars were very faint (12th magnitude) and to gain a useful spectrogram using the Cassegrain focus required four hours exposure time on average.[7]

Of all those on the nighttime staff in 1923 Adrian van Maanen was the only one other than Humason working in both stellar and nebular research, and Humason was involved in nearly every type of investigation that was underway at the time. With each new challenge, Humason was honing his

Fig. 6.1 Cowboy rancher turned gentleman astronomer - Milton Humason at the MWO offices ca 1925. (Courtesy, Ann Humason Bernt family)

techniques for ever fainter objects and as his skill grew so did the demands for his assistance. He was also being asked to work on longer research programs which combined his skill at the telescopes with the time-consuming reduction of data for a variety of surveys.

After publishing nine papers in 1922 Humason would publish only nine more through 1928. Two of these were the discovery of the 22nd nova near the nucleus of M31[8] and a note on the star Boss 1985 that was thought to "show the very interesting type of spectrum characteristic of the three M-type stars W Cephei, Boss 5650 and HD 42474," in relation to the Kapteyn Selected Areas project with Adams and Joy.[9]

Later reclassified as a VV Cephei variable, this eclipsing binary system is an M-class supergiant with a hot, blue B-type companion. For at least part of its orbit the supergiant fills its Roche lobe and takes on the appearance of a teardrop with the tapered end pointing at its companion star. As material escapes this lobe it is drawn to the companion, forming a ring around it and the excitation of the ionized gas increases the system's brightness. The concept of a star filling a gravitationally limiting surface and losing mass was predicted by the 19th century French astronomer and physicist Édouard Roche. Detailed work on various types of accreting binaries had to await the development of x-ray astronomy decades later.

Humason's work on Boss 1985 came during his long survey programs with Adams, Joy and Merrill. He and Merrill published their findings based on the study of 90 B-type stars displaying emission lines (and known as Be-type) in 1924 along with Cora Burwell in the Computing Division.[10] Many of the stars compiled for this paper are no longer classified as being this type but the work is nevertheless an indication of the sheer number of stars the large reflectors at Mount Wilson made available for study.

The other large project on Humason's agenda was the absolute magnitude and parallax program on M-type stars in the Kapteyn Selected Areas with Adams and Joy, photographed using the 100-inch. In their report in the Astrophysical Journal (ApJ) in August 1926 they listed 410 stars giving magnitudes and spectral types, comparisons between spectroscopic and trigonometric parallaxes, and magnitudes with prior published works. Twenty-eight of them had absolute magnitudes brighter than −1.0 and their concentration toward the plane of the Milky Way was reminiscent of "super-giant" stars of other spectral types. The paper was read at the 37th meeting of the AAS in Philadelphia in December 1926.[11]

The term supergiant had recently been coined by the English American Cecilia Payne (later Payne-Gaposchkin) in the 1925 dissertation which earned her the first Ph.D. awarded to a woman at Harvard College Observatory. Harlow Shapley had established the graduate program in astronomy soon after becoming director. The first student was Adelaide Ames, co-author of the *Shapley-Ames Catalogue of Bright Galaxies* in 1932.[12] Payne was the second. She had completed her studies at Cambridge in England but was not awarded a degree because at the time the school wasn't granting degrees to women. Her dissertation *Stellar Atmospheres: A Contribution to the*

Observational Study of High Temperature in the Reversing Layers of Stars reached a few groundbreaking conclusions, chief among them being that stars were composed almost entirely of hydrogen and helium.[13] In 1962 the Russian American astronomer Otto Struve called Payne's work "undoubtedly the most brilliant Ph.D. thesis ever written in astronomy."[14]

Despite her obvious practical and theoretical prowess, Payne-Gaposchkin spent most of her career battling gender discrimination at Harvard even while accumulating a large number of honors. She earned a lower middle-class wage, her work at the observatory was deemed unofficial and remained unacknowledged until 1938, and the courses she taught were not recorded before 1945. After three decades she was finally awarded a full professorship in 1956 and later the first female Chair of the Department of Astronomy, which she held until retiring as Professor Emeritus in 1966.[15]

Throughout her career, Cecilia Payne-Gaposchkin maintained a close relationship with some of the members of the Mount Wilson and Palomar Observatories including Hubble, who described her as "the best man at Harvard," a dig at Shapley who he knew would catch wind of the slight from her.[16]

The year 1923 was favorable for the Humason clan. The solar assistant and observatory photographer Joseph Hickox left Mount Wilson for life in the valley and Milton's brother Lewis Humason was appointed in his place.[17] Hickox's time in the valley must have made him lonely for the life on the mountain because he was back at the observatory two years later and would remain there until 1961 as one of the most-liked members of the organization.

The Humason brothers had signed up for the draft on the same day in 1917 but only Lewis, a mechanic working at an auto shop, was drafted. He spent the war as part of an army artillery battalion in Europe.[18] His work prior to and during the war implied Lewis would be well-suited to the complex mechanics of the coelostat mirrors, spectrographs and gratings in the solar department. Although his career at MWO lasted only five years, the romance of the mountain captured him just as it had his older brother. Shortly after joining the observatory, he met Beatrice Mayberry, a member of the computing division who got a degree in physics and astronomy from Stanford in 1919. She had been hired the following year, and when she met Lewis, she was working closely with George Hale on his sunspot cycle program. Her maternal grandfather was George Wing, inventor of Wing Function Symbols, a system of grammatical symbols used to teach sentence structure to deaf and mute people, and the Mayberry family had been instrumental in the development of the Alhambra township as well.[19] Just as had Milton and Helen 15 years earlier, Lewis and Beatrice fell in love with Mount Wilson. They were married February 11th, 1925.[20] Their only child, Harry Wing Humason was born on Christmas Eve of that year and the family would remain on the mountain until Lewis's resignation on January 1st, 1928.[21]

Beatrice Mayberry was a serious scientist who also found the going tough as a woman in a field dominated by men. The timing of her first and only joint reports published as a member of the observatory in 1924 and 1925 is curious, coming after Milton Humason was established as a member of the staff. The articles were published with Humason's former mentor Seth Nicholson. A teacher's teacher, Nicholson brought Mayberry in with him on a report that updated his analysis of Jupiter's ninth moon Sinope in 1924,[22] and a report on sunspot activity for the year that appeared in the February issue of the PASP in 1925.[23] In fact, Nicholson also oversaw a joint report on spectroheliograms with Lewis Humason that appeared in 1926.[24]

Beatrice's betrothal to Lewis brought her career as an astronomer to an abrupt end, as at that time married women were not allowed to work on the mountain. She was obliged to resign on February 1st, 1925, just ten days before her wedding day. She left the organization disgruntled and carried her disgust with her for many years.[25] But she was sought after for her skill in applied mathematics and would later work as a physicist in the Underwater Ordnance Department of the Missile Systems Division at the Naval Ordnance Test Station in Pasadena. Her son Harry also worked there as the head of Production Design Engineering.[26]

Neither Milton Humason nor any other member of the staff, regardless of stature or level of respect inside the organization, and no matter his sympathy for her situation, could have changed Beatrice Mayberry's fate. Her story provides a sad commentary on the slow pace of our common social evolution through the 20th century.

* * *

For Edwin Hubble, the 10 years between 1921 and 1931 were phenomenal by any standard of scientific discovery. His publications make up a tour de force of astronomical advancement. Left virtually alone in the field of nebular research, he would go on to essentially plan and develop the observatory's nebular department. By the end of the decade, he would submit conclusive evidence explaining the luminescence in various types of nebulae, establish the existence of external galaxies, create a classification system for galaxies that would be used by future generations of astronomers, and of course make the definitive practical discovery of universal expansion. Ironically, that last advance, the one for which he is singularly known, required the most collaboration, chiefly in the form of Milton Humason.

Knut Lundmark, who would become another of Hubble's disputatious antagonists, spent a year at Mount Wilson studying the spiral nebulae in association with the Observatory of Uppsala University in Sweden where Anders Celsius and Anders Jonas Ångström dwelled in the 18th and 19th centuries, respectively.[27] A wide field astrographic triplet (one for visual observations, and one each for blue and visual photography) with a six-inch Zeiss objective lens had been installed in 1914 but offered little efficacy in deeper fields to the Swedish born astronomer whose main interest, like Hubble's, was in nebular research. This likely prompted Lundmark's later request for a position at Mount Wilson, but this was ignored by Adams in order to obviate a conflict of interest with Hubble. Lundmark's year at Mount Wilson finished on May 1st, 1923, leaving Hubble with free reign over the nebulae once again.

Hubble's work was progressing steadily, yielding insights into nebular luminescence, motion, structure and composition. Of key interest was the resolution of stars, particularly in spirals near enough for their arms to be photographed for investigations of their intrinsic stellar makeup.

On the 5th of October 1923 he entered the 100-inch dome for a night's work during an observing run to photograph M31 in search of novae and other related stars. He had pulled a box of Seed 30 plates, named after their inventor, Miles Seed, an English-born American whose invention so revolutionized photographic plate technology that in 1902 his company was purchased by Eastman Kodak. Hubble prepped one side of each of the plates with the silver gelatin emulsion that when dry would easily record the light from M31 with less than an hour's exposure. With the telescope aimed at the Andromeda Nebula he placed a plate in the holder and opened the shutter to permit the nebular light to reach the emulsion. When the time was up, he slid the shutter closed and removed the plate from the holder. One by one, he exposed the plates as he worked the controls of the telescope, platform and dome, expertly guiding the telescopic camera across the sky until the approaching sunrise ended his run.

In accordance with the guidelines set up by the observatory, Hubble marked the rear of the plates just above the dried emulsion using a black marker pen. On one of them he wrote "H335H M31 45 S-30 Seeing 3+ Oct. 5, 1923 HA at end 2:08 E" to signify the Hooker (or Hundred-inch) telescope; plate number 335; the observer, Hubble; the object, M31; an exposure time of 45 minutes on a Seed 30 plate; the seeing 3+ (pretty good); the date; and the hour angle at the end of his run being 2 hours 8 minutes east.

Several days later Hubble began blinking his plates at the MWO offices in Pasadena, comparing them with earlier plates taken by other members of the staff. He was primarily hunting for novae, a popular pursuit at the time, and marked several on plate H335H. Then something magical happened.

As he compared one of his assumed novae on the plate against images of the same star on earlier plates, he realized that the star's brightness did not just grow over time. Rather, its brightness grew and then diminished and then grew again. The star must be a variable. Perhaps still disbelieving his luck at what he had found, Hubble took out his red marker and crossed out the "N" he had written to indicate the star as a nova and below it he wrote in all caps, "VAR!" (Fig. 6.2).

Of all of Hubble's discoveries, this one was unquestionably the purest. A moment of such surprise and elation that it might only be relatable to that of an archaeologist who happens upon an ancient fossil while sweeping away the sanded earth to reveal an entirely new prehistoric species. Waxing mythological, he may have imagined Cepheus, in whose constellation you will remember the first Cepheid variable star was discovered, smiling as he peered over Cassiopeia's shoulder toward the head of mighty Pegasus in where the constellation M31 resides. Taking nothing for granted, Hubble spent several days measuring the brightness curve for the star in question based on the time of year, the telescope used and the seeing, weighing these and other factors to calculate a period between the star's peak brightness of 31.415 days.[28] The shape of the light curve indicated the star to be a Cepheid variable.

In the months to come, Hubble began a deep study of various nearby galaxies seeking other Cepheid candidates and computing their distances using the period-distance formulas for Cepheids developed by Hertzsprung and Shapley. What he found astonished him. The spiral nebulae resided far outside the limits of Shapley's proposed universe. Henrietta Leavitt would have been proud to know that her work had facilitated definitive proof that galaxies existed beyond the Milky Way.

In view of the immensity of his discovery and its probable impact on the field, Hubble understandably took his time to gather further evidence to corroborate the discovery. The MWO contribution to the CIWY in December 1924 focused first and foremost on Hale's Pasadena solar laboratory, which was being created in conjunction with contributions from Carnegie and the optical and mechanical shops on Santa Barbara Street. The main items to feature were the failed mission to observe the total solar eclipse using an interferometer at Point Loma, details of Hale's sunspot cycle program and Charles St. John's confirmation of the validity of general relativity based on shifting

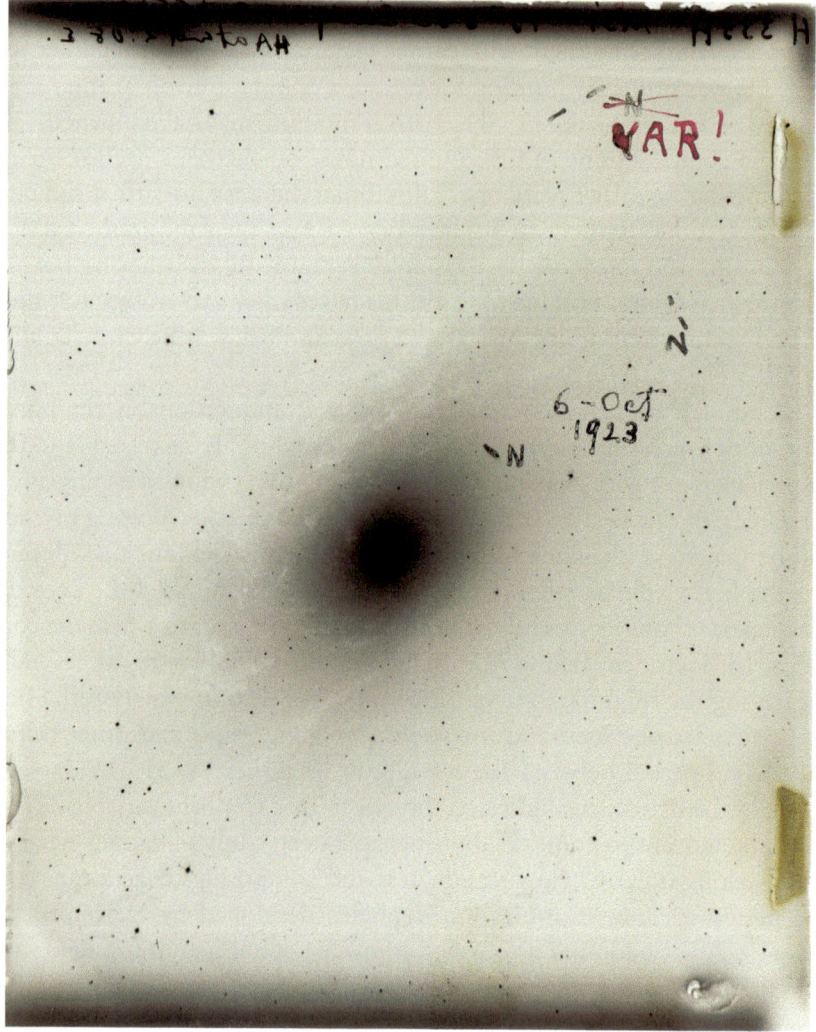

Fig. 6.2 Edwin Hubble's plate of M31 from October 1923 that helped establish the evidence for external galaxies. (Courtesy of Carnegie Institution for Science)

iron lines in the spectra of the Sun's atmosphere. Adams's note about Hubble's discovery comes in the fifth paragraph, in the outline of work for the past year:

> Among other results of interest in this [stellar] field [is] the discovery by Hubble of several variable stars in the great spiral nebula of Andromeda, from a study of which the distance of the nebula probably can be determined with considerable precision…[29]

Although the note is inconclusive, reading it would have caused some handwringing for Lundmark and others in the field whose interests and desires for professional and public acclaim were closely tied to its discovery. Harlow Shapley wouldn't have to wait that long.

In a letter to Shapley on February 19th, 1924, Hubble disclosed his discovery of Cepheid variables in M31, the "characteristics" of which were "unmistakable." With help from Seares in photometrics he was able to discern a "pv [photovisual magnitude] of 17.6" and a distance "something over 300,000 parsecs."[30]

This would put M31 roughly a million light years away from Earth (the modern figure is roughly 2.5 million light years) and Shapley immediately understood the implications in his former MWO colleague's gloating letter. The snake bitten astronomer handed Hubble's letter to Cecilia Payne, who was in his office, saying, "Here is the letter that has destroyed my universe." In later paying homage to her late director, she defended Shapley, whose "private response" and "public comments were less than enthusiastic," and said his "actions spoke louder than his words. From that day he planned a gigantic assault on the problem of galaxies," with the result the "emphasis of Harvard astronomy shifted abruptly."[31]

The circumstances that led to Edwin Hubble's marriage to Grace Leib were a blend of the romantic and the macabre. The degree to which the two had fallen for each other in the early stages of their relationship is lost to history but based on their virtual inseparability in marriage one may reasonably conclude there was more than a passing interest at the very least. In the early stages, however, Grace's marital status presented the star-crossed couple with a seemingly unsurmountable obstacle. But this would soon be removed by a terrible tragedy. While Earl Leib was inspecting a coal mine near Sacramento in June 1921, he was overcome by gas fumes and fell to his death down a mineshaft.[32]

Not wishing to appear unseemly Hubble, the self-styled English gentleman did little more than send letters of condolence and support for the remainder of the year. But by the spring of 1922 he was calling regularly on Grace at the Burke family home where she had moved after the death of her husband. Edwin needed a story that would prove his ability to provide for Grace in the eyes of her father. The proof of external galaxies would make him a scientist of note.

Edwin and Grace were married on February 26th, 1924, in a small, closed ceremony, witnessed by Grace's parents and sister. They spent most of the next three months touring Europe, from Ireland to England, then on to France and Italy before returning to Pasadena in May. Grace found them a small one-bedroom apartment near Caltech. For the first time in his life Edwin was utterly happy and content.[33]

Word of Hubble's Cepheid discovery in spiral nebulae and their awesome distances swept through the offices in Pasadena and up the dusty road to the summit of Mount Wilson quicker than a thorn stuck mule.

Two of the plates Hubble had used to make his discovery were the very ones that Milton Humason had shown to Shapley a few years prior. They still bore the telltale smudge marks of Shapley's handkerchief (and indeed may still). As it became clearer that the plates were among those used by Hubble to deduce the distance to M31, deciding the debate over their nature in favor of the island universe theory, it gave Humason pause to contemplate what might have been. More than his failed attempt to find Pluto, his near miss on the Cepheids, thwarted by his former mentor and friend Shapley, had to weigh heavily on the conscience of the ambitious cowboy astronomer.

After his promotion in the fall of 1922, Milton and Helen had moved Billy back down the mountain and into a two-bedroom home at 1034 North Hudson Avenue, a couple miles north of Caltech. William and Laura Humason were a 10-minute drive to the house from their home on Lakewood Place in the wealthy Oak Knoll neighborhood. The beautiful new Huntington Library with its sprawling botanical gardens was also located nearby. The two most important factors were the reputable school district and being sufficiently close to the Santa Barbara Street offices for Milton to ride his bike, which he did at a notoriously high rate of speed, to and from work when he wasn't on the mountain.[34]

Led by the sons of Joseph Witmer, the Witmers had by now re-established themselves in the community and were again contributing to the development of Los Angeles and the surrounding areas. William Witmer was carrying on the family business while his brother David Witmer was on his way to becoming a prominent architect designing modern homes and apartment complexes, some of which were occupied by various members of the nuclear and extended Witmer family including the Humasons.[35]

Milton's sister, Virginia, had been surprised by a visit from Billy Suermondt, the friend of a boarding school acquaintance, who came to LA on a tour of the United States from Holland (Netherlands). Taking his friend's advice, the dashing young Dutch businessman called on Ginny and "they fell in love on first sight." It didn't take long for him to sweep the beautiful American girl off her feet, down the aisle and onto a ship bound for a new life in the Dutch countryside.[36]

For Milton Humason, the conspicuousness with which his name was appearing in the Carnegie yearbooks carried with it a sense of surreal joy and amazement intermingled with a very real sense of anxious dread. His formal association with the observatory was more than two years old and although he had contributed much to the work of others, he was frustrated by his inability to land a substantial discovery or line of research which he could call his own. His most recent missed opportunity did nothing to buttress his flagging self-confidence.

His research in 1924 included a lengthy search for "very red stars" in the open cluster M37 in Auriga, like the one "discovered by [Edvard] von Zeipel" at Uppsala Observatory in 1921, using highly color sensitive "panchromatic plates and a red filter." The outcome of this search was that stars of this nature and magnitude were "exceedingly rare."[37]

The following year Humason spent some significant time investigating the orbit and other characteristics of the unusually massive OB binary star system HD 163181, which he had observed while hunting Merrill's Be stars. Study of its spectrum using the 10-inch in 1920 had revealed bright hydrogen lines characteristic of the B-type stars and they had classified it as a Be-type spectroscopic binary. Working with Seth Nicholson for his research on this peculiar star he determined its velocity to be +400 km/sec and variable with a mass almost 35 times that of the Sun and a period of just over 12 days.[38] In 1939 Cecilia Payne-Gaposchkin reported this system, also listed as V543 Scorpii, to be an eclipsing binary.[39]

Unusual, even remarkable though Humason's report on this spectrographic binary was, it was hardly groundbreaking. Now more than seven years into his education and three into his career in astronomy he was beginning to settle into the routine upon which the majority of careers in astronomy and astrophysics rest: continuously mining the depths of space for answers to tiny mysteries and clues that over time contribute a broader understanding. The work required competence in not just astrophotography and spectroscopy, but photometrics and a familiarity with parallactic studies of stars and nebulae, facilities for which Humason is seldom given credit despite the breadth of his contributions.

Since his report with Nicholson in 1919, Humason had published 27 papers on a variety of phenomena from variable stars to novae and radial velocities of faint stars and nebulae, 19 of them in the five years since he was promoted to the staff. Considering the educational invisibility from which he came, his rise to becoming the most sought after spectroscopist on the staff of the most prominent observatory in the world was truly extraordinary.

But Humason still had reservations about his plain-spokenness and depended on those around him for help writing and editing his reports to Carnegie, the PSAP and the ApJ. In reality he was a gifted astronomer with an uninhibited approach to both his education in the field and his research and, of course, a most formidable and creative practical observer. "He became," in the words of Carl Sagan, "the virtuoso of the 100-inch telescope."[40]

By 1927 Humason was completely at home at the controls of the 100-inch reflector and its various spectrographs. He had spent the better part of five years at both its Newtonian and Cassegrain focus making direct photographs and spectra for several ongoing research programs and along the way he had discovered three novae and three planetary nebulae.

In this regard at least, he was carving out a niche for himself in direct photography and spectroscopic research, particularly of very faint stars. What the work lacked in formative power it more than made up for in sheer prodigious output with one or two interesting lines of inquiry and interpretation and one Herculean feat of mastery for the astronomy world to chew on as well. The publication of three major works in that year illustrates this point.

The first was the classification of giant and dwarf stars with Adams and Joy, a titanic survey of stars in the Boss Catalogue, some from the Kapteyn Selected Areas and also the Perseus Cluster that he started in the fall of 1922 just after his appointment to the staff. The report on M-type giants that congregate in the core of the galaxy, their brightness and distribution compared with dwarfs of the same type, represented hundreds of hours of effort both at the telescope and in the lab. Allan Sandage later noted that "Humason was a strong part of that 4,653-star catalogue. He made the observations and also made the reductions."[41]

The faintest giant in the study had an absolute magnitude of +0.7. What was remarkable was the prominent star Vega in Lyra is +0.58 and 25 light years away and the giants in the 1927 study are at distances as far as 7,500 light years. The brightest dwarf star was recorded as +6.9, an order of magnitude beyond the limit of naked eye seeing from Earth. The report covered 312 M-type giants and 100 dwarfs, a significant contribution to the overall mission of the observatory to understand the evolution of the stars.[42]

The second significant report, *Note on Very Cool Stars*, was published with Paul Merrill on one of Merrill's favorite subjects, late-stage giant long-period variables. It pointed to an apparent temperature minimum for this type of star "slightly above 2000 degrees Celsius." On this, Adams wrote in the observatory's contribution to the CIWY for that year:

Cooler stars are not known in spite of the fact that observational methods are apparently competent to reveal them…[There is] strong [evidence] that a real discontinuity in brightness exists in the giant sequence at a temperature just below that of the[se] variables. The fact that the coolest known stars are all long-period variables may indicate physical instability at their temperatures. Still cooler stars, if they exist, may resemble long-period variables at [their] minimum [brightness].[43]

Red supergiants are among the rarest in the giant sequence and are known to be nearing the end of their lives. They consumed their remaining hydrogen during the giant phase and are now fusing helium into carbon and even heavier elements. Their diameters are vast, in some cases many times that of the Earth's orbit around the Sun. Once they build up iron in their core, the core collapses, and the release of gravitational energy results in a supernova.

Although they were unable to detect any stars cooler than these, Merrill and Humason suggested in their note that possibly "more complete observations of the future will disclose cooler and fainter dwarfs than any in the present lists" to create "a smooth continuation of the series as now known."[44]

This suggestion proved prescient. Eventually infrared technology revealed the existence of brown dwarf stars (Class L, T, and Y) that are cooler. Although formed in the same way as other stars, brown dwarfs are often referred to as "failed stars" because they are not hot enough to have started the fusion of hydrogen into helium. In fact, they are comparable in diameter to the planet Jupiter with roughly 15 to 75 times its mass.

The third and perhaps most important of the reports Humason published that year came from the work with Merrill on red supergiant stars and was far less significant in terms of its impact on the field of stellar evolutionary research. What it did, however, was highlight his inordinate skill in obtaining clear and readable spectra of extremely faint objects, which was becoming the stuff of legend.

"By far the faintest star, intrinsically, known" at that time was a dim red dwarf named Wolf 359 for its position in a list of stars published by the German astronomer Max Wolf in 1919.[45] A superlative astrophotographer, Wolf catalogued over 1,500 low luminosity stars during his career as director of the Heidelberg-Königstuhl State Observatory at the University of Heidelberg. The high proper motions of these stars meant they were relatively nearby, and their faintness meant they were of low intrinsic luminosity and therefore interesting targets for research.[46]

Wolf 359 had an absolute photographic magnitude of +18.5, three magnitudes fainter than the next faintest star known at the time. Or to put it another way it was 25,000 times fainter than Sirius, the brightest star in the sky. For this reason, it was thought to be too faint for its radial velocity to be accurately measured.

The star was added to Adrian van Maanen's parallax program and Humason was asked to see if he could get a good spectrum of it. He made his way to the summit on the 21st of April 1927 and prepared his run at the primary focus. After preparing the telescope Tom Nelson, the assistant, opened the dome at 5:30 p.m. Problems with the mirror and poor seeing led Humason to call off the trial search after a few hours having taken only one dubious plate. The condition of the mirror was listed as "very poor" in the logbook for that night.

The next night the seeing and the mirror improved slightly, but not enough to acquire a good spectrum of such a fantastically faint object. Humason managed to obtain two more plates and developed them to help him gauge the amount of time that would be needed for a good exposure of the spectrum and to select a suitable guide star for tracking.

On the final night, April 23rd, he and Nelson opened the dome at the same time hoping to nab the elusive quarry. This time the sky was listed as "clear & calm" and the mirror's condition as "fair". Humason spent a little over nine hours at the controls, simultaneously controlling the movement of the telescope, the dome and the Cassegrain platform exposed to the cold mountain air as he peered through the guide scope, steering the giant reflector, expertly predicting the periodic error in the drive clock, and seamlessly easing the telescope back to remain squarely on his target.[47]

When he was finished, he had two plates, at least one of which captured a clear spectrum showing two bright hydrogen emission lines. Measurement of its spectrum revealed a radial velocity of –90 kps. The report noted that no attempt was being made to ascertain the star's spectroscopic parallax because the curve for deriving absolute magnitude from its type and the intensity of the lines in its spectrum did not yet exist. Humason was literally setting the curve[48] (Fig. 6.3).

Word of his feat in nabbing a useable spectrum for Wolf 359 spread like wildfire through the observatory. The skill, stamina and resolve to obtain a plate like that was unheard of at the time. Regardless of his misgivings about his education and his worldly sophistication, Humason had become the undisputed king of the most technologically advanced telescope in the world.

*   *   *

While Milton Humason was winning friends and fighting his inner demons, Edwin Hubble was going to war with several members of the astronomy community.

Fig. 6.3 A master at work. Milton Humason at the Cassegrain spectrograph of the 100-inch telescope on Mount Wilson. (Courtesy, Ann Humason Bernt family)

There were lesser foes like Vesto Slipher, the acting director at Lowell Observatory and president of the IAU Commission on Nebulae and Star Clusters. In the latter role, Slipher was merely acting on principle when he delayed bringing Hubble's nebular classification to the table for a vote to adopt it as the new preferred system. His trepidation was certainly understandable. To have forced Hubble's plan into the field prior to a thorough discussion within the group and achieving consensus on its effectiveness would've been irresponsible and ethically untenable given his position.

For Hubble the delay was frustrating. He was determined to dominate nebular research and getting his classification scheme into publication first was a crucial step to achieving that goal. He decided to take matters into his own hands, publishing his plan for classification in *A General Study of Diffuse Galactic Nebulae* in the October issue of the ApJ.[49]

Hubble must have reckoned that after a few minor adjustments based on feedback from the IAU panel and responses to the report, acceptance of his system would be a formality. There was no written rule or guideline forbidding him from publishing his ideas for review by the field, just as there was no reason the field was obligated to accept his system, in part or in whole. But his impatience was noted by Slipher and everyone else on the panel.

When no vote on ratifying his new system had come to pass by 1924 Hubble was starting to chafe. He knew that he was going to make a splash with his report on external galaxies, in preparation for publication the following year, and his political instincts were as engaged as his ego was now.

Remembering his success in the final stages of the Rhodes Scholarship he began to press the issue, showing off his classification proposal to some of the men with influence in the field such as Hale, Russell and the English astronomer James Jeans. Russell and Jeans liked the new scheme and decided to promote it as part of the commission on nebulae at the next IAU meeting in Cambridge in July of 1925. Hubble then wrote Slipher informing him that Hale had given permission to publish his classification, which by then had been revised to simplify some of its details. In his February letter to Slipher, Hubble wrote, "I would prefer that it go through the committee if that is feasible within a reasonably short time." Having no desire to be usurped by the irascible Mount Wilson astronomer, Slipher sent the copies of Hubble's notes to the committee members asking for their feedback.[50]

As the subjective analyses came in, Hubble probably wished he had just gone ahead and published the new scheme. The most favorable of the reviews came from Solon Bailey at Harvard, who generally thought it as complete as could be expected in view of the relatively scant knowledge of the nebulae at that time. His main objection was to the nomenclature that Hubble had created to describe the chart.[51]

In formulating his system Hubble had decided on three basic groups of nebulae, which he arranged in the shape of tuning fork starting with the elliptical nebulae (gaseous clouds apparently devoid of resolved stars) that ranged numerically from E0 for round objects like planetaries to E7 for the most elongated objects. They formed the stem of the tuning fork. At the point where the nebulae took on a more pinwheel-like shape he switched to the "S" designation with sub-categories for the prongs of the tuning fork. There were "Spirals" and "Barred Spirals" to differentiate between those with and those without a bar-like band of light spanning their centers. Sa, Sb, and Sc were the spirals on one prong of the fork while SBa, SBb and SBc were the barred spirals on the other. Objects that showed development toward becoming spirals but lacked evidence of spiral arms were designated S0 and placed where the stem of the fork divided[52] (Fig. 6.4).

Hubble's scheme was meant to depict the evolution of a galaxy from the early elliptical stage with little or no stellar development through to a fledgling spiral and continuing until it reached a final stage, depending on its trajectory. He had developed the idea based on his discussions with James Jeans. It was this notion of nebular evolution that raised the greatest exception among the committee members. Even if he was correct, there was no empirical evidence to support the idea and it seemed an incredibly large leap to assume that evolution occurred, let alone in the arrangement Hubble was suggesting. To many on the panel, all of whom were aware of the Cepheid

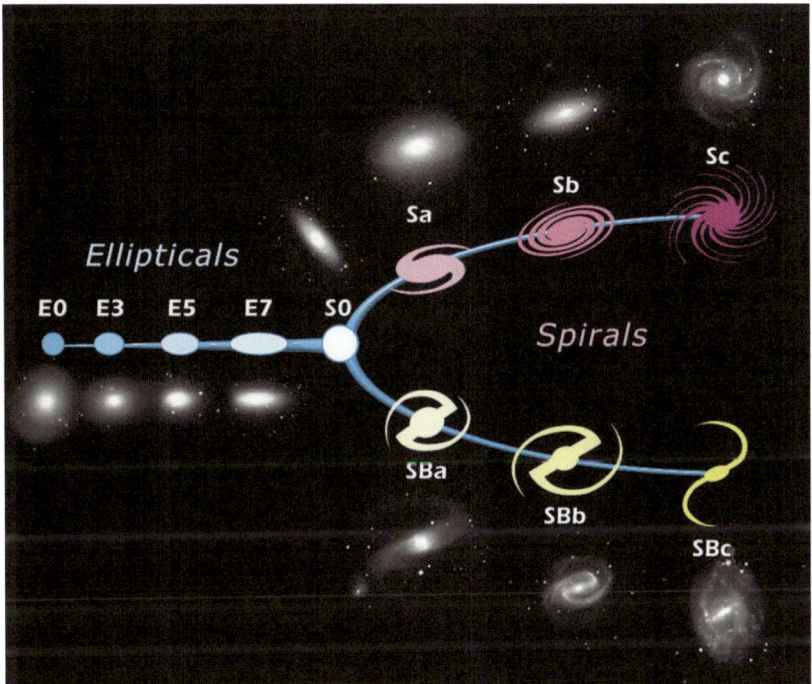

Fig. 6.4   The Hubble galaxy classification scheme. (NASA/ESA)

discovery by now, this Shapleyesque leap overstepped acceptable limits of presumption.

Deepening Hubble's resentment at the committee's resistance, Slipher had accepted a request from Harlow Shapley to join the commission as the resident expert on star clusters. Although he had made a highly successful start at Harvard, Hubble's letter had left an open psychological wound at Shapley's core. The two had last met in person as the newlywed Hubbles traveled to the east coast on their way out of the country to tour Europe. Although he was cordial and professional both in person and correspondence, and generous in aiding Hubble's Cepheid research when he could. Shapley continued to resist the Mount Wilson astronomer's findings even as the evidence in support of external galaxies mounted. There remained the issue of the internal motions that Shapley's great friend and colleague Adrian van Maanen was measuring at Mount Wilson, but increasingly Shapley was being obliged to accept Hubble's findings.

Although Hubble had disclosed the Cepheid in M31 he had yet to publish his distance measurements and, by extension, the revelation that the nebulae were external galaxies. He was being urged by Jeans and Russell to make his revelations public. His paper *NGC 6822, A Remote Stellar System*, published in the ApJ in December 1925 shocked the astronomy community as it revealed "the first object definitely assigned to a region outside the galactic system."[53]

The secret was finally out. As Nicholas Mayall would later observe, "Hubble did for the universe what Shapley did for the Galaxy, and what Galileo did for the solar system…[He] opened up a greater volume of space."[54]

The year before, Shapley had written Hubble with a suggestion regarding Hubble's use of the word "nongalactic nebulae" to describe those far away from the plane of the Milky Way. In Shapley's opinion the word "nebula" ought to be used to designate inner-galactic dust clouds and "galaxy" used to designate the external ones.[55]

Although the details were debatable, Shapley had a good case and almost everyone with knowledge of the matter agreed with this strategy. But Hubble would have none of it. This seemingly small issue of semantics wasn't too small for the prideful Mount Wilson man, who wanted ultimate creative power over the domain of nebular research. His ascendency in the nebular field led Hubble to believe this was his call to make. After consultation with Hale, Adams and others, he settled on the term "extra-galactic nebulae." This terminology remained at MWO until after Hubble's death, but with the passage of time "galaxy" gained ever greater acceptable outside of Pasadena.

Despite Hubble's maneuvering and the weight of the organization he had behind him, Slipher's committee remained skeptical of some of the specifics of Hubble's classification scheme. Fourteen years Hubble's senior, the well-established leader of the committee was not going to be pushed around by the upstart astronomer. He led the committee in voting not to endorse *any* new classification system for nebulae in 1925.

Hubble took this decision as a personal rebuke, and in response he published a revised version of the scheme in a 48-page report *Extra-Galactic Nebulae* in the December 1926 issue of the ApJ. Still bearing its signature evolutionary feel this report laid out the root of the descriptive system still in use to this day, but its evolutionary aspect was never adopted.[56]

However, before Hubble could submit his paper he was hit by a bombshell when Knut Lundmark published his own classification system *A Preliminary Classification of Nebulae* in a journal for the Royal Swedish Academy of Sciences.[57]

Hubble and Lundmark had worked together while Lundmark was at Mount Wilson as a visiting astronomer from June 1922 to May of the following year. They had even published a joint report on the nova Z Centauri in 1922.[58]

After digesting Lundmark's classification scheme Hubble, no doubt feeling the sting of being beaten to the punch, sent an angry letter to Slipher decrying his rival's classification as "practically identical" to his and complaining that in presenting his evidence Lundmark had "calmly ignored [my] existence" in claiming the classification as his own. Hubble was charging his former collaborator with plagiarizing his work and complaining to the Lowell director that Lundmark had stolen the idea while at the IAU meeting in 1925.[59]

For Hubble, who had published his first report on the matter in 1922, albeit with a far less evolved scheme, the matter was intensely personal. After expressing his displeasure to Slipher, he wrote to Lundmark demanding an explanation and promising he would "take considerable pleasure in calling constant and emphatic attention…to your curious ideas of ethics" in the future.[60]

Lundmark was under no restrictions in putting his ideas in print and his decision to do so merely mirrored the precedent previously set by the American. He was free to compare and contrast his ideas with Hubble's (or anyone else's for that matter) and let the science community decide which was better. Nevertheless, Hubble could hardly contain himself.

Hubble's scorn betrayed an underlying pressure that still exists today to be the first to print or establish any new idea or discovery, one that permeates the realm of invention and sometimes results in scandal. The dispute in the latter part of the 19th century between Alexander Graham Bell and Elisha Gray over the rights to the patent for the telephone is an example of this. As Steven

Johnson wrote in his wonderful book *How We Got to Now: Six Innovations that Made the Modern World,* "Most discoveries become imaginable at a very specific moment in history, after which point multiple people start to imagine them."[61] In most cases it is the person who first presents an idea in print or by demonstration who ultimately lays claim to it.

In the case of the nebular classification scheme the facts are somewhat hazy but support Lundmark's overall argument. In rebutting Hubble's accusations, the Swedish astronomer acknowledged being at the Cambridge meeting but denied being part of the Committee on Nebulae and Star Clusters where Hubble alleged that he had stolen his approach. Lundmark claimed he hadn't yet been appointed to the committee and had no access to the papers being presented. He said the first he heard about Hubble's concurrent classification idea was in the American's angry letter in the fall of 1926. But whether or not Lundmark attended the committee is irrelevant for two reasons. One, his scheme focused on stellar concentrations and suggested "Magellanic" galaxy types. In contrast, Hubble's centered on evolutionary structure and bore none of Lundmark's key points. Two, an unpublished paper by Lundmark from 1922 reveals that he had been working on a classification system at that time.[62]

By publishing what he thought was a well-reasoned approach to identifying nebulae and galaxies, Lundmark had inadvertently stepped on the Mount Wilson prima donna's foot, and Hubble was not about to take it lightly. It should further be noted that no fewer than a dozen other astronomers, including Slipher, Curtis and Bailey were working on a way to classify different types of nebulae, all of which fell along vaguely similar lines.

It would be Hubble's growing dominance of the field that led to his ultimate victory in claiming the classification system for himself. By the mid-1930s his classification system for galaxies was almost universally adopted. However, it is easy to see how this idea was spawned within a climate of reason and adaptation that would likely have yielded a system of that type with or without Edwin Hubble.

To Hubble's credit, his system was both reasonable and sufficiently general to allow for future variations. In particular, the French astronomer Gérard de Vaucouleurs incorporated two separate lenticular galaxy types leading to spirals and barred spirals, plus a variety of intermediate spirals and irregular galaxies.

The evolutionary approach Hubble set out in his diagram is still largely theoretical. The time scale of galactic evolution and the quantitative scale in terms of numbers of galaxies is too great to ensure the accuracy of the divisions or their origins and anomalies. Although a more general understanding of galaxy and universal formation has been gleaned through decades of astrophysical research using ever-improving technology and techniques, many mysteries remain.

Hubble was leaning hard on James Jeans' theoretical work on the development of stars in the arms of spirals. Jeans was best known for his work in theoretical physics, especially on the effects of gravity on interstellar gas clouds in the formation of stars. His later support of the Steady State theory (discussed in a later chapter as a rival to the Big Bang) and the theory that planets condensed from matter that was drawn out of the Sun by near collisions with other stars proved to be less compelling.

Broadly speaking, this period in Hubble's life points to the moment at which he began to unravel as an influencer and leader among his peers, both at home and abroad. Not that he could be blamed for getting angry at what he saw as people trying to steal his thunder. But in opting for open combat, insinuation, insubordination, moral tyranny and out and out bullying he was setting a foundation which would damage his legacy among some inside and outside the field and give rise to future professional embarrassment.

The great Russian theater actor, director and practical theorist Konstantin Stanislavski famously said, "Love art in yourself, and not yourself in art." It would be easy to conclude that as his status grew, Edwin Hubble was starting to show himself as a preening narcissist more interested in fame than the less glitzy romance of a life spent in research. The reality is far more complex, revealing at once the subconscious mind at work behind the countless hours of study, the fundamental discovery and press-seeking, and the reflections of a youth dominated by silent rebellion against paternal imperiousness and neglect. At best, his increasing tendency toward abstinence, introspection and vaingloriousness was unfortunate.

His ongoing feud with Adrian van Maanen does not reflect well in this light, either. In forming his conclusions van Maanen relied on photometric and parallactic measurements. It had long been suspected that these techniques were insufficient for providing substantial proof of motion, distance, or any other standards of measure for objects apart from those relatively nearby. But van Maanen believed the nebulae *were* relatively close and therefore measurable by the techniques he had learned and perfected. Working from this hypothesis he allowed his beliefs to inform his analysis, rather than analyzing his data and letting the results inform his beliefs. As the years went on and evidence mounted against his perceived rotations in spirals the stubborn Mount Wilson photometrist continued to insist that he was right and that his conclusions would eventually be borne out as technology improved. He would carry this belief to the grave, but not before (as we shall see later) his dust up with Hubble came to a head.

On the second floor of the offices of the Carnegie Observatories in Pasadena stands the van Maanen comparator. It was on this machine that van Maanen spent years working on what he was sure were internal motions in nebulae, motions which remained elusive even as the quality of direct photographs improved. On the front of the machine is the note which he wrote many years ago warning would-be researchers to keep their hands off:

Do not use this Stereocomparator without consulting. A. van Maanen.[63]

The machine and its calibrations, presumably made by van Maanen himself, was his to use and his alone. But by 1925 few at the observatory could be bothered with it, preferring to measure objects in the sky using Cepheid variables and supernovae as standard candles and the spectroscopic parallax technique developed by Adams. Although he was well liked within the ranks at MWO, no one would corroborate the Dutch astronomer's measurements of internal motions in nebulae.

With nearly the entire astronomy community on his side Hubble could reasonably have ignored van Maanen, but Harlow Shapley was one of van Maanen's few supporters in the field and it is easy to imagine that every time Hubble passed van Maanen in the halls of the observatory what he saw was a two-headed galactic monster, Shapley's enormous head on one side, speaking a mile a minute, and van Maanen's on the other, nodding in agreement while smoking his pipe.

Mayall recalled being "pretty much a middleman" between the two enemy combatants while at Mount Wilson in the early 1930s. Placed in charge of the observing schedules for the large reflectors, he noted that Hubble and van Maanen "would completely ignore one another" and Mayall would "shuttle between them to find out what nights they wanted." He was often assisted in this chore by Humason who "had become tired of [the] situation."[64]

Hubble's overt distain for van Maanen boiled over into his other relationships at Mount Wilson. Paul Merrill, perhaps the most conservative member of the staff had befriended Hubble in the early days as the two second generation astronomers were getting acquainted, but their relationship later soured. For one thing, Merrill was of the most devout Catholic persuasion, something that Hubble had long since fervently expunged from his life. More to the point, Merrill was a spectroscopist who valued reason and integrity over showboating and what he regarded as unsubstantiated conclusions. The distance scale to M31 and other galaxies in 1926 was the event that drove a permanent wedge between Merrill and Hubble. Hubble's absolute magnitudes for the brightest stars in his galaxies was three magnitudes brighter than those for the same stars as measured in the spectral classification program for M-type stars, most of which had been measured by Humason. Believing the spectroscopic parallax measurements to be more precise (which they were), Merrill decided Hubble's measurements must be in error and therefore sided with van Maanen in the kerfuffle over internal motions. In Hubble's view, no friend of van Maanen's could be a friend of his and so Merrill simply responded in kind.

Van Maanen's refusal to budge on the "motions" question so irked Hubble that the tension between them remained well into the 1930s. The most oft told story of the confrontational nature of their relationship revolved around a week at the observatory when they both had observing runs. As was customary, van Maanen, the 100-inch telescope operator that week, would have been seated at the head of the table for dinner at the monastery. It had become the tradition that the 100-inch operator sit at the head of the table and lead the conversation with the 60-inch operator to one side and the solar telescope operators on the other.

Hubble prepared in the afternoon as he was inclined to do, reading up on arcane subject matter intending to present himself as an authority on the topic. It was another of his methods for gaining an intellectual advantage over his peers. Knowing van Maanen would be seated at the head of the table, Hubble stole into the dining room ahead of the others and switched the name plates to place his at the head of the table and van Maanen's in the usual spot for the 60-inch operator. He took his seat and watched gleefully as van Maanen's face blanched on entering the small dining room. At over six feet, Hubble was an intimidating presence. He knew his bullying would not be resisted by the Dutchman.[65] It was an obscene show of disrespect that should have been beneath a man of Hubble's education, experience and adopted English manners but one he could not resist. That no one made a show of support for van Maanen is telling. The Cepheid revolution had cemented Hubble's public notoriety and, as everyone (except van Maanen) knew, he was correct in his assertion of external galaxies.

The animosity between Merrill and Hubble persisted to Hubble's death but his attitude toward Shapley and van Maanen evolved (if you can call it that) from out and out disdain to one of pejorative condescension. He eventually patched things up with Lundmark in the early 1950s on a visit by the Swedish astronomer to the 200-inch Hale telescope on Mount Palomar.

At the close of 1927 Edwin Hubble and Milton Humason had reached what might have been the pinnacles of their individual careers, Hubble as a visionary observer and Humason as a virtuoso photographer and spectroscopist. Hubble had opened the eyes of the public to a universe of galaxies, each with billions of stars, the extent of which was unknown, while Humason had contributed to research on a multitude of questions concerning stellar evolution stretching the boundaries of collected data and the limits of the best photographic equipment of the day.

However, the merging orbits of these two rising stars of the astronomy world were about to enter a new phase that would unite them in the most difficult challenge yet conceived in astronomy and test their acumen, creativity, and resolve. Something that had emerged from Albert Einstein's general theory of relativity was about to send them on the most incredible journey of their lives.

References

1. *U.S. Timeline – The 1920s*, (americasbesthistory.com), Retrieved June 5th, 2019, https://americasbesthistory.com/abhtimeline1920.html
2. Letter from Edwin Hubble to V.M. Slipher, Lowell Observatory, February 23, 1922
3. Walter S. Adams, Contributions to the Mount Wilson Observatory, (CIWY 1922-1923), No. 22, pg. 181
4. *Ibid.*
5. Walter S. Adams, Contributions to the Mount Wilson Observatory, (CIWY 1923-1924), No. 23, pg. 81
6. Walter S. Adams, Contributions to the Mount Wilson Observatory, (CIWY 1922-1923), No. 22, pgs. 193-194
7. *Ibid.* pgs. 200-205
8. M. L. Humason, *Nova No. 22 in the Andromeda Nebula,* (PASP 1923), Vol. 35, No. 204, pg. 124
9. W.S. Adams, A.H. Joy, M.L. Humason, *An Additional Star of the W Cephei Type of Spectrum,* (PASP 1925), Vol. 37, No. 161, pgs. 161-162
10. Paul W. Merrill, Milton L. Humason, Cora G. Burwell, *Discovery and Observations of Stars of Class Be,* (ApJ 1925), Vol. 61, pgs. 389-417
11. W.S. Adams, A.H. Joy, M. L. Humason, *Absolute Magnitudes and Parallaxes of 412 M-Type Stars,* (PASP 1926), Vol. 38, No. 257, To be read at the 37th Meeting of the American Astronomical Society, December 28-30, 1926, Philadelphia, Pennsylvania
12. Harlow Shapley, Adelaide Ames, *A Survey of the External Galaxies Brighter Than the Thirteenth Magnitude,* (AHCO 1932), Vol. 88, pgs. 41-76
13. Cecilia Helena Payne, *Stellar Atmospheres: A Contribution to the Observational Study of High Temperature in the Reversing Layers of Stars (Ph.D. Thesis),* (Harvard University Press 1925)
14. Richard Williams, *This Month in Physics History: January 1, 1925: Cecilia Payne-Gaposchkin and the Day the Universe Changed,* (APS News January 2015), Vol. 24

No. 1, Retrieved July 28th, 2019, https://www.aps.org/publications/apsnews/201501/physicshistory.cfm

15. Cecilia Payne-Gaposchkin, *Cecilia Payne-Gaposchkin: An Autobiography and Other Recollections,* (Cambridge University Press 1996), pgs. 219-227

16. *Ibid.* pg. 194

17. Walter S. Adams, Contributions to the Mount Wilson Observatory, (CIWY 1922-1923), No. 22, pg. 183

18. Lewis Howard Humason Draft Registration Card, Los Angeles, California, Precinct 329, June 5th, 1917, Retrieved August 14th, 2006, www.ancestry.com

19. V. Gerald Iaquinta, *Beatrice Mayberry Humason,* (Los Angeles County Biographies 2013) http://freepages.rootsweb.com/~npmelton/genealogy/lahuma.htm

20. Index to Marriages for Los Angeles, Feb. 11, 1925, Mayberry, Beatrice W. and Lewis H. Humason

21. Harry Wing Humason WWII Draft Registration Card, 1943

22. S.B. Nicholson, Beatrice W. Mayberry, *Ephemeris of Jupiter's Ninth Satellite,* (PASP 1924), Vol. 36, No. 211, pg. 143

23. Seth B. Nicholson, Beatrice W. Mayberry, *Sun-Spot Activity During 1924,* (PASP 1925), Vol. 37, No. 215, pg. 34

24. S.B. Nicholson, L.H. Humason, *Spectroheliograms Using the Iron Lines Λ3720, Λ3735, Λ3860 (Abstract),* (PASP 1926), Vol. 38, No. 224, pg. 263

25. Interview with Harry Wing Humason, 2010

26. V. Gerald Iaquinta, *Beatrice Mayberry Humason,* (Los Angeles County Biographies 2013) http://freepages.rootsweb.com/~npmelton/genealogy/lahuma.htm

27. Walter S. Adams, Contributions to the Mount Wilson Observatory, (CIWY 1922-1923), No. 22, pg. 184

28. EPH writes to astronomer Harlow Shapley, 1924-02-19, mssHUB 611

29. Walter S. Adams, Contributions to the Mount Wilson Observatory, (CIWY 1923-1924), No. 23, pg. 85

30. EPH writes to astronomer Harlow Shapley, 1924-02-19, mssHUB 611

31. Cecilia Payne-Gaposchkin, *Cecilia Payne-Gaposchkin: An Autobiography and Other Recollections,* (Cambridge University Press 1996), pgs. 209-210

32. Gale E. Christianson, *Edwin Hubble: Mariner of the Nebulae,* (University of Chicago Press 1995), pg. 165

33. *Ibid.* pgs. 165-169

34. Records of the Pasadena City Directory, 1924-1932

35. Virginia Linden Comer, *In Victorian Los Angeles, The Witmers of Crown Hill,* (Talbot Press 1988), pg. 91

36. Letter from Truus Suermondt to Ann Humason Bernt, Sunday, Jan. 27th, 2013

37. Walter S. Adams, Contributions to the Mount Wilson Observatory, (CIWY 1923-1924), No. 23, pg. 96

38. Milton L. Humason and Seth B. Nicholson, *H.D. 163181, A Spectroscopic Binary,* (ApJ 1928), pgs. 341-346

39. J.B. Hutchings, *Spectroscopy of the Massive Eclipsing Binary HD 163181*, (PASP 1975), Vol. 87, No. 516, pg. 245

40. Carl Sagan and Ann Druyan, *Cosmos 10: The Edge of Forever*, Documentary, November 30, 1980

41. Interview of Allan Sandage by Bert Shapiro on 1977 February 8, Niels Bohr Library & Archives, American Institute of Physics, College Park, MD USA, www.aip.org/history-programs/niels-bohr-library/oral-histories/32867

42. W.S. Adams, A.H. Joy and M.L. Humason, *The Absolute Magnitudes and Parallaxes of 410 Stars of Type M*, (ApJ 1926), Vol. 64, pgs. 225-242

43. Walter S. Adams, Contributions to the Mount Wilson Observatory, (CIWY 1926-1927), No. 26, pg. 123

44. Paul W. Merrill and Milton L. Humason, *Note on Very Cool Stars*, (PASP 1927), Vol. 39, 230, pg. 199

45. A van Maanen, J.A. Brown and M.L. Humason, *A Star of Extremely Low Luminosity*, (PASP 1927), Vol. 39, No. 229, pg. 174

46. *Obituary Notices: Associates: Wolf, Max*, (MNRAS 1933), Vol. 93, pgs. 236-238

47. Mount Wilson Observatory logbooks. The Huntington Library, San Marino, CA., 100-inch telescope, April 1927, pg. 223

48. A van Maanen, J.A. Brown and M.L. Humason, *A Star of Extremely Low Luminosity*, (PASP 1927), Vol. 39, No. 229, pg. 174

49. E.P. Hubble, *A General Study of Diffuse Galactic Nebulae*, (ApJ 1922), Vol. 56, pgs. 162-199

50. Gale E. Christianson, *Edwin Hubble: Mariner of the Nebulae*, (University of Chicago Press 1995), pg. 174

51. *Ibid.* pgs. 174-175

52. R. Hart and R. Berendzen, (JHA 1971), Vol. 2, pgs. 109-119

53. E.P. Hubble, *NGC 6822, A Remote Stellar System*, (ApJ 1925), Vol. 62, pgs. 409-433

54. Interview of Nicholas Mayall by Bert Shapiro on 1977 February 13, Niels Bohr Library & Archives, American Institute of Physics, College Park, MD USA, www.aip.org/history-programs/niels-bohr-library/oral-histories/4766

55. Gale E. Christianson, *Edwin Hubble: Mariner of the Nebulae*, (University of Chicago Press 1995), pg. 175

56. Edwin Hubble, *Extra-Galactic Nebulae*, (ApJ 1926), Vol. 64, pgs. 321-369

57. K. Lundmark, *A Preliminary Classification of Nebulae*, Arkiv för Mathematik, Astronomi öch Fysik, Vol. 19, No. 8, 1926

58. E.P. Hubble and K. Lundmark, *Nova Z Centauri (1985) and N.G.C. 5253*, (PASP 1922), Vol. 34, No. 201, pgs. 292-293

59. Hubble, Edwin, "Edwin Hubble to V. M. Slipher, June 22, 1926, on sub-committee appointments and potential plagiarism," *Lowell Observatory Archives*, accessed February 23, 2021, https://collectionslowellobservatory.omeka.net/items/show/982.

60. Gale E. Christianson, *Edwin Hubble: Mariner of the Nebulae,* (University of Chicago Press 1995), pg. 176
61. Steven Johnson, *How We Got to Now: Six Innovations That Made the Modern World,* (Viking 2018)
62. P. Teerikorpi, *Lundmark's Unpublished 1922 Nebula Classification,* (JHA 1989), Vol. 20, No. 3, Oct. pgs. 165-170
63. Van Maanen's Stereocomparator was viewed by the author on a tour of the observatory offices with Ann Humason Bernt and her family in 2017 on the occasion of the 100th anniversary of the 100-inch Hooker telescope
64. Interview of Nicholas Mayall by Bert Shapiro on 1977 February 13, Niels Bohr Library & Archives, American Institute of Physics, College Park, MD USA, www.aip.org/history-programs/niels-bohr-library/oral-histories/4766
65. Interview with Don Nicholson, 2006

7

Eclipsing Binaries (1928-1929)

As Edwin Hubble and Milton Humason reach new heights in their respective disciplines, advances in theoretical physics and in telescope and photographic technology are starting to coalesce around them, planting the seed which will ultimately bring their unlikely partnership to bear on the question of universal expansion. Analyses of the Einstein's general relativity by Schwarzschild, de Sitter, Friedmann and Lemaître raise the question of whether the universe is static, as Einstein presumed, or might be expanding or contracting. With the most powerful telescope in the world at his disposal, Hubble need only find a partner with skill and endurance enough to bring down spectra of increasingly distant and faint galaxies. Advances in cameras, filters and emulsions have increased the ability for the highly skilled Humason to give this a try but he quickly realizes that digging deeper into the universe will require much more technological development. The episode will test the emotional and intestinal fortitude of both Hubble and Humason, the professional acumen of the Mount Wilson director and its founder, and the operational limits of the state-of-the-art apparatus, almost bringing the entire mission down before it can really get started.

The run of events that led to the practical discovery by Hubble and Humason of universal expansion began with the publication of the first of Albert Einstein's theoretical papers on relativity in the early part of the 20th century. As seminal works go, you couldn't do much better than his twin masterpieces, the second of which was conceived and refined during an extensive 10-year period of mathematical deliberation and presented before the Prussian Academy on November 18th, 1915.[1]

© Springer Nature Switzerland AG 2021   **265**
R. Voller, *Hubble, Humason and the Big Bang*, Springer Praxis Books,
https://doi.org/10.1007/978-3-030-82181-4_7

The first of these papers, *On the Electrodynamics of Moving Bodies* (Chapter 1), was published in September of 1905, about the time Edwin Hubble was entering his junior year of high school and Milton Humason was taking up more or less permanent residence on Mount Wilson.

For two centuries since Isaac Newton revealed the laws or gravity (Chapter 1) the world believed that space was infinite and unchanging, and that time was linear. Einstein's mathematical and theoretical development of the mechanics of what he called "spacetime" rewrote Newtonian mechanics stating that space and time were not linear and not subject to the relative perspective of an observer. The remaining question in Einstein's imaginative mind concerned gravity. Newton's famous math described how gravity worked on celestial bodies in terms of "action at a distance," but neither he nor anyone since had been able to explain what gravity actually was.

Eager to figure this out, Einstein had what he later called "the happiest thought of [his] life." While thinking about gravity in 1907 he realized that a person falling from a building should not feel their own weight. Gravity and acceleration were equivalent. He concluded that space must be a kind of fabric that was warped, or curved, by large celestial bodies so that smaller bodies trapped within that warped space would continually travel around them like the Moon orbiting the Earth.[2]

It was a brilliant insight but not so easily proved in mathematical terms. It took Einstein eight years and a couple of failed attempts before he came up with the right solution. This was theoretical physics at levels beyond all but a few scientists in the world, notably David Hilbert, who had been working on his own ideas involving relativity at the time.[3]

The field equations of the Theory of General Relativity that Einstein published in 1916 could be used to understand the relationship between space, time and gravity. The paper he published led to an explosion of theoretical and experimental work to test the new theory and explore its possible uses.

Within a year the German physicist Karl Schwarzschild is thought to have used the field equations to predict that a star could collapse under its own gravity to produce a black hole (although he didn't use that term) while serving in the German army during the Great War. Unfortunately, he died in the trenches on the Russian front the following year.[4]

In 1917 Einstein applied the field equations to a model of the structure and state of the universe and realized it predicted either expansion or contraction. At that time, astronomers were of the opinion that the universe was static. On this basis, and because an infinite and eternal universe matched his personal vision of the cosmos, Einstein added his now famous cosmological constant to the field equations to achieve a static solution. The same year, Dutch astronomer Willem de Sitter devised his own nuanced, non-static version of the universe from general relativity.

Further progress was largely stifled by the war and its aftermath but in May of 1919 the English astronomer Arthur Eddington, another early master of relativity, proved Einstein's prediction that gravity would cause starlight to "bend" when passing by a massive celestial body. These observations were made by an expedition to the Atlantic Island of Principe to photograph a total solar eclipse.[5] A solar eclipse expedition led by Mount Wilson Observatory in the same year was foiled by cloud cover.

In 1922 a Soviet physicist named Alexander Friedmann boldly suggested that contrary to Einstein's invention of the cosmological constant to hold universe static the equations indicated it must be dynamic, either expanding or contracting. The idea might have gained traction within the science community more rapidly except for Friedmann's untimely death three years later at age 37.[6]

While Edwin Hubble was making his distance measurements to M31 and other nearby galaxies in 1924, Knut Lundmark was studying Slipher's radial velocities of galaxies and Shapley's velocities for globular cluster in search of verification of an idea by de Sitter that sought to explain redshifts not as recessional velocities but as the slowing down of time across great distances, a quirk of his model of relativity known as the de Sitter effect. The Swedish astronomer would certainly have benefitted from Hubble's distances, published two years later, and not from Shapley's Milky Way cluster velocities which had no bearing on large scale cosmology. Even with Hubble's measurements, Lundmark's inclusion of the cluster velocities in his research would have doomed his approach to the expansion of the universe problem.[7]

In 1925 Hubble published his initial papers on galaxy distances from Cepheid variables and the following year he added absolute magnitude measurements in a broader paper (the one that caused the rift with the spectroscopists at Mount Wilson). The 1926 paper caught the eye of the Belgian Catholic priest and physics professor Georges Lemaître. Lemaître was aware of Friedmann's use of the field equations and while reading Hubble's report he had a remarkable epiphany. Pairing Hubble's distance measurements with Slipher's radial velocities for various galaxies the Belgian physicist published a paper in 1927 in which he formulated a linear velocity-distance relationship.[8]

In parallel, Howard Percy Robertson, a promising assistant professor of mathematics at Caltech had the same insight. On the basis of the limited data, both Lemaître and Robertson calculated a velocity-to-distance ratio of +550 kps per million parsecs. Robertson's paper was published in 1928.[9]

By July, when Hubble made his way to Leiden to take part in the IAU Commission on Nebulae and Star Clusters, which he was to chair, the physics community was awash with speculation about the implications of the theoretical determination of a velocity-distance relationship for galaxies. After the conference, Edwin and Grace toured Europe, which was slowly recovering from the ravages of war. In France he made a couple of stops to deliver lectures on the state of nebular research. It was during one of these stops that he learned of the predicted universal expansion from some of the French astronomers and physicists who had read the recent article by Lemaître.

Interested in the possibility of confirming Lemaître's conclusions experimentally, upon his return to Pasadena he met with Robertson to discuss this prediction in detail.[10]

Eager to add this new effort to his nebular research program, Hubble next met with Walter Adams to get the director's opinion on which of the spectroscopists might be right for the job. Hubble would handle the photometrics, but he needed someone to play the part of Slipher, nabbing spectra of faint galaxies to determine their redshifts. If the velocities of galaxies were directly correlated with their distances (obtained from his own measurements of their absolute magnitudes) then the expansion ratio was linear and uniform.

Although a simple idea in principle, getting spectra of the faintest galaxies yet measured would require painstaking – possibly even dangerous – hours at the telescope, an intimate knowledge of telescope and photographic equipment, and copious amounts of spirit and stamina. When Hubble wondered aloud who a good candidate for the assignment might be, Adams knew there was only one man on the mountain he could recommend. Milton Humason was, in Adams's view, "the finest observer Mount Wilson has ever had."[11]

The plate archives at the Carnegie Observatories in Pasadena contain a set of remarkable photographic plates of M101, a spiral known as the Pinwheel Galaxy. They are remarkable not for their spectacular clarity or the features they reveal but for the manner in which they were obtained. The plates were taken by Milton Humason nearly a hundred years ago while crossing the meridian at the Newtonian focus of the 100-inch telescope while orientated at a declination of +54 degrees with the cage facing northward. It was almost an impossible shot and demanded not only great skill but heaps of nerve.

The 100-inch (Fig. 4.7) rotates north-south in a closed yoke that rotates east-west on its axis with the aid of two massive bearings. The south bearing is set low near the observing floor while the north one rests on a high pier above the floor. The bearings were filled with mercury to help them move smoothly, and the imaginary line between their center points was referred to as the meridian. The Newtonian platform rides on rails attached to the dome at the slit doors and the astronomer observing at this focus sits or stands on the platform, keeping his eye on the guide star while coordinating the movement of the telescope and dome. With the front of the telescope (the "cage" carrying the secondary mirror) pointing south, the act of crossing the meridian poses no threat at the Newtonian focus, but if the cage is pointing north, the observer must raise the platform high enough on its dome rails so that the bottom of the platform does not strike the north bearing, breaking its seal and provoking a major "mercury event." The official declination limit for the telescope tube itself is +63 degrees but the actual northern working limit for a meridian passage at the Newtonian focus is less, with the precise amount depending on the skill, the moxie, and the steady hand of the observer.

To obtain the plates of M101, Humason had to lie prone on the platform, easing it into position just barely above the bearing housing with the giant telescope tube positioned just over his head. From this position he was somehow able to keep his eye on the guide scope, keep the telescope and dome rotating in unison, and keep his heart out of his throat as he carefully moved across the meridian in near total darkness.[12]

This amazing feat, which entered the folklore of the observatory, illustrated Humason's understanding and comfort level with the 100-inch as well as his skill as an observer, much of which he attained before 1928. Although the bulk of his work was in stellar photography, spectroscopy and photometry, he also had some experience nabbing velocities for nebulae, most notably M101 and Hubble's famous NGC 6822.[13]

This then, was the Humason who met with Hubble in the latter's office in the summer of 1928.

At this point, the almost diametric opposition between them was very real. Humason was first and foremost a spectroscopist, part of the crew that had opposed Hubble's photometric magnitudes for M31 and hence his distance measurements. He was a close friend of Paul Merrill, with whom he had studied Be-type stars and long-period variables. What is more, he was a friend and admirer of both Harlow Shapley and Adrian van Maanen and a frequent party guest of the latter.

Unaccustomed to the frequency of visits to the observatory offices by members of the press, the spectroscopists were somewhat envious of Hubble's growing fame. His scientific success and leading man looks had made him a media darling and he was not one to miss an opportunity for self-promotion. His feuds with Lundmark and van Maanen were well known within the ranks of the observatory and the director had been irritated by Hubble's decision to take an inordinately long vacation in Europe with Grace.[14]

Given Hubble's increased seclusion, aloofness and other affectations, in 1928 there was ample cause for Humason to have steered well clear of Mount Wilson's brightest star. But prompted by Adams and being of generally amiable character he probably felt compelled to explore Hubble's idea on behalf of the organization that had supported his development as an astronomer. He also appears to have enjoyed a challenge and Hubble's program of work would clearly push the limits.

Hubble likely saw Humason, an inveterate gambler, rake, rebel and raconteur, as little more than a crude, albeit skilled necessity. The ability of Humason to get on with several of Hubble's arch enemies would also have given him pause at the onset. But, like Humason, the endorsement of Adams was sufficient to convince Hubble to try out the former cowboy and muleskinner. Without someone capable of digging out spectra of faint galaxies Hubble simply couldn't get far, and he knew it.

It was therefore inevitable that Hubble and Humason would combine to probe the deeper reaches of space, but no sooner had they begun than the venture seemed doomed to failure.

As mentioned earlier, the highest velocity measured by Slipher was 1,800 kps for NGC 584 in Cetus, but that was at the limit of his spectrograph on the telescope at Lowell Observatory. If Lemaître's linear velocity relationship was valid, fainter galaxies would have much higher velocities. To test this Hubble and Humason chose NGC 7619 and NGC 7626, a pair of faint elliptical galaxies in Pegasus.

On the morning of Monday, August 6th, 1928, Milton said goodbye to his wife and son and stepped out the door to his waiting bicycle. Soon he was pedaling along at breakneck speed barreling down the street on his way to the observatory offices to board the waiting auto shuttle that would take him and his fellow observers to the summit of Mount Wilson.

On arrival to the observatory grounds, he checked into his room at the monastery. After touching base with Nelson about the set up for their observing run, he would've milled around the library or chatted with others on the back porch overlooking the mountains and the valley before retiring for a short nap. As 100-inch operator he sat at the head of the table for dinner. He would have been quizzed about his plans by curious fellow observers. Finally, as the Sun was setting in the western sky, he headed to the dome to begin the run.

Given the late summer heat, Nelson didn't open the dome slit doors until 5:30 p.m., and even then, they decided to sit for a couple hours to allow the interior temperature to equalize with the outside air. This precaution was only mildly successful because the warm air made the mirror's condition only "fair" on that first night.[15]

What is recorded next is a series of remarkable installments in the 100-inch logbook for that week. Humason chose "Spect #6" for his attack, a small-dispersion spectrograph that was his go-to instrument for nabbing spectra of faint objects at the time. It used a 24-inch collimating lens, two prisms and a 3-inch camera. The key was to give enough exposure time to capture a spectrum that was both wide enough and clear enough to eliminate an excessive probable error in his analysis of the displaced absorption lines that he was targeting. The smaller the probable error in his velocity measurements, the stronger would be the case for their being real.[16]

Next to the instrumentation column the astronomer would state the number of plates he exposed on a given night. On this night, for the first time in the long history of the Mount Wilson logbooks, the words "Begin one" were written. The faintness of the nebula meant that Humason was working not in plates per night, which was the norm at this and every other observatory in the world, but in nights per plate. The next night the word "Continued" was written in the column. On the next two nights there was a single set of quotation marks. Then the word "finis" on August 10th signaled the end of the five-day run.

In an effort to get the mirror into better condition, Humason and Nelson waited to open the dome until around 7:15 p.m. for the rest of the week, started to observe at 9 p.m. and finished at around 4 a.m. The total observing time for the five-day run was over thirty-six hours, all for one spectrum of the faintest object ever recorded for the purpose. The average viewing time for a given night was seven hours. In that time, as related by the other entries in the book during that week, an observer might expect to bring down as many as a dozen or more plates for direct photography depending on the magnitude and number of targets.[17]

It would be difficult for a modern observer to comprehend the difficulty and extreme stress involved in chasing a single object across the sky for hours on end, night after night. With automation, computer-guided telescopes and digital data retrieval astronomers can key their target's coordinates into the telescope's guidance system (which synchronizes the movement of the instrument and dome automatically) and simply wait for the data to come in. The hard work of running the telescope is mostly removed from the equation, allowing the modern astronomer more time and energy for data reduction and analysis.

Humason was a pioneer of ultra-deep space spectroscopy and direct photography at the end of the era in which the work of setting up the photographic transfer system and finding, tracking and capturing the target was the job of the astronomer alone. By the introduction of the 200-inch Hale telescope the computer age was underway and by the 1960s Humason was remarking on the advantages his colleagues were already enjoying from technological improvements.[18] He was one of the last of a rare breed of observer who worked at a time when technology was developing in ways that would vastly alter physical research.

This work was not only difficult, but it was also dangerous. Imagine yourself observing a single very faint star through a camera on a tripod installed on a narrow 10-meter diving platform above an empty pool for five nights in a row, through good skies and bad, braced against the cold wind. A false move or a moment of lost consciousness and you could go over the edge. And all the time you have to keep the crosshairs on the guide star, as otherwise your image will be blurred and useless. But this was a challenge that Humason simply couldn't resist.

At the end of that first run an exhausted Humason went to the darkroom to develop the tiny plate bearing the spectrum. The absorption lines, he later reported, were of "very poor quality" and "wide and diffuse" but his well-trained eye recognized that the displacement toward the red end of the spectrum was far greater than any he had previously seen. When he later measured the plate, it gave a velocity of 3,828 kps, which was more than twice the fastest speed Slipher had measured.[19]

It was the kind of discovery that would raise the hackles on the back of your neck if you were in Humason's place. On returning to the observatory offices on Santa Barbara Street, he gave the plate to Elisabeth MacCormack in the computing division for a second opinion on the spectral displacement. Her measurement was close enough to confirm the extremely large radial velocity.

When he heard the news Hubble was understandably elated, but Humason was already preparing for another run with a longer exposure. The equipment on hand was not adequate to take good quality exposures of these moderately distant galaxies without extremely long exposures. Nevertheless, Humason thought another run might yield a more accurate result.

Five weeks later, Humason felt he was ready to give it another go and made his way up the mountain on September 14th. By that time in the year the sky was getting dark sooner and the ambient temperature correspondingly dropped more quickly, which kept the mirror in pretty fair condition for the weeklong run. Once again, with Nelson at the control desk, Humason guided the telescope in the direction of Pegasus and began his run. Cloud cover disrupted the first night, allowing an actual exposure time of only two hours between 10:30 p.m. and 12:30 a.m. Fortunately the weather, the sky and the mirror improved through the remainder of the run and Humason was able to make a total run of 47 hours over six nights. With Nelson cheering him on, they set a nightly goal of opening the dome at 6:30 p.m. and observing for nine hours until closing the dome at 4 a.m. As the sky started to brighten in the early morning of September 20th a weary Humason walked with Nelson from the 100-inch dome over the bridge way, past the 60-inch done, and down the trail to the monastery.[20]

This time Hubble was on the mountain for a run on the Hooker telescope to get some direct photos of the region and others to be used in the development of their analysis. When he heard that the longer run had provided a better exposure and a rough recessional velocity of 3,754 kps he was satisfied they were on to something.[21] He could plot these new measurements along with distances derived from absolute magnitudes for these and the galaxies measured by Slipher to form the basis of the velocity-distance relationship.

The results were encouraging and made Hubble eager for more data, but there was one big problem. Humason had been broken by the experience. While Hubble had spent two nights making 16 direct photographic plates of relatively wide fields Humason had spent eleven nights, over 80 hours, chasing a tiny point of light across the sky to obtain just two spectra. The last run had been too much. Humason went down the mountain and into a sick bed and didn't return to the summit for six months.

Many years later the retired Mount Wilson spectroscopist would cast this experience as not "a very happy kind of work" – the understated sentiments of a survivor who had lived to fight another day.[22] His first deep space galaxy, while definitely fainter than any that he or anyone else had targeted before, was only scratching the surface of the cosmos and at this point nobody knew how far out it would extend. All that Humason knew was he simply could not go any farther. The work was physically and mentally demanding. He had to fight off sleep-inducing fatigue, muscle cramps, aching joints and sore, swollen eyes. To continue posed what he perceived to be a clear and present danger to life and limb. History would later prove that he had reason to be concerned.

During its long history only two men were ever seriously injured in accidents related to the 100-inch. One, mentioned earlier in the book, was the high steel man who died during its construction. The other was Alfred Joy, the Bruce Medalist and longtime member of the Mount Wilson staff, who lost his balance crossing the narrow gangway between the ladder and the Cassegrain platform. He plummeted 30 feet to the concrete floor. Gravely injured and unable to be carried down to a waiting vehicle, he was secured to a stretcher and lowered to the ground floor of the dome using the crane that was there to move the mirror and cages. It took six months for the 60-year-old to recover from multiple internal and external injuries, but he eventually returned to the mountain.[23]

Reflecting on Joy's unfortunate accident from our vantage point in history (his fall occurred in the 1940s) it is easy to understand the peril even a seasoned observer like Humason might have felt in continuing with Hubble. His wariness caused him to balk at the idea of moving forward on the project, just as the partnership made its highly auspicious debut. If the two men were suspicious of each other at this point, Humason's decision to quit would do nothing to bring them closer. In full view of the prize and not quite able to reach it, Hubble's driving ambition would not stand for Humason, who he saw as a subordinate, simply deciding to walk away. However, it was Humason's prerogative. He was assisting Hubble, but he wasn't Hubble's assistant.

In his otherwise captivating book, *Mariner of the Nebulae,* relating the life and times of Edwin Hubble, Gale E. Christianson writes of Humason in highly reductive terms relative to Hubble, referring to an "educationally impoverished Humason" as Hubble's "assistant," "perched, like a monkey, on the small Cassegrain platform." He states eloquently how the "pair hoisted sail and set a course for far-flung waters, Humason in the crow's nest and Hubble at the helm." The passage makes for a pretty read but requires some reconstruction out of deference for both men and the respectful partnership that they were eventually able to forge.[24]

First of all, Humason was by the late 1920s no longer "educationally impoverished," which is one of the most remarkable things about his rise. He had come to Mount Wilson as a zero on a knowledge scale of one to ten and by the time he and Hubble started working together he was very nearly a ten.

Humason was what would be referred to in sports terms as a walk-on, earning his right to stand with the men on the mountain and, in fact, creating his own special place among them. Like others in Humason's rare position he proved to be one of the most productive and prolific men at the observatory, a characteristic common to those who find their calling, their passion, and their purpose in life irrespective of their upbringing or background. As regards the second point, far from being in the "crow's nest" Humason was very definitely at the wheel of the great "Mariner's" ship. Hubble may have known where he hoped to go but without Humason at the helm there was no way in hell he was going to get there.

Christianson's use of metaphor does reflect Hubble's overall sense of superiority over Humason as a researcher (and possibly a man) at the onset of their project. But Hubble was in no way constitutionally, experientially, mechanically, or psychologically equipped to do what he was asking Humason to do, and he knew it.

After taking the initial challenge, Humason was now all too aware that to continue he would have to be either crazy or stupid, or perhaps both. Having published five papers, the previous year and being involved in ongoing projects with Seares, Merrill, Adams, and Joy he had plenty of other work. So Humason merely stepped away.

Hubble was apoplectic, a characteristic which was finding him increasingly at odds with some in the astronomy world. He complained to Adams, who for once took notice of his predicament. Adams was as excited about the implications of Hubble's findings as anyone, and his correspondences with Carnegie president John C. Merriam were already brimming with anticipatory glee over what might be a fourth dramatic discovery in the observatory's 25-year history. But if Humason was balking at going forward with the work, Adams knew that he must come up with something special to convince the reticent spectroscopist that it was a quest worth pursuing.

The honorary director was the ace up Adams's sleeve. On learning of the extraordinary radial velocity for NGC 7619, George Hale became so animated he had to be restrained. A visit with him just might be the key to convincing Humason to continue the work. In an act of desperation, Adams went to Hale to respectfully ask his longtime friend and boss to step into the fray between Hubble and Humason.

Hale sent an invitation to Humason asking him to visit his home and take a look at his solar laboratory. Surprised by the invitation from the man he regarded as highly as anyone he had known, Humason accepted. Hale led his guest on an extensive tour of his study and solar lab, telling him tales of his first forays into the development of the spectroheliograph and walking him through the workings of the device. As a curious observer and student of telescope mechanics and optics Humason had spent many hours learning the operation of the various solar telescopes on Mount Wilson. During an interlude as the two sat and talked, Hale disclosed to Humason that he had been in grave error not promoting him to the staff when initially prompted by Shapley and Adams years before. It was clear to everyone now that Humason was deserving of his place. Then Hale asked Humason what it would take to get him to resume work on the deep space nebular program.[25]

To have received such praise from Hale must have been a moment of solemn pride for Milton Humason. It is hard to imagine him not recommitting himself to the work again in that moment, but there were technical challenges to overcome if he or anyone else were to attempt radial velocity measurements of remote galaxies. The camera in use with the slit spectrograph on the Hooker telescope had a 3-inch focal length and a single 60-degree prism. The dispersion of 3.21 mm offered an eight-fold reduction from slit to plate that was more than sufficient for most of the work being done at the observatory, but it fell well short in both clarity and speed for Hubble's project. A shorter focus would be required to move forward.[26]

Hale understood immediately the problem Humason was facing and told him he would start work on a solution to the camera and optics required to improve the effectiveness of the spectrograph. In the meantime, he wondered if the embattled spectroscopist would stay with the project, which Hale believed was poised to make a history-making breakthrough. Humason agreed to continue if the apparatus was improved. Adams, with help from Hale, had found a way to keep the expansion train on the tracks.

In the meantime, there was sufficient evidence for a compelling paper suggesting the plausibility of the distance-velocity relationship to the astronomy community at large.

Hubble decided to publish his initial paper on his own, either because he wanted to set the record independently or out of a distrust in Humason's resolve, or both. Humason later insisted that the idea for the program had been Hubble's alone and that he only participated at Hubble's request and therefore claimed no dominion over the practical discovery of the expansion.

Still, Hubble had a choice in publishing alone or jointly and opted for the former. In his defense, of the 46 galaxies for which Slipher had provided velocities Hubble had obtained reliable distances for 24 and the other 22 were part of the photometric work that lay ahead. However, Humason would provide velocities of NGC 6822, NGC 224 (M31), NGC 221 (a dwarf galaxy satellite of M31) and NGC 5457 (M101) using the Mount Wilson reflector as a check against the previous measurements. Hubble also included Humason's velocity of NGC 7619 in suggesting the linear relationship held out at least for that one target, while announcing Humason was launching an investigation to determine "velocities of the most distant nebulae that can be observed with confidence."[27]

As has been pointed out Hubble was untrusting of many of those around him and bent on cementing his legacy, while Humason was immensely deferential, disinclined to defend his own interest and inherently reluctant to seek out, let alone enter the limelight.

For his part, Humason had finally found a course of research he could carve out entirely for himself so long as an improved camera was made available in the near future. He was essentially on his own in extending Slipher's original list of galaxy velocities out as far as the 100-inch and spectrograph would go. Depending on depth and density this represented hundreds if not thousands of potential targets, more than enough to occupy him for the rest of his career. Hubble and Humason submitted independent papers to the National Academy of Sciences on January 17th, 1929, and on March 15th they were published back-to-back.

The Large Radial Velocity of NGC 7619 by Humason came first. He explained that the measurement had been made in support of a program being pursued by "Mr. Hubble," gave a description of the procedure and presented the remarkable velocity which so dramatically extended Slipher's results. He finished by saying he hoped that velocities for more galaxies would "soon" be forthcoming.[28]

A Relation Between Distance and Radial Velocity Among Extra-Galactic Nebulae by Hubble explained the linear velocity-distance program based largely on solar motions and suggested several possible solutions for the phenomenon. He had correlated the distances of the 24 galaxies in the survey with their velocities and derived a "proposed" relative "K term" of 500 kps per million parsecs. This implied that distances to galaxies could now be determined spectroscopically from their radial velocities. If the relationship remained true for the vast majority of galaxies, then it implied the universe was expanding. Alternatively, in the de Sitter model the relationship might derive from the "general curvature of space." But as Hubble stated, the role of his paper was to introduce "a first approximation" of the relationship between redshift and distance.[29]

Allan Sandage suggested in his book, *Centennial History of the Carnegie Institution of Washington: Volume I The Mount Wilson Observatory*, that Hubble's determination of the expansion was made entirely on his own, without prior knowledge of the work of Einstein, Friedmann or Lemaître. On page 502: "in 1929 Hubble had no knowledge of Friedmann or the Einstein equations except that de Sitter had predicted a static model" or knowledge of "the prior predictions…by Lemaître and Robertson."[30]

With due respect to the late Dr. Sandage, the idea that Hubble had no knowledge of the theories, or their implications doesn't hold up to scrutiny. Sandage himself suggested that Robertson told him that he had discussed his derivations of Einstein and Friedmann's work with Hubble in 1928.[31] Humason remarked in an interview many years later that Hubble had returned from the conference in Leiden in August of that year excited to test the idea (presumably of the possible expansion) and asked Humason if he "would try and check that out."[32] Hubble was even photographed with Lemaître (Fig. 7.1) on a tour of the east coast with Grace on their way to Europe in 1925. Whether or not the two discussed the Belgian's theory isn't known but there can be little doubt that Hubble knew of him in 1928.

Taken together with Hubble's general curiosity and determination to dominate the field of nebular research as well as his correspondence and knowledge of de Sitter's predictions, it is hard to believe he had no knowledge of the possible velocity-distance relationship prior to starting the project with

Fig. 7.1 Edwin and Grace Hubble (center) stand for a photograph on a visit to Harvard College Observatory in 1925 where Georges Lemaitre (second from left) was working as a visiting researcher from St. Edmund's College, Cambridge, UK. (Image from Edwin Hubble Papers, Huntington Library, San Marino, California)

Humason. Sandage would appear to base his opinion on serving as Hubble's assistant from 1949 to 1953, during which time Hubble made no reference to either the physicists or their theories. But Sandage knew as well as anyone that Hubble was an immensely private and complex person, and it hardly defies the imagination to suppose he simply kept that information to himself: an act meant to preserve and protect his legacy.[33]

None of this is intended to detract from the practical discovery itself, which required the initiative and competence to pull together work in a way which substantiated the expansion for the majority of the science community (with at least one incredible exception). For the most part, the general public couldn't wrap its head around the theories put forth in general relativity and its various predictions. But concrete evidence that the universe was in a state of expansion would spread like wildfire through the press and catapult Hubble's name to the pinnacle of the science world and alter perceptions in ways that would impact on almost every aspect of life for decades to come. Like Einstein, Hubble would become an object of obsession, praised, revered, and ridiculed privately and publicly for years to come.

References

1. A.J. Kox, Martin J. Klein and Robert Schulmann (editors), *The Collected Papers of Albert Einstein, Volume 6: The Berlin Years: Writings, 1914-1917,* (Princeton University Press 1996), pg. 243
2. A.J. Kox, Martin J. Klein and Robert Schulmann (editors), *The Collected Papers of Albert Einstein, Volume 7: The Berlin Years: Writings, 1918-1921,* (Princeton University Press 1996), pg. 136
3. Daniel Kennefick, *Was Einstein the First to Discover General Relativity,* (Princeton University Press March 09, 2020), https://press.princeton.edu/ideas/was-einstein-the-first-todiscover-general-relativity
4. Abraham Zelmanov, *Biography of Karl Schwarzschild (1873-1916),* (The Abraham Zelmanov Journal 2008), Vol. 1, 2008, pg. xvii
5. F.W. Dyson, A.S. Eddington, C. Davidson, *A Determination of the Deflection of Light by The Sun's Gravitational Field, From Observations Made at the Total Eclipse of May 29, 1919,* (Philosophical Transactions of the Royal Society of London 1920), Series
6. Simon Singh, *Big Bang: The Origin of the Universe,* (Harper Perennial 2005), pgs. 149-155
7. Allan Sandage, *Centennial History of the Carnegie Institution of Washington: Volume 1 The Mount Wilson Observatory,* (Cambridge University Press 2004), pg. 503
8. Abbé G. Lemaître, *A Homogenous Universe of Constant Mass and Increasing Radius Accounting for the Radial Velocities of Extra-Galactic Nebulae,* (Annales de la Société scientifique de Bruxelles 1927), Translated for the MNRAS 1931, Vol. 91, pgs. 483-490
9. Allan Sandage, *Centennial History of the Carnegie Institution of Washington: Volume 1 The Mount Wilson Observatory,* (Cambridge University Press 2004), pg. 501
10. *Ibid.*
11. Letter from Walter S. Adams to Milton L. Humason, June 29th, 1950
12. Allan Sandage, *Centennial History of the Carnegie Institution of Washington: Volume 1 The Mount Wilson Observatory,* (Cambridge University Press 2004), pgs. 377-380
13. Milton L. Humason, *Radial Velocities in Two Nebulae,* (PSAP 1927), Vol. 39, No. 231, pgs. 317-318
14. Gale E. Christianson, *Edwin Hubble: Mariner of the Nebulae,* (University of Chicago Press 1995), pgs. 222-223
15. Mount Wilson Observatory logbooks. The Huntington Library, San Marino, CA., 100-inch telescope, August 6-10, 1928
16. Milton L. Humason, *The Large Radial Velocity of N.G.C. 7619,* (PNAS March 15, 1929), Vol. 15, No. 3, pgs. 167-168

17. Mount Wilson Observatory logbooks. The Huntington Library, San Marino, CA., 100-inch telescope, August 6-10, 1928
18. Interview of Milton Humason by Bert Shapiro in circa 1965, Niels Bohr Library & Archives, American Institute of Physics, College Park, MD USA, www.aip.org/history-programs/niels-bohr-library/oral-histories/4686
19. Milton L. Humason, *The Large Radial Velocity of N.G.C. 7619*, (PNAS March 15, 1929), Vol. 15, No. 3, pgs. 167-168
20. Mount Wilson Observatory logbooks. The Huntington Library, San Marino, CA., 100-inch telescope, Sept. 14-19, 1928
21. Milton L. Humason, *The Large Radial Velocity of N.G.C. 7619*, (PNAS March 15, 1929), Vol. 15, No. 3, pgs. 167-168
22. Interview of Milton Humason by Bert Shapiro in circa 1965, Niels Bohr Library & Archives, American Institute of Physics, College Park, MD USA, www.aip.org/history-programs/niels-bohr-library/oral-histories/4686
23. Allan Sandage, *Centennial History of the Carnegie Institution of Washington: Volume 1 The Mount Wilson Observatory*, (Cambridge University Press 2004), pg. 260
24. Gale E. Christianson, *Edwin Hubble: Mariner of the Nebulae*, (University of Chicago Press 1995), pgs. 191-192
25. Interview with Don Nicholson, 2006
26. Milton L. Humason, *The Rayton Short-Focus Spectrographic Objective*, (ApJ 1930), Vol. 71, pg. 353
27. Edwin Hubble, *A Relation Between Distance and Radial Velocity Among Extra-Galactic Nebulae*, (PNAS 1929), Vol. 15, No. 3, pgs. 168-173
28. Milton L. Humason, *The Large Radial Velocity of N.G.C. 7619*, (PNAS March 15, 1929), Vol. 15, No. 3, pgs. 167-168
29. Edwin Hubble, *A Relation Between Distance and Radial Velocity Among Extra-Galactic Nebulae*, (PNAS 1929), Vol. 15, No. 3, pgs. 168-173
30. Allan Sandage, *Centennial History of the Carnegie Institution of Washington: Volume 1 The Mount Wilson Observatory*, (Cambridge University Press 2004), pg. 502
31. *Ibid.* pg. 501
32. Interview of Milton Humason by Bert Shapiro in circa 1965, Niels Bohr Library & Archives, American Institute of Physics, College Park, MD USA, www.aip.org/history-programs/niels-bohr-library/oral-histories/4686
33. Allan Sandage, *Centennial History of the Carnegie Institution of Washington: Volume 1 The Mount Wilson Observatory*, (Cambridge University Press 2004), pg. 501

8

Bang! Goes The Universe (1929-1931)

The Carnegie Institution's funding stabilizes the Mount Wilson Observatory as the world sinks into economic depression. Hale makes good on his promise to Humason by developing an improved camera that is many times faster than any previously produced and the misfit duo of Hubble and Humason embark on their mission to prove universal expansion. On combining Slipher's earlier spectral results with their own, Hubble and Humason jointly report their first conclusive evidence in 1931, with Einstein and the rest of the scientific world eager for the news. The groundbreaking paper brings the German physicist to Pasadena, where he tours the observatory and consults with astronomers and physicists at Caltech on the cosmological implications of this work.

The Wall Street Crash on October 29th, 1929, sent American stocks into a freefall that the market would not entirely recover from for a quarter of a century. The effects of this "Black Tuesday" rippled until by the beginning of 1930 most of the world was suffering the effects of low consumer confidence, low tax revenue and profits and degradation of gross domestic product and international trade.

Soon one in four Americans would be out of work while the new president, Herbert Hoover, who had defeated Calvin Coolidge in a landslide election months before, struggled in the face of an unfriendly Congress to pass legislation designed to help pull the country out of its financial collapse. Public trust in local and federal government had begun to erode years before, when fallout from corruption within the Harding administration fueled anti-government sentiment.

© Springer Nature Switzerland AG 2021

R. Voller, *Hubble, Humason and the Big Bang*, Springer Praxis Books,

https://doi.org/10.1007/978-3-030-82181-4_8

Among the groups that capitalized on public unrest was the mob, whose bosses stoked suspicion of government corruption while robbing taxpayer's blind. It wasn't until the St. Valentine's Day Massacre on February 14th, 1929, when seven members of Bugs Moran's crime gang were gunned down in broad daylight at a warehouse in Chicago by members of Al Capone's notorious gang, the Outfit, that public support for the mob began to wane.

The first Academy Awards were held on May 16th at the Hollywood Roosevelt Hotel. The entire ceremony, in effect a private dinner attended by 270 filmmakers and actors, hosted by Douglas Fairbanks sr., lasted a mere 15 minutes and presented 12 awards. Guests paid $5 and the best picture award went to the 1927 silent film *Wings*, which was set in the Great War and starred Clara Bow and Charles "Buddy" Rogers with a small role played by Hollywood newcomer Gary Cooper.

Galvin Manufacturing Corporation introduced the first car radio in the US in Chicago, combining the words "Motor" for motorcar and "ola" for Victrola to produce the Motorola brand name.

In June Bell Laboratories introduced color television and in October the retail giant J.C. Penney opened its 1,000th store in Milford, Delaware. By that time the chain had stores in 48 states and gross earnings of $190,000,000 ($2.83B in 2020).

Adding to the misery of struggling families, the worst influenza epidemic in a decade broke out in the winter of 1928-1929, and the public was still reeling from its effects in the spring of 1930.

In the face of destabilizing events such as these, new products like Bib-Label Lithiated Lemon-Lime Soda (7-Up), with its mood-enhancing drug lithium citrate, rapidly took root in the public sphere.

Despite the economic doldrums, Europe was making its own advances technologically. German Zeppelin airships were achieving round-the-world flights and the first telephone booths were being introduced to streets in England.

The decline in international commerce further soured the mood of the German public, which had been struggling under the weight of reparations since the Treaty of Versailles in 1919. The Young Plan signed by Congress in August of 1929 reset the target for German repayment of war debts for damages caused at just over 112 billion marks to be paid over six decades. But the plan would disintegrate by 1932 with Germany paying only about 20 billion of the 132 billion marks decided at Versailles. The subversion of the German public and economy had given rise to the Nazi Party and Adolf Hitler who by the end of the decade was just beginning to take hold of the country.

Life at Mount Wilson carried on almost unaltered by the economic hardship being faced by much of the country. Carnegie's vast financial backing allowed operations to continue unabated apart from a degree of belt-tightening imposed by the always fiscally conscious Walter Adams.

Regardless of whether Edwin Hubble was aware of the name Lemaître, by the spring of 1929 the ordained Belgian physicist was well aware of the astronomers at Mount Wilson. Hubble's practical establishment of the predicted expansion coupled with Humason's latest velocities resonated with Lemaître, and he began to think through their implications. If the expansion was real, then the steps in the process should be reversible until all of the energy in the universe was shrunk to a single point or as it was sometimes called, a "cosmic egg." He had suggested the plausibility of this idea in 1927 in the paper *l'Hypothèse de l'atome primitif (Hypothesis of the Primeval Atom)*, giving the first description of the concept that would later come to be known as the Big Bang.

The reviews from the pinnacle of the physics world were mixed. Albert Einstein, who had read up on his work, wryly suggested to Lemaître at a conference in Brussels that year, "your calculations are correct, but your physics is atrocious."[1]

Arthur Eddington, who was having Lemaître's paper translated for publication by the Royal Astronomical Society (RAS) and who had invited the Belgian physicist to a meeting for the British Association in London in 1931, declared repugnant the proposal that the universe had a beginning because it smacked of Creationism.

Without implying a religious correlation, Lemaître responded in an article in the science periodical Nature arguing that if the laws of quantum theory were to be believed then such a beginning was not all that farfetched. He "conceive[d] the beginning of the universe in the form of a unique atom, the atomic weight of which is the total mass of the universe." If "energy…is distributed in discrete quanta," he wrote, "ever increasing," then as "we go back in…time we must find fewer and fewer quanta, until we find all the energy of the universe packed…in a unique quantum." Here, he continued, "the notions of space and time would altogether fail to have any meaning," and "would only begin to have a sensible meaning when the original quantum had been divided into a sufficient number of quanta." In this vital event, Lemaître was suggesting, "the beginning of the world happened a little before the beginning of space and time."[2]

His 1929 paper offered a moment of revenge for Hubble, who was still holding a grudge against Lundmark after their run-in about the nebular classification system. As part of his discussion Hubble recalled Lundmark's ill-fated 1925 paper which sought evidence for the de Sitter effect. In his conclusions the Swede had predicted an upper limit of 2,000 kps for redshift velocities. Yet Humason had measured a velocity considerably higher than this for NGC 7619.[3]

Harlow Shapley reacted to Hubble's report of a relationship between the redshifts and distances of galaxies by writing in the July 15th issue of the PNAS that he had anticipated the apparent expansion fully a decade ahead of Hubble's 1929 paper:

> I pointed out some ten years ago the systematic recession of these objects on both sides of our Galaxy and the fact that "The speed of the spiral nebulae is dependent to some extent upon apparent brightness, indicating the relation of speed to distance, or possibly to mass."[4]

Hubble basically ignored this due to the Harvard director's longstanding insistence that the Milky Way was the entire universe. What use would such a relationship have been in a single galactic system? Evidently Shapley was still losing sleep over missed opportunities.

While MWO's experts created a faster spectrographic camera for the 100-inch Humason continued to measure spectra within the limits of the existing apparatus to support Hubble's work. In addition to corroborating velocities measured prior to 1929 he was obtaining some new ones. Meanwhile, Hubble was measuring magnitudes photometrically for these same galaxies and charting their combined data.

The scientific frenzy caused by Hubble and Humason's work had finally given George Hale the means to sell the idea of a new, even larger telescope to possible benefactors. This time the honorary director had convinced the International Education Board to provide the funds through Caltech "for the construction of a 200-inch reflecting telescope." The award was announced on October 18th, 1928 and was later chronicled in the opening to the 1929 Mount Wilson Contributions. Hale's latest vision, spawned through conversations in 1927, was to be jointly conceived, designed, constructed and operated by staff members of both Mount Wilson and Caltech. Importantly, the project had the support of the CIW, which underwrote the project as a whole. The plan would provide staff and visiting astronomers from Mount Wilson to continue key investigations with an improved telescope while also creating a research facility for graduate level astronomers and astrophysicists at the university.

Hooker telescope designer, Francis Pease had called for a 300-inch primary mirror, but Hale realized that a diameter of 200 inches was more sensible in both scale and cost during the lingering and painful depression.

As part of the coordinated effort, the CIW agreed to support the salaries for some of the staff at Mount Wilson to help with the designing and building the enormous new instrument and finding a site suitable for a facility to house it. Dr. John A. Anderson, who was involved in the development of the new camera to aid in deep space spectroscopic investigations on the 100-inch, became executive officer in charge of design and construction. An advisory board was established by the two institutions with Hale as chairman and including Adams, Seares and Hubble, with the latter serving as Secretary of the Advisory Committee.[5]

This appointment obliged Hubble to take a lead role in the development of the apparatus to accompany the new telescope. A seven-page report *Proposed Equipment for the Prime Focus of the 200-inch* laid out in detail the nature of investigations likely to be undertaken by the new telescope and the equipment that investigators would require to conduct them.

Specifying the main categories of research as (1) Direct Photography, (2) Photographic Photometry, (3) Spectrography and (4) Physical Measurements, Hubble then identified the necessary supporting instruments in each discipline for discussion and submission to the policy committee and subsequently the construction committee. The list included upgrades to conventional equipment which called for various plate sizes and bases with modernized guidance, collimating and field rotation controls, industry standard thin Wratten filters by Kodak and limited absorption (Zeiss) Jena glass color filters. Anticipating a relatively large improvement in magnification an enlarging camera for planetary photography was listed, as were field correctors to deal with off-axis comatic aberrations caused by common focus errors in the parabolic (Why, George? Why?!) primary mirror. Finally, a jiggle camera was included for built-up or layered images of various objects and field surveys.

In addition to these items, Humason's intended research into the depths of the universe in pursuit of ever greater redshift velocities would require specially fitted slit and ultraviolet slitless spectrographs. Hubble further suggested the creation of a special spectrograph with no slit to be adapted from a zero corrector for very large fields. And for better definition in enlarging objects at moderate distances (for the time) he suggested a corrective lens with a focal ratio of f/5.

With the help of Nicholson and others, an extremely sensitive photoelectric cell could be developed, as could radiometric and polarizing equipment. These antecedents of modern photovoltaic cells were relatively new and were making advances in the determination of magnitudes of very distant objects with ever-increasing precision.[6]

Given the new instrument and the combined organizations behind it, vested as they were in its creation, there was every reason to believe the 200-inch's predicted completion date of 1938 would be achieved with few, if any, interruptions. But as Hubble was drawing up plans for the most modern telescope the world had ever seen, no one could have predicted that the emerging nationalism rippling through Europe would engulf the world in the most horrific war in history.

Instead of incurring the expense of pouring and annealing the glass disk in Europe and shipping it across the Atlantic, as was the case for the 100-inch, Hale and Adams decided to exploit advancements in American engineering and hired the General Electric Company in New York to design and manufacture the enormous blank for the primary mirror under the supervision of Dr. Elihu Thomson of G.E.[7]

Selecting the best location for the new telescope was an important consideration for the development committee. During the spring of 1929 Hubble and Humason, accompanied by Ferdinand Ellerman, made a number of trips into the hills south of Los Angeles and in Arizona, using small telescopes to undertake studies of the seeing conditions. In his role as Secretary of the Advisory Committee, Hubble was likely overseeing these experiments and taking notes. Humason would have been in charge of set-up and observations. Ellerman, a skilled mechanic, would have looked after the instruments. Humason and Ellerman carried these investigations forward the following year[8] (Fig. 8.1).

In addition to providing possible clues to the perfect location for Hale and company to construct their next great observatory and research center, these excursions likely played a role in evolving the working relationship and trust between the arch-conservative democrat Hubble and the progressive republican Humason. Although few in number, the mornings and early afternoons between observing and documentation were often given over to hiking and fishing, common pastimes of most of the men at Mount Wilson. Hubble and Humason were both accomplished fly-fishermen.

In 1930 Lowell Observatory announced the discovery of the now dwarf planet Pluto by Clyde Tombaugh using a 13-inch astrograph and blink comparator.[9] Pouring over plates taken at the 10-inch and 60-inch telescopes on Mount Wilson by Humason on his unsuccessful search for the planet in the winter of 1919-1920, Seth Nicholson was able to locate it near the position predicted by W.H. Pickering and Percival Lowell.

Fig. 8.1 A small telescope set up on Palomar Mountain during a location scout ca 1930. (Courtesy, Ann Humason Bernt family)

Assisting Nicholson was Nicholas Mayall, a first-year graduate student at the University of California at Berkeley who was taking time off to obtain practical experience at Mount Wilson prior to going back to "do his doctoral thesis." With all the energy and excitement at the observatory at that time, Mayall decided to stay on from 1929 to 1931 and soak up everything he could from Nicholson and later Hubble and Humason.[10] Mayall was therefore treated to master classes with Hubble in nebular photometry and classification, Humason in astrophotography and spectroscopy, Merrill in stellar spectroscopy and Nicholson in the broader field of astrophysics. It was an enviable education and set the table for his work with Hubble and Humason in future years. During this time Mayall and Humason became close friends.

On this occasion however, he was not doing Humason any favors. Mayall spent several evenings in March and May at the 60-inch exposing new plates of the region to assist him and Nicholson in their search. Once they located Pluto on their new plates it was easier to locate it on the plates that Humason took years before, a fuzzy freeloader at the edge of his images. A full analysis by Nicholson and Mayall called *Positions, Orbit, and Mass of Pluto* in the ApJ in January 1931 gave an orbital period of 247.70 years (currently 247.94) and a mean distance of 39.457AU (currently 39.482) and predicted the perihelion of September 1989 with an orbital eccentricity at 0.2485.[11]

At that time, everyone assumed that the ninth planet must be massive enough to have a perturbing effect on the orbits of Neptune and Uranus because Lowell based his prediction on an analysis of their motions. According to Lowell it would have seven times the mass of Earth. Nicholson and Mayall calculated a mass similar to that of Earth.

The Dutch American astronomer Gerard Kuiper was able to reduce Pluto's size to about that of Mars using the 200-inch in 1948-1949 with the aid of Humason, whose assistance, Kuiper said was "so vital that he may well be considered co-author of the principal result" of the paper.[12] This may have offered a small bit of redemption for the legendary Mount Wilson spectroscopist.

In 1976 the existence of methane ice was suggested by measurements of Pluto's albedo. This meant it was extremely bright for its size, and therefore must be closer to 1 percent of Earth's mass. The discovery of the moon Charon by James Christie in 1978 led to further a reduction of its mass to its current value of 0.0022 Earth masses.

Finally, at a meeting of the IAU in 2006 the planet was downgraded to dwarf status. According to IAU guidelines at that time an object must meet three criteria to be counted as a planet. One, it must be in orbit around the Sun. Pluto, cleared that hurdle. Two, it must have sufficient mass to assume a nearly round shape. The round, beautiful, multi-colored surface of Pluto revealed by the New Horizons mission in 2015 clears that hurdle as well. Three, its orbit must be free of other objects of similar mass. Pluto is known to travel within the Kuiper belt, which is loaded with similarly sized objects, some even larger, so it doesn't clear that hurdle and is therefore considered a dwarf planet.[13]

Humason likely took the news of Pluto's discovery in stride, having long since resigned himself to the sidelines in the search. The work he was doing with Hubble was by this time offering plenty of excitement and more press attention than the camera-shy spectroscopist was comfortable with anyway. Key to his increased enthusiasm was the fact that the camera Hale had promised for his push to extend the redshift record was ready for trial.

In May 1930 the results of his analysis of the "world's fastest" camera appeared in the ApJ. *The Rayton Short-Focus Spectrographic Objective* detailed its attributes as "an eight-fold enlargement of a microscope objective, with a focal length of 32 mm and an aperture of 50 mm or a focal ratio of F/0.6."[14]

Wilbur Bramley Rayton was a second generation American from a New York farming family and a graduate of Syracuse University in 1906. He married after college and moved to Rochester, NY, where he went to work for Bausch and Lomb Optical Company. There he worked his way from computer to assembler to optical engineer. He taught optics at the Institute of Optics at the University of Rochester from 1929 to 1931 and served as president of the Optical Society of America from 1933 to 1934.[15]

According to Humason, this masterpiece delivered "a gain of 50 percent in speed over the fastest short-focus camera previously used" (19 times the reduction from slit to plate). The result was a well-defined spectrum with clear narrow lines. Observing with the new camera and accompanying spectrograph with a 24-inch collimator and two 60-degree prisms Humason would have twice the speed and could go after the spectra of galaxies that would have been "hopeless" previously.

As a test, Humason selected one of the faintest known galaxies in a cluster in Ursa Major. The exposure time for a five-day run in September was 45 hours. That was roughly the same duration as for the galaxy which nearly forced him to quit the project a year earlier, but the velocity was three times greater at 11,500 kps. Applying Hubble's ratio for velocity to distance this implied a distance of "twenty-three million parsecs, or seventy-five million light years."

Absorption lines in the spectra of galaxies were much more diffuse and fewer in number than those in stellar spectra due to the convolution of light of large stellar systems and their small angular diameter. The H and K lines of (singly ionized) calcium in the near-ultraviolet end of the spectrum were very dependable. For his laboratory comparison Humason used helium, whose lines were similar in number and spacing.[16]

Despite the improvement in camera speed, Humason knew that his involvement in the project with Hubble would require still longer hours of exposure time as the galaxies he sought grew fainter and more difficult to target. Just how faint or how far neither he nor anyone else knew for sure but Hale had made good on his promise to deliver the fast camera and Humason wasn't going to back out now. He was curious to find out whether Lemaître's prediction of a linear relationship would hold up.

Finally, as the date for their next publication approached the partnership of Hubble and Humason was improving, and a mutual respect was growing. This latter point was brought home when, to Humason's undoubted surprise, Hubble suggested that they make their next article a joint effort. For all these reasons, Humason's heart was back in the game and the dogged determination and competitive spirit was returning to the deep space pioneer.

He set out to perform his latest act of observational lunacy with the 100-inch in March 1930. The target, in a cluster of galaxies in Leo was easily the faintest he had yet attempted. To play it safe he booked nine consecutive nights to allow sufficient time to expose the image of its spectra.

Starting March 24th, Humason spent seven nights averaging almost nine hours per night at the telescope, chasing a tiny point of light across partly clouded skies with a frigid wind lashing at him. Protected from the elements by the bear skin coat Helen had bought for him, he guided the great reflector with unprecedented skill. The dome was closed all night on the last two nights of the month while a storm blew over, dumping rain and sleet on the mountain. He and Nelson went back to work on April 1st for the final run before removing the plate at 4:15 a.m. on April 2nd.

Mayall was on the mountain that week working at the 60-inch and Humason invited the younger protégé to the dark room to watch him develop the tiny plate. Once the plate was "hypoed clear" Humason "held it up…in front of the light box." Humason's reaction was imprinted on Mayall's mind for all time. "My God, Nick, this is a big shift!" The telltale "H and K lines…were shifted way over to hell-and-gone from where they should've been." According to Mayall, Humason's increasingly large redshifts were subject to skepticism from his fellow spectroscopists, a matter that was testing his already wavering self-esteem. Gaping transfixed at this awe-inspiring plate, Mayall told his friend that if there were "any doubting Thomases" remaining in the department, "you won't have any after they see this."[17]

With what looked to be an enormous new redshift in hand, Humason was in the mood to "celebrate" and invited Mayall to his dorm for a nip of his famous elixir. Mayall "had one" and his host "had several." In the morning they had "breakfast at the monastery," then Mayall stood by the telephone while Humason called down to Hubble in Pasadena with the news. On hearing what his partner had nabbed, Hubble said, "Milt, you are now using the 100-inch telescope the way it should be used."[18]

His most precise calculation (which typically involved a relatively large probable error) revealed the incredible recessional velocity of 19,700 kps, more than five times that he had obtained for NGC 7619 two years earlier. The velocity corresponded to a tenfold increase in distance to over one hundred million light years (Fig. 8.2).

As in 1929, Humason led with a paper detailing his exploits in the extragalactic realm. *Apparent Velocity-Shifts in the Spectra of Faint Nebulae* laid out his analysis of 46 galaxies all of which were shifted increasingly to the red end of the spectrum. The exposure time for the brightest one was only 13 hours. As usual, after Humason photographed a spectrum, it was measured by both him and a computer (in this case Elizabeth MacCormack). In what would become a theme, Humason underscored the fact that all the faint galaxies thus

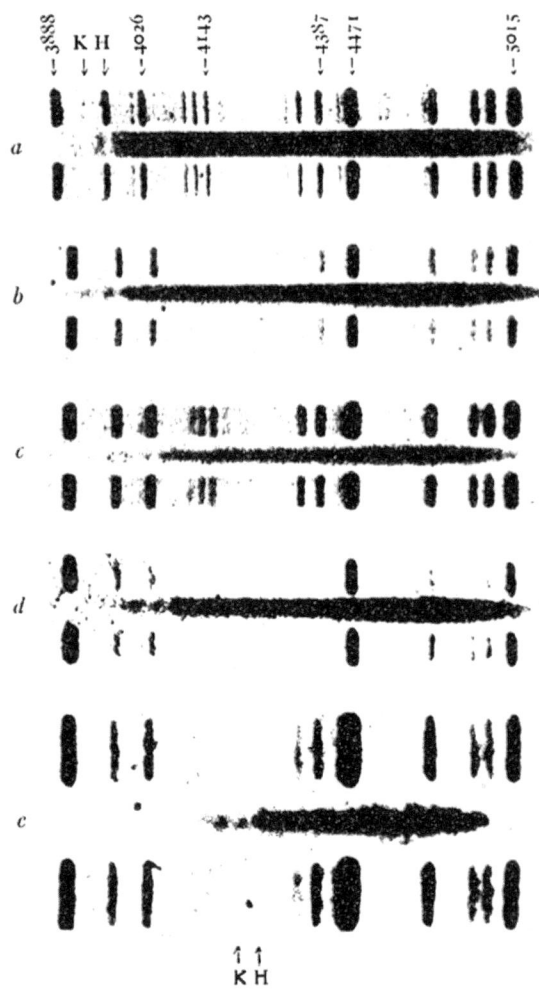

Fig. 8.2 The "big shift" from Humason's 1931 paper showing a) normal position for the H and K lines; b) NGC 221 vel. = -185 kps; c) NGC 385 vel. = +4,900 kps d) NGC 4884 vel. = + 6,700 kps and e) The bright nebula in Leo vel. = +19,700 kps. Humason indicated that the enlargement for a – d is 28 times the original negative, and that e was enlarged 47 times at half the dispersion of the first four.

far measured matched Hubble's linear ratio for velocity and distance, suggesting the expansion was real.[19]

One unique feature of this report was Humason's mention of galaxy groups which differ from clusters in that their spectra are similarly redshifted. Pioneering at the outer edges of the known universe with the world's most advanced observing equipment, Humason was virtually the only astronomer who could potentially discover this relationship at the time.[20] Thus, we see that Hubble's ambitious program to better understand galactic evolution was revealing unanticipated layers, an attribute of research in the field that continues to this day in modern cosmology.

Immediately following Humason's report in the ApJ came the jointly authored paper, *The Velocity-Distance Relation Among Extra-Galactic Nebulae*. In it Hubble revealed a distaste for taking too long a view of the data. Having stared through the eyepiece of the 100-inch long enough to have realized the depth of faint light still out of reach, both he and Humason likely understood that they were only scratching the surface. Hubble pointed out that only 57 percent of the galaxies in the report had resolved stars in them to aid the verification of magnitudes and distances. In addition to Cepheids there were "irregular variables, helium stars…and P Cygni stars." Unable to resolve stars in the fainter galaxies, he was obliged to derive a mean magnitude formulation. He lamented the large degree of uncertainty due to "the lack of a well-established system of nebular photometry, free from serious systematic errors," and he and Humason recognized that new standards would be necessary, stemming from non-photographic means (such as spectral type) in the pursuit of greater precision in calculating distances.[21]

If Hubble's 1925 paper on external galaxies had opened the door to the broader universe, Hubble and Humason's report in March of 1931 switched on the lights. As Allan Sandage would later write, their remarkable first joint paper "posed…problems that would occupy observational cosmologists for the next 70 years"[22] (Fig. 8.3).

As those in the field began to grapple with issues surrounding the observations, testing, equipment and techniques required to probe deeper into the vast new universe, most of the world was yearning to hear what Albert Einstein thought of the report by the Mount Wilson duo. But even Einstein wasn't sure what it all meant.

Visiting the United States for three months through February 1931 the world-renowned scientist spent much of his time in Pasadena touring the observatory and talking shop with staff at MWO and Caltech. By now Einstein had become extremely interested in the work of Hubble and Humason and he came to Pasadena to view their findings and discuss their theoretical implications. Mayall remembered that the German theoretical physicist "was given an office right across the hall from Hubble," and how strange it was to see "the great man…sitting at his desk" in the observatory offices.[23]

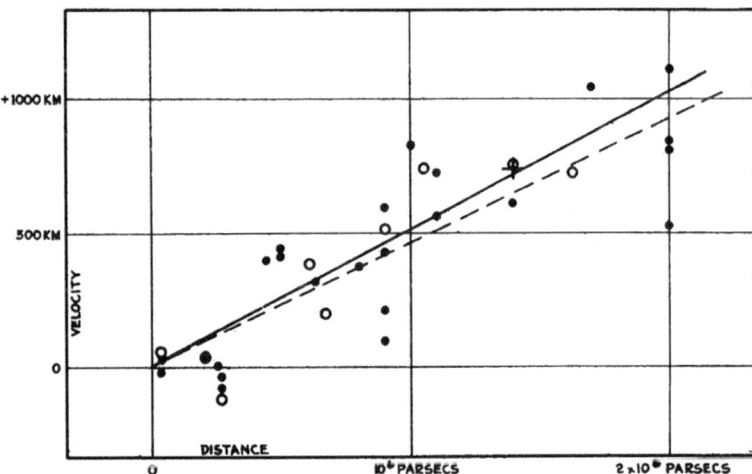

Fig. 8.3 Hubble's chart from the joint 1931 paper indicating the velocity-distance relationship for 24 individual nebulae and nebulae in groups.

On reviewing the evidence Einstein later told the New York Times, "New observations by Hubble and Humason concerning the redshift of light in distant nebulae make the presumptions near [i.e., *make* it appear likely] that the general structure of the universe is not static." When asked about the future of his cosmological constant he commented, "The redshift of the distant nebulae have smashed my old construction like a hammer blow." But he was not sure what the correct solution was to the "Doppler effect," as he referred to the velocities of galaxies, or the science surrounding them.[24]

During his visit, Einstein posed for a photograph with members of the Mount Wilson and Lick Observatories inside the Hale Library at the observatory offices in Pasadena. In the picture Hubble and Humason stood side-by-side at right, beneath the likeness of George Ellery Hale whose vision and persistence had made their work possible. To Hubble's left stood Charles St. John, who in his later years was working on solar solutions to general relativity, Albert Michelson, Einstein (chalk in hand), Lick Observatory director William Campbell and Walter Adams (Fig. 8.4).

Einstein was animated by the presence of Michelson, who he considered to be a titan of physics and who, as a native of Poland, spoke fluent German. At the time, Michelson was reprising his light velocity experiment with an improved design that was expected to yield a more accurate measurement of the universal speed limit.

Fig. 8.4 Einstein and fellow scientists (From left) Milton Humason; Edwin Hubble; Charles St. John; Albert Michelson; Albert Einstein; W.W. Campbell (Lick Obs.); Walter Adams gathered for a photo inside the Hale Library at the Mount Wilson Observatory offices in 1931. (Image courtesy of Observatories of the Carnegie Institution for Science Collection at the Huntington Library, San Marino, California)

One evening Einstein was driven to the property of James Irvine jr., a level plot of land southwest of Irvine (Fig. 8.5). There a 36-inch corrugated steel tube had been assembled over a mile-long stretch by a team of workers using a crane to hoist the heavy 60-foot-long sections of the tube into place. Michelson's design called for air-tight connections and installing inside the tube the optics, mirror supports, slow-motion controls, vacuum pump connections and all other elements of the apparatus. The pump was capable of moving 350 cubic feet of air per minute to a pressure of between two and three millimeters of mercury (approximately five hundred times less than the standard air pressure at sea level). After testing, the joints received a coating of rubber paint to further seal the structure. For best results Michelson designed his new experiment to time the beam of light from its source at the far end of the tube over ten laps, back and forth along the tube. A motorized wheel at one end of the tube featured 32 finely polished and lacquered mirrors set at precise angle. A system of remote controls manipulated the various elements to fine tune the measurement.[25]

Fig. 8.5 Building and area adjacent to Albert Michelson's speed of light experiment, Irvine, California. (Image courtesy of the Observatories of the Carnegie Institution for Science Collection at the Huntington Library, San Marino, California)

The entire assemblage was surveyed by the commander of the US Coast and Geodetic Survey who derived a value of 1,594,259 plus or minus 0.47 mm (5230.76509 feet). Francis Pease then reduced this measurement to the distances between the instruments for a final measurement of 15,999,680.0 mm across the 10-mile path (9.94174 miles).[26]

The measurements were begun on February 19th and were to continue intermittently for several months to determine a good median velocity for light. Einstein watched as Michelson and Pease revved the mirrored wheel up to 730 revolutions per second and then opened a slit to allow light from a powerful arc to shine through. The light flashed off one of the rotating mirrors through the mile-long tube, where it reflected off a flat mirror and back up the tube, reflecting off another of the fastened mirrors and so on until the test was over in a small fraction of a second.

This exercise in scientific measurement at the very foundations of nature was no doubt a highlight of Einstein's stay, impressing him with the scientific might and industriousness of his American counterparts.

Albert Michelson died after an illness in May at age 78 but the impression he had made on Einstein was evident in the latter's praise of his upon hearing of his death:

In carrying out physical experiments Dr. Michelson was an artist of the highest type. His invention of the interferometer, his celebrated attempt to discover the Earth's motion through ether, which was a precursor of the enunciation of the principle of relativity, the diffraction method for determining the diameters of stars, the proof by purely optical methods of the rotation of the Earth, and, finally, the wonderfully exact measurements of the velocity of light – all these feats of experimental skill will be admired as long as mankind concerns itself with physics.[27]

I had the good fortune to become acquainted with him personally before his death, and I asked him why he took such infinite and never-ending pains to measure constants so accurately. He replied that he did it because it amused him. That characterized the man and his work. On one hand, physics to him was an art; on the other, it was pure sport.[28]

Another physicist Einstein was most eager to meet was Richard Tolman at Caltech. He had published two of several recent papers (1929, 1930) on possible dynamic solutions to the cosmic model; the others being Lemaître (1927, 1931), Eddington (1930) and de Sitter (1930).[29]

One of the leaders in physics of the day, Tolman had done important work in developing the relationship between relativity and quantum theory. One idea was a solution that would incorporate Einstein's cosmological constant to yield a static universe of constant radius and matter density yet permit the perceived expansion that Hubble and Humason were measuring. Tolman's 1930 paper derived a solution for the apparent expansion "with a continuous transformation of matter into radiation" allowing for the redshifts of galaxies without a discernible growth in radius.[30]

For this to work however, the universe would require the creation of matter to keep the level of matter density constant as de Sitter had also suggested in his 1930 paper. Einstein found the possibility intriguing.[31] In essence Einstein was considering a steady state model. He likely scrawled the draft of the idea (only recently discovered, on American paper) while sailing with his wife Elsa on the S.S. Deutschland bound for Hamburg, Germany. However, he never used a term to account for the creation of matter in his latest version of the universe and apparently abandoned the idea altogether. Yet as will be discussed later, it would be explored independently nearly two decades later by Fred Hoyle.

Meanwhile, after returning to Berlin in April, Einstein spent a long weekend working on another possible solution which in his view showed more promise. The new model was distinct in that it marked the first occasion on which he put aside his notion that the universe must be static. The evidence amassed at Mount Wilson was sufficient to convince him that his field equations implied a universe that was far more dynamic than he was comfortable acknowledging. His former wariness of expansion was replaced now by a distaste for the so-called primeval atom of Lemaître's hypothesis and he was keen to find a way around it.

On the Cosmological Problem of the General Theory of Relativity was Einstein's first formal attempt to find a solution that took into account expansion. As always, he set out to identify the simplest plausible solution. His thesis was in keeping with the tenets of the 13th and 14th century Bavarian philosopher William of Ockham that sought the simplest of a set of solutions. Einstein was frustrated by the notions of a beginning to spacetime and an ever-expanding universe, both of which used relativity to alter his own perception of spacetime. There was just too much uncertainty in their conclusions. His latest idea was to present a model that fit within the bounds of the field equations while accounting for Friedmann's proposed (and increasingly material) expansion. And to circumvent Lemaître he created a model in which the universe first expands and then contracts.[32] Whether he intended this to occur in a cyclical manner isn't indicated exactly within his analysis, but the idea has been thought of as being Einstein's attempt to find a solution for an oscillating universe. What is clear is that the findings of Hubble and Humason were occupying his thoughts on universal structure.

Much has been said and written about the "discovery" of the expanding universe in the decades since the first formal and practical case was made for it in 1931 at Mount Wilson. Edwin Hubble, who is portrayed variably as an unruly, arrogant, self-indulgent and camera hungry egotist, certainly has his detractors. The arguments of those who can't see their way past the personality of the man hold little value in this discussion, however.

Hubble is seen in some circles as a media prop for an undeniably momentous discovery whose origins predated his arrival on the scene by years, if not actually decades. Still others give him credit for the practical discovery of universal expansion, together with those who suggested it in theory first. Without the real discovery, it must be said, the theory was really rather hollow. The full picture of Hubble's place in the history of the development of Big Bang cosmological theory is as complex as the man himself.

First and foremost, there can be no doubt that the theoretical case was well-established before Hubble and Humason started their mission to confirm it in practical terms. Einstein himself had discovered the possibility in the field equation as soon as he developed them in 1917 but swept it under the rug mathematically. From there a steady stream of physicists had developed the idea and Lemaître explored it explicitly in his 1927 paper.

It has been argued in this book that the claim Hubble was unaware of these theoretical developments prior to 1929 is almost impossible, although the evidence to judge it one way or another is largely circumstantial.

Hubble's immersion in the field of nebular research started formally in 1915 as a student at Yerkes, which included scouring scholarly texts on the subject. How else would he have found the work of Lundmark (and others) in European journals? The accounts of those who worked closely with Hubble, like Humason and Sandage, seem to substantiate the idea that he knew at least some of the theory when he began his assault on universal expansion with Humason in 1928.

Sandage's depiction of Hubble making the great discovery independently, like a stellar archaeologist brushing away layers of stardust until he serendipitously exposed the bones of the ancient universe rings hollow. It seems probable that Hubble knew of the theory and that he formed his analysis from a different angle (namely a study of the kinematic problem of solar motions correlated with galactic velocities) to enable him to promote the discovery on his own terms. This takes nothing away from Hubble's discovery and subsequent work on expansion, however. There can be little doubt that the 1931 report offers the first formal and concrete evidence in support of Big Bang cosmology, although many astronomers had made suggestions in the years prior to that.

Vesto Slipher, the undoubted leader of nebular spectroscopy in the early part of the 20th century and an influencer of both Humason and Hubble, was personally more interested in the velocities in their own right than any possible correlation with distance. The mission of the Lowell Observatory was more centered on planetary research, particularly on Mars and after 1916 Slipher spent a good portion of his time in directing the observatory and focusing on public outreach.

Knut Lundmark had made an extensive attack on the problem but made fatal connection errors in his analysis by including Shapley's globular clusters and imposing a "speed limit" on spectral redshifts that doomed his quest.

Shapley's own claim to a prior suggestion of the phenomenon simply didn't hold water because increasing nebular velocities would be incongruous in his Milky Way universe.

Perhaps the most obvious evidence in support of Hubble is the arrival of Albert Einstein in Pasadena in 1931. He came with one overriding problem in mind, namely, to understand the implications of Hubble and Humason's evidence in connection with relativistic theory. Certainly, the naysayers wouldn't suggest Einstein was drawn there out of a compulsion to create a scene for the press.

Hubble rightly published his report in 1931 with Humason, without whom he could not have made the discovery and definitely could never have pushed into the depths of space to further corroborate his evidence. Humason was inventing new strategies to track galaxies many times fainter than could be seen at the eyepiece of the big reflector in order to capture their spectra. Had Humason pressed Hubble on the issue of partial claim to the discovery, Hubble would probably have acquiesced. But Humason continued to defer to his teammate on the issue both in private and in print throughout his long tenure and in retirement.

At the end of the day, Hubble deserves credit for the momentous discovery. In doing so, he called upon theoretical and experimental evidence years, decades and even centuries in the making. Given his pivotal involvement, partial credit ought also to go to Humason even though the legendary cowboy astronomer didn't attest to it himself.

As is common in the advancement of science, Hubble was standing on the shoulders of those who made practical and theoretical inroads on a multitude of problems that eventually led to its discovery. When the technology was available to advance the idea, he was in the right place, of the right mind, education and approach to make it. These facts are revealed by his arrival at the mecca of astronomical research just as the most powerful telescope in history was going into regular rotation and as the most influential theoretical physicist in modern history was revolutionizing our perception of time and space. The challenges this new viewpoint presented to the whole of scientific research mystified everyone, including Einstein, who in the face of the evidence against his static universe became simply another gawking admirer (although an exceedingly educated one) of the mysteries of space. Given such uncertainty, Hubble's reaction to his own data comes as little surprise, as we shall see later.

Major development of the grounds at Mount Wilson peaked well before 1931 because by that time most of the technological infrastructure – structural, electrical and mechanical – had been installed. One of the events that signaled this fact was the retirement on January 1st of Merritt Dowd, the man largely responsible for the electrical infrastructure. As was written of the 23-year run of the stalwart engineer in the Mount Wilson contribution to the CIWY for that year, his contribution had "been of immense value to the Observatory. Much of the electrical switchboard equipment of Mount Wilson and in the Pasadena laboratories was designed and built by him and the efficiency of the operating plant on the mountain is in large part due to his unfailing skill and ability."[33]

The respect Milton Humason felt for his father-in-law would, to say the least, have been profound. Although he never spoke of it, he likely owed his incredibly good fortune to the decision by Dowd in 1917 to throw his name into the mix to assume the janitorial position at the observatory.

In 1931 the Humasons were celebrating another family milestone. Billy was graduating high school and about to launch his freshman year at Caltech. To support their son's education, the family moved to the cozy Oak Knoll neighborhood near to the campus in order to allow him to walk to and from class. William and Laura Humason also moved nearby.[34]

Milton's extended observing runs on the mountain had left Helen to manage the home and raise Billy largely on her own. As he got older, Billy's busy schedule with school and extracurricular activities kept her busy and distracted during her husband's absences. In the quiet times, she tended her garden and spent hours combing local and national newspapers for articles mentioning Milton, which she kept in a scrapbook to commemorate his career as family lore.[35]

When he wasn't on the mountain, Humason spent his days in his office on Santa Barbara Street analyzing data for his ongoing projects. These included not only the galaxy program with Hubble but also his program with Merrill's on Be-type stars and the classification of stars in the Kapteyn Selected Areas, which by now accounted for something like 90 percent of the catalogued stars down to the 11th magnitude.

His confidence as an observer was still undermined by his inhibition, especially when asked to write or speak publicly, which he was being asked to do increasingly often. His propensity to "say things too simply," as he would later put it, made him feel ill at ease in the company of others who often gathered at Caltech or at the Hubble residence to discuss cosmological matters.[36]

Nevertheless, Humason was revered now, not just for his earthy charm and easy-going collegiality but for his style and skill as an observer. The spectroscopists, who distrusted Hubble, held Humason as one of their own and appreciated him for being able to work with the stardom-seeking Hubble in spite of their very real differences in ideology and behavior.

As his respect and admiration of Hubble increased through the 1930s, "Humason would explain Hubble to the staff," Sandage would later say. "Humason was Hubble's confidant. He was essentially the only person that Hubble ever talked with, and only occasionally he would become less than formal."[37]

By this time, Hubble was pursuing his research and his fame with equal zeal, distilling the riddles of the cosmos for a phalanx of reporters who often called on him for comment. The need to legitimize his decision to betray his father's wishes was likely in his rear-view mirror, but his obsession with the spotlight and almost Daliesque fashioning of his persona, whether innate or learned, remained with him.

He was nevertheless a most serious and dedicated scientist. Over the years his methods would sometimes confound his colleagues for what they lacked in depth and precision, but his genius for finding the holes in research in the field, his openness and objectivity about his findings, and the caution he displayed in the wake of the revelations that they brought forward were remarkably effective qualities.

To be sure, Hubble was tenacious in defending his territory. He was fiercely competitive but just as loyal and respectful to the few people he felt he could trust. By this time Milton Humason had become one of those rare birds, no doubt aided to some degree by his ingrained, almost molecular aversion to the spotlight (Fig. 8.6).

While Humason appeared in just one photo with Albert Einstein during his long visit to Pasadena, Hubble followed the famous physicist around from office to summit, ever alert for a photo opportunity. This was warranted given the occasion. Humason, whose contributions were so vital to Hubble's success, likely kept out of the way to allow Hubble to bask in the light that he himself abhorred. Humason would not have been starstruck by Einstein, although he revered the German for his work and exchanged letters professionally with him after the latter's return to Germany.

In the wake of the landmark 1931 paper, candid photos of Hubble became increasingly scarce. Like Dalí and other mavens of self-promotion he had learned the benefits of a well-cultivated public image. A photo taken of him at the time illustrates this point. Hubble sits at his desk in a suit and tie cradling an enlargement of a galaxy between the fingers of his left hand and his elbow. His fingers gently clutch the edge of the picture while he points his pencil suggestively to a spot on the outer region of the galaxy, as if his work was interrupted for the photo (Fig. 8.7).

There are moments in history when great ambition is met with dismal failure or soaring victory. There are few better exemplars of the latter than the life of Edwin Hubble. By this time the 42-year-old astronomer had made not one but two discoveries, either one of which would've cemented his name in the annals of science. Both discoveries had been predicted for many years (or even centuries in the case of the external galaxies) but the evidence he (and Humason) had presented catapulted him to world acclaim. Comfortable in

Fig. 8.6 Edwin Hubble (Sitting with his back against post) is interviewed by a member of the press as Milton Humason (Standing) stands in the background ca 1930. (Image courtesy of Caltech Archives)

his dual role as an important man of science and media darling he was prepared to press on in the development of his research while safeguarding his

Fig. 8.7 Edwin Hubble poses for a photograph in his office at the Mount Wilson Observatory offices in Pasadena, California ca 1931. (Image from the Edwin Hubble Papers, Huntington Library, San Marino, California)

claim to history. He could now be sure of Humason's participation, which would be so vital to his success.

References

1. Simon Singh, *Big Bang: The Origin of the Universe*, (Harper Perennial 2005), pg. 160
2. G. Lemaître, *The Beginning of the World from the Point of View of Quantum Theory*, (Nature May 1931), Vol. 127, Issue 3210, pg. 706
3. Allan Sandage, *Centennial History of the Carnegie Institution of Washington: Volume 1 The Mount Wilson Observatory*, (Cambridge University Press 2004), pg. 503
4. Harlow Shapley, *Note on the Velocities and Magnitudes of External Galaxies*, (PNAS 1929), Vol. 15, pg. 565
5. Walter S. Adams, Contributions to the Mount Wilson Observatory, (CIWY 1928-1929), No. 28, pgs. 101-102
6. Edwin Hubble, *Proposed equipment for the prime focus of the 200-inch*, mssHUB 48

7. Walter S. Adams, Contributions to the Mount Wilson Observatory, (CIWY 1929-1930), No. 29, pg. 136

8. *Ibid.* pg. 136

9. R.G. Aitken, *The Discovery, at the Lowell Observatory, of A Body That May Be A Trans Neptunian Planet,* (PASP 1930), Vol. 42, No. 246, pgs. 105-107

10. Interview of Nicholas Mayall by Bert Shapiro on 1977 February 13, Niels Bohr Library & Archives, American Institute of Physics, College Park, MD USA, www.aip.org/history-programs/niels-bohr-library/oral-histories/4766

11. Seth B. Nicholson and Nicholas U. Mayall, *Positions, Orbit, and Mass of Pluto,* (ApJ January 1931), Vol. 73, No. 1, pgs. 1-12

12. Gerard P. Kuiper, *The Diameter of Pluto,* (PASP August 1950), Vol. 62, No. 367, pgs. 133-137

13. IAU 2006 General Assembly: Result of the IAU Resolution votes

14. Milton L. Humason, *The Rayton Short-Focus Spectrographic Objective,* (ApJ 1930), Vol. 71, pg. 351

15. Wilbur Bramley Rayton biography, OSA Living History pages, June 12th, 2013, Retrieved September 4th, 2020, https://www.osa.org/en-us/history/biographies/bios/wilbur-bramleyrayton/

16. Milton L. Humason, *The Rayton Short-Focus Spectrographic Objective,* (ApJ 1930), Vol. 71, pgs. 351-356

17. Interview of Nicholas Mayall by Bert Shapiro on 1977 February 13, Niels Bohr Library & Archives, American Institute of Physics, College Park, MD USA, www.aip.org/history-programs/niels-bohr-library/oral-histories/4766

18. *Ibid.*

19. Milton L. Humason, *Apparent Velocity Shifts in the Spectra of Faint Nebulae,* (ApJ 1931), Vol. 74, pgs. 35-42

20. Allan Sandage, *Centennial History of the Carnegie Institution of Washington: Volume 1 The Mount Wilson Observatory,* (Cambridge University Press 2004), pg. 508

21. Edwin Hubble and Milton L. Humason, *The Velocity-Distance Relation Among Extra-Galactic Nebulae,* (ApJ 1931), Vol. 74, pgs. 43-80

22. Allan Sandage, *Centennial History of the Carnegie Institution of Washington: Volume 1 The Mount Wilson Observatory,* (Cambridge University Press 2004), pg. 507

23. Interview of Nicholas Mayall by Bert Shapiro on 1977 February 13, Niels Bohr Library & Archives, American Institute of Physics, College Park, MD USA, www.aip.org/history-programs/niels-bohr-library/oral-histories/4766

24. Duncan Aikkan, *With Einstein at Pasadena: Informal Clinics on Cosmos,* (NY Times Feb. 8, 1931)

25. Walter S. Adams, Contributions to the Mount Wilson Observatory, (CIWY 1929-1930), No. 29, pg. 179-180

26. Walter S. Adams, Contributions to the Mount Wilson Observatory, (CIWY 1930-1931), No. 30, pgs. 218-219

27. NY Times, Thursday, May 14, 1931, *Tribute to Michelson*

28. NY Times, Saturday, July 18, 1931, *Einstein Pays Tribute to Prof. Michelson*

29. C. O'Raifeartaigh and B. McCann, *Einstein's cosmic model of 1931 revisited: an analysis and translation of a forgotten model of the universe,* (Waterford Institute of Technology 2013)

30. R.C. Tolman, *More Complete Discussion of the Time-Dependence of the Non-Static Line Element for the Universe,* (PNAS 1930), Vol. 16, pg. 409

31. Cormac O'Raifeartaigh, Brendan McCann, Werner Nahm and Simon Mitton, *Einstein's steady state theory: an abandoned model of the cosmos,* (Waterford Institute of Technology 2014), pg. 5

32. C. O'Raifeartaigh and B. McCann, *Einstein's cosmic model of 1931 revisited: an analysis and translation of a forgotten model of the universe,* (Waterford Institute of Technology 2013)

33. Walter S. Adams, Contributions to the Mount Wilson Observatory, (CIWY 1930-1931), No. 30, Pg. 221

34. Pasadena City Directory for 1933-1935, Milton Humason 26 Oak Knoll Gardens Dr.

35. Interview with Ann Humason Bernt, 2006

36. Interview of Milton Humason by Bert Shapiro in circa 1965, Niels Bohr Library & Archives, American Institute of Physics, College Park, MD USA, www.aip.org/history-programs/niels-bohr-library/oral-histories/4686

37. Interview of Allan Sandage by Bert Shapiro on 1977 February 8, Niels Bohr Library & Archives, American Institute of Physics, College Park, MD USA, www.aip.org/history-programs/niels-bohr-library/oral-histories/32867

9

Grading Reality On A Curve (1932-1934)

After the announcement of proof that the universe appears to be expanding, Edwin Hubble begins his count of thousands of galaxies and their distribution in the expectation of finding a determination of de Sitter's predicted curvature of space. Milton Humason continues his deep space investigations to provide further evidence of the expansion through the recording of ever greater radial velocities. Comparing Humason's measurements for 35 isolated galaxies with Hubble's magnitude measurements and data in the *Shapley-Ames Catalogue of Bright Galaxies* they compile data for a second report on the expansion. In 1933 Hubble produces a blockbuster report on the distribution of galaxies that will ultimately lead to his greatest failure as an astronomer. His extended leaves of absence and hostility toward those who disagree with him continues to build mistrust and animosity within the leadership at the Mount Wilson Observatory and the Carnegie Institute of Washington. Hubble and Humason issue their second major report on the velocity-distance relationship in 1934. Although he has earned world acclaim, questions about the holes in Hubble's method begin to bubble to the surface, propelled largely by the newly arrived Walter Baade.

The appearance of an expanding universe, revealed for the first time in full practical terms in the joint paper by Hubble and Humason, brought three of the most influential theoretical physicists on the implications of general relativity to Mount Wilson in the years following its publication: Albert Einstein in 1931, Willem de Sitter in 1932 and Georges Lemaître in 1933. Their contributions to cosmology continue to drive the field to this day, but it was de Sitter who would have the greatest influence on Edwin Hubble.

© Springer Nature Switzerland AG 2021
R. Voller, *Hubble, Humason and the Big Bang*, Springer Praxis Books,
https://doi.org/10.1007/978-3-030-82181-4_9

Fig. 9.1 Willem de Sitter with his wife, Eleonora, outside the Kapteyn cottage on Mount Wilson ca 1932. (Courtesy, Ann Humason Bernt family)

Willem de Sitter (Fig. 9.1) was born in the Netherlands in 1872 and served as director of Leiden Observatory from 1919 until his death in 1934. A master mathematician he was one of the first to fully grasp relativistic theory.

An early indication of this insight came in 1913 when de Sitter used a simple thought experiment to resolve the conflict between the "ballistic light" theory of the Swiss physicist Walther Ritz and the "constant velocity of light" theory of Einstein and Lorentz. In *A Proof of the Constancy of the Velocity of Light*, de Sitter focused on how an observer would view the light from a distant system of two stars orbiting a common center of mass.

Special relativity held that the speed of light was constant in all frames of reference to the observer, whereas the ballistic theory of light said that starlight was propelled from the star according to its motion and velocity.

What de Sitter showed was that if ballistic theory was correct, the light from a star in a binary system receding in its orbit from the point of view of an observer would be moving more slowly than the light from the same star while it was approaching the observer in its orbit. The overlapping light approaching the observer at different velocities would give rise to strange anomalies such as seeing the same star in two different positions at once, which was inconsistent with the observations of binary systems.[1]

It was this sort of astute reasoning that endeared the Dutch astronomer to Einstein, and the two started to correspond in 1915 shortly after the latter published his paper on general relativity.

In August of 1916 de Sitter published the first of three papers in the Monthly Notices of the Royal Astronomical Society (MNRAS) describing in detail the "Astronomical Consequences" of gravitation on universal dynamics, a tour de force of relativistic interpretation. By the publication of the third paper in November 1917 (a decade before the arrival of Lemaître on the scene) de Sitter's interpretation of general relativity differed in stark ways from Einstein's.[2]

By 1917 Einstein realized that introducing gravity to the theory of relativity implied a collapsing system. But astronomers believed the universe to be static, so he introduced the cosmological constant to zero out this contraction and maintain a static, spherical universe of finite matter.

But de Sitter suggested a perfectly plausible counter explanation wherein the universe was shaped like a hyperboloid with a positive cosmological constant that expanded space into infinity. In a universe of finite matter, this expansion meant that sometime in a distant future there would cease to be any discernible matter at all. To form his argument, de Sitter pulled matter out of the equation and focused on the inertial aspects of relativity. As an analogy, if we think of the inner stem of an hourglass whose curve is smooth between top and bottom, then by taking sections of that hyperboloidal shape and staging the timeline from bottom to top we can imagine a spherical universe that is first contracting and then expanding with the passage of time.[3]

Although Einstein grudgingly entertained de Sitter's mathematical interpretation of the field equations, he rejected a universe that was devoid of matter. Without concrete evidence leading in one direction or the other, they called a truce and agreed to disagree.

When Lemaître entered the field, he simply merged these theories, suggesting Einstein's cosmology was unstable. After an epic eruption disrupted an incredibly dense composite entity that held the entire mass of the universe the cosmological constant would take over. The result would be a funnel shaped universe whose past looked much like Einstein's and whose future resembled de Sitter's spacetime devoid of matter. It was by this mathematical reasoning the Belgian physicist hit upon the idea of the "primeval atom" reported in a 1927 paper.

The publication of the linear relationship between velocity and distance for galaxies by Hubble and Humason convinced Lemaître he was on to something. Einstein and de Sitter though, remained skeptical. After corresponding, the two legendary physicists published a paper in 1932 called *On the Relation Between the Expansion and the Mean Density of the Universe* where they suggested ideas for the future of observational research. Although the evidence against Einstein's static universe was incontrovertible it by no means implied "a positive curvature of three-dimensional space," the only direct evidence to date "being the mean density and the expansion." Humason's redshifts, they agreed, appeared to be certain while Hubble's measured nebular distances were still "very uncertain." What remained to be answered, they added, was "whether it is possible to represent the observed facts without introducing a curvature at all." Density was dependent upon the "assumed masses" of the galaxies and their relative distances. Furthermore, it was unclear whether or not "all the material mass in the universe [is] concentrated in the nebulae." They concluded saying that these issues would be resolved only when sufficiently precise data was available to indicate the sign and magnitude of the curvature.[4]

At this point Einstein withdrew from the scientific and media circus that surrounded the subject of universal expansion. Instead, he set his sights on an ultimately futile attempt to unify relativity and quantum mechanics.

Willem de Sitter died two years after the publication of the paper with Einstein, having published 15 papers on the expansion since Hubble's 1929 report on the appearance of the linear ratio of velocity to distance for galaxies. Unlike Einstein, de Sitter became obsessed with what the new evidence meant for relativity. He could hardly have expected to become embroiled in a scientific dust up with his former protégé, Edwin Hubble.

In May of 1930 de Sitter published an expansive report *On the Magnitudes, Diameters and Distances of the Extragalactic Nebulae and Their Apparent Radial Velocities* in the Bulletin of the Astronomical Institutes of the Netherlands (BAIN) in which he attempted to coalesce the practical and theoretical evidence on expansion into a broad discussion of the facts that were more or less understood and those that remained uncertain. The paper was a model of scientific reasoning and expression, displaying de Sitter's comprehensive understanding of Einstein's theory. On the basis of Hubble's 1929 report on the expansion de Sitter concluded that the evidence appeared to show a static universe was not possible, and cited Lemaître's 1927 paper (which he had just heard of via Eddington) as an "ingenious solution."[5] Upon publication, de Sitter sent a copy of the report to Hubble in Pasadena, where he was intending to visit in a few months, probably thinking it would be met with a measure of appreciation and conversation. What he received was a suspicious, albeit respectful rebuke from Hubble over what the latter perceived to be the former's attempt to take credit for the discovery of expansion.

Allan Sandage would later refer to de Sitter's failure to cite Hubble as the "discoverer" specifically as "callous," pointing in particular to this casual remark by de Sitter:

> It has been remarked by several astronomers that there appears to be a linear correlation between the radial velocities and the distances.[6]

Sandage's motives are unclear, but a thorough read of de Sitter's report reveals little or nothing of a callous nature or anything that would lead one to believe he was trying to steal Hubble's thunder.

The title of the paper is the first indication of its contents, a comprehensive discussion of the theoretical and practical evidence. At no point did de Sitter lay claim to a discovery, despite being one of the first to draw attention to the theoretical connection between general relativity and expansion.

Hubble himself related this to his Dutch mentor in his angry letter of reply several weeks later, saying the idea had been "in the air for years."[7]

De Sitter even draws a correlation between the formulaic conclusions in the paper with Hubble's from the groundbreaking 1929 report, the title and contents of which were meant explicitly to highlight the fact that a discovery had been made. In contrast, de Sitter's paper does none of this, even drawing more attention to Lemaître in the end than to his own work. Never mind that no one with a view to coopting the discovery would make a statement like the one above, which seems to promote the essence of the development leading to Hubble's practical breakthrough.

Hubble's stated concern was the "formulation, testing and confirmation" of the linear velocity-distance relationship be considered as "a Mount Wilson advance" and that he was "deeply concerned in its recognition as such."[8] In other words, he believed he alone was responsible for the discovery, and he wanted de Sitter (and everyone else) to shout it from the mountain tops.

Whatever de Sitter thought of Hubble's outburst, it did not take him long to placate the Pasadena prima donna. In a letter dated September 23rd Hubble wrote that he and Humason were "both deeply sensible of [de Sitter's] gracious appreciation of the papers on velocities and distances of nebulae." Hubble revered de Sitter and even his "angry" letter from August of that year began with "Many thanks for the Bulletins from the Observatory at Leiden…"[9] Any remaining uncertainty was put to rest during de Sitter's visit.

Despite his incredible success, Hubble remained irascible in the face of perceived slights throughout the 1930s. In these disputes Hubble may have felt victorious in battle, but he was losing the war of collegiality. To many in the field who witnessed his insubordination, and his combativeness in private and in the press, Hubble was an astronomer by trade and a boxer by nature.

Routinely monitoring the work, publications and social competence of all members of the staff at the observatory, Adams was becoming increasingly exasperated with Hubble's quarrelsome and self-serving attitude. Nonetheless his popularity was an excellent public relations tool, and Adams was conscious of the need to promote the facility as leading the world in astronomical research. He raised Hubble's salary to the third highest level on the staff behind himself and Seares. As Humason probed deeper into space and required ever-longer exposures to nab usable spectra Adams awarded lengthy runs on the 100-inch, rather to the annoyance of other members of the night staff seeking to advance their own research programs.

By 1934 Hubble's notoriety and wide-ranging knowledge of history and current affairs had earned him entry into the starry world of Hollywood and literature. Aldous Huxley and Douglas Fairbanks sr. were among those he called friends, and he and Grace were frequent guests at celebrity parties. He was becoming a celebrity in his own right and was relishing every moment of it.[10]

For Hubble the ultimate peace came in his daily routine, walking to and from the office, conferring with Humason and others in his department or within the halls of Caltech, and telling Grace about his day. He was at his best in researching, writing and lecturing on the subject of the cosmos. His polish and command of the material made him a highly sought-after lecturer at Carnegie, Princeton, the RAS and just about anywhere that astronomy was studied. He was developing a theatrical sense of story that captivated audiences.

The events transpiring within the astronomy world were taking place against a backdrop of unprecedented economic decline in a world clinging to hope for a way out of the morass. Mistrust of the global political and business establishment was virtually universal, and few signs of a brighter future were evident.

As economic depression gripped the United States, two massive construction projects were completed, the San Francisco Bay Toll Bridge in 1929[11] and the Golden Gate Bridge in 1937 which jointly served to link Oakland, San Francisco and Marin County. These projects had created thousands of jobs but did little to help those struggling in the middle part of the country.[12]

In 1932 the economic news reached its nadir with the Dow Jones Industrial Average at 41.22, the lowest level of the depression, and 23 percent of American's were out of work.

President Herbert Hoover, the former Commerce Secretary under Calvin Coolidge and a darling of progressives for his work as the country's "food czar" under Woodrow Wilson during the Great War, struggled to reverse the effects of the downturn. Increasing numbers of people filled the shantytowns known as Hooverville's as ever more defaulted on home mortgage loans. The establishment of the Reconstruction Finance Corporation to bolster local governments and financial institutions, the Home Loan Bank Act and Glass-Steagall Act to provide backing for national banking systems through the Federal Reserve, and the Emergency Relief Construction Act for the advancement of public works projects were all introduced in the last year of Hoover's presidency.[13]

It was too little too late for Hoover whose lackluster performance in attempting to avert economic meltdown led to his crushing defeat in the election of 1932 by the governor of New York, Franklin Delano Roosevelt.

Following one of the most stirring presidential speeches in the nation's history which produced the historic line, "the only thing we have to fear is fear itself," Roosevelt launched into a 100-day offensive on the economy that included the extension and refinement of Hoover's policies. This set a precedent for all future presidents in the early months of their administrations.[14]

In Europe the worldwide depression was itself manifesting in different and sometimes more sinister ways.

In the Soviet Union, famine resulting from poorly administered and executed planning under Joseph Stalin's Collectivization policy directly or indirectly caused the deaths of up to twelve million citizens of Ukrainian S.S.R. There remains dispute among academics as to Stalin's intentions regarding this situation. Inside Ukraine it is referred to as Holodomor or "death by famine" and many believe it was perpetrated as a genocide on the Ukrainian people. Under Collectivism, landholders were stripped of their land and made to work for peasant wages. Those who opposed these, and other new policies of the Soviet government were exiled to Siberia. Those trying to escape the regions beset by famine were summarily shot.[15]

In Germany, 30 percent of the workforce was unemployed, and populism was rising as the Nazi Party gained a majority in two party elections while the

Austrian-born and recently naturalized German citizen Adolf Hitler jockeyed for the Chancellorship of the new Reich.[16]

A failed attempt on the life of Japanese emperor Hirohito became the pretext for conflict with China in late January that led to the second Sino-Japanese War five years later. Acting as arbiters of peace, the League of Nations attempted mediation.

Meanwhile Pope Pius XI was meeting at the Vatican with the Italian dictator Benito Mussolini, whose seizure of Italy in the 1920s had served to inspire other strongmen across the continent.

Elsewhere pacifists and progressively minded individuals were pushing back against an increasing wave of nationalism in various countries.

In India, the barrister turned activist Mohandas Gandhi was arrested for his involvement in marches against the British imposition of a salt tax and other non-violent uprisings and incarcerated in Poona prison, but he was confident that British rule was coming to an end.[17]

To many millions of people around the world, the USA continued to shine as a beacon of hope and a symbol of freedom. Hattie Caraway of Arkansas was elected the country's first female Senator and female track and field star, Babe Didrikson dominated at the 1932 Summer Olympic Games in Los Angeles (Didrikson later won 10 major championships as a member of the LPGA).

The world mourned the kidnapping and death of Charles Lindbergh jr., the infant son of American flying legend Charles Lindbergh and his wife and fellow flyer Anne Morrow Lindbergh. The couple's 10-week ordeal captured the hearts of millions worldwide before ending tragically with the youngster's body being found near the family home in Hopewell, New Jersey.[18]

On a brighter note, in May 1932 Amelia Earhart took off in her Lockheed Vega 5B solo from Newfoundland aiming for Paris to emulate Lindbergh achievement five years earlier. Although strong headwinds obliged her to set down in Northern Ireland after a few minutes short of 15 hours her flight was another advance for women's equality.[19]

For those who could afford it, there were small wonders at play in the American cinema. *Tarzan the Ape Man* opened on March 25th starring Olympic gold medal swimmer Johnny Weissmuller in the title role alongside Maureen O'Sullivan, and the Disney animated dog Goofy made his screen debut opposite Mickey Mouse in the short titled, *Mickey's Revue*.[20] The reopening of the Palace Theater in New York City as a cinema signaled the end of the age of vaudeville.[21]

If you had a radio, you could tune into Buck Roger's in the 25th Century or listen while Babe Ruth famously called his homerun against the Chicago Cubs in the fifth inning in the third game of the 1932 World Series.

Also, the Radio City Music Hall in New York City opened for its first performance on December 27th proving once again that with enough will and determination one "cancan" accomplish just about anything.[22]

After suffering for the better part of two years with an undiagnosed spinal disorder, 77-year-old Eastman Kodak founder George Eastman committed suicide with a gunshot to his heart. The entrepreneur and philanthropist had helped popularize photography through the development of "roll film" cameras that could be used by the masses. For years the research laboratory at Kodak had contributed greatly to the work at MWO, as well as other areas of science and scientific facilities through the development of highly sensitive photographic plates. Kodak plates capable of producing exposures into the red and infrared parts of the spectrum were being used by virtually every member of the spectroscopy department at the observatory. His suicide note read simply, "To my friends: my work is done. Why wait?"[23]

While the Hubbles were celebrating the publication of their friend Aldous Huxley's new novel *Brave New World* (a dystopian futuristic look at a genetically altered human state where technological advances are supposed to produce a utopia) and the Humasons were facilitating their son Billy's college career at Caltech, two momentous events occurred in 1932 with implications for the future of Mount Wilson Observatory.

The first was the arrival in October 1931 of Walter Baade as one of the new members of staff. When the brilliant Hamburg Observatory astrophysicist visited Pasadena in 1927 Adams realized he would make a valuable member of the Mount Wilson team, but this took some time to arrange.[24]

A master of both physics and observational astronomy, Baade was part of a new group that included Olin Wilson and Rudolph Minkowski who bridged the gap between the old observationally oriented astronomers and the new theoretically based ones who would start to join the ranks at Mount Wilson and Caltech after the Second World War. He came with a desire to use the 100-inch for his own photometric research of galaxies and star clusters. By that time, he was already assembling the evidence for what (13 years later) would be his groundbreaking discovery of stellar populations.[25] He was also eager to study the effect of interstellar dust on the magnitude and color of starlight and the measurement of distances of stars. His research would end up shrinking Shapley's "universe" and enlarging Hubble's.

An inveterate practical joker who walked with a noticeable limp due to a hip defect at birth, Baade made an immediate impact at Mount Wilson and at Caltech where he began conversations with several physicists and a collaboration with Fritz Zwicky, a Swiss mathematician who was making some interesting investigations on the nature of matter.

It wasn't long before Baade was complaining about the inordinate amount of time given to the Hubble program, which depended on obtaining data and then repeating the exposure run to corroborate evidence for the deep space nebulae. The largest proportion of the most precious observing on the 100-inch (when moonlight was not present to mask faint sources) was being allocated to Humason's deep space spectroscopy.

Hubble was eager to investigate the "outstanding feature" from his 1929 paper, namely "the possibility that the velocity-distance relation may represent the de Sitter effect." The increasing redshift in spectra of faint galaxies might be related to expansion by a "general curvature of space" rather than the primordial eruption of condensed matter. If time slowed with increasing distance along the curve of space, an observer would perceive a passage of time.[26] He hoped that by surveying a large number of galaxies to be able to extrapolate from their magnitudes and distribution something approximating the curvature of space.

Baade objected on the grounds that there was still too much research to be done on the evolution of the galaxies. As he often repeated to Sandage and others, Baade believed that the key to understanding universal structure lay within the dynamics of the galaxies. This caution was warranted. It was heavy material for even the most detailed of researchers, but Hubble, perhaps overestimating his abilities in the wake of his recent success, felt he could accomplish the task and with it garner a fourth important contribution to science.

The second significant occurrence of that year was the publication of the *Shapley-Ames Catalogue of Bright Galaxies* (SAC) by Harvard Observatory director Harlow Shapley and Adelaide Ames, a young astronomer working as his research assistant.[27]

One of the most promising in a sea of promising women in the field, Adelaide Ames was a member of the American Astronomical Society (AAS) and best friend of Cecilia Payne-Gaposchkin. Ames received her undergraduate degree from Vassar College in 1922 and her graduate degree in astronomy at Radcliffe in 1924. Like her mentor, Shapley, she had been interested in becoming a journalist but after looking in vain for work she accepted a research assistant post at Harvard. Tragically, she drowned while on holiday when her canoe capsized on Squam Lake south of the White Mountains in central New Hampshire in June of 1932. The daughter of US Army Colonel T.L. Ames, she was buried at Arlington National Cemetery.[28]

The SAC included 1,249 bright galaxies taken from the NGC and its Index Catalogue (IC), which were the combined efforts of William, Caroline and John Herschel in the 18th and 19th centuries and later work by the Danish-born astronomer John Dreyer (Chapter 1).

In an interesting sign of Shapley's capitulation to Hubble's victory in the external galaxy competition the SAC listed not only the position, brightness and size of a galaxy but also its Hubble classification. Hubble had begun to use his own classification for the nebulae in his initial report on the expansion. The only possible snub from Shapley was his use of the word "galaxy" which Hubble refused to accept, preferring "extra-galactic nebula." Hubble in turn returned the snub, paying Shapley a backhanded compliment meant to diminish his groundbreaking work on galactic structure in the 1910s telling those around him that the new publication was the best work the Harvard director ever did. In their text, Shapley and Ames asserted that galaxies were not evenly distributed between the northern and southern hemispheres. Along with the NGC, the SAC became one of the most trusted and widely used resources for astronomical research.[29]

Among the first to use the new catalogue were Hubble and Humason. With the reports of his enormous redshifts, everyone was anxious to see what new depths of space Humason would come down from the mountain with after his runs at the giant reflector. By that year the largest redshift he felt sure enough of to report to Carnegie was 24,000 kps for a cluster in the constellation of Gemini. The larger plan was now in place. Using the SAC as a guide, Hubble and Humason would fill in the gaps in their velocity-distance relation to further corroborate the evidence of expansion. Feeling the momentousness of the occasion, Humason was eager to continue his breathtaking probing until he reached the limit of the 100-inch's capability.

Hubble, meanwhile, had counted some six thousand nebulae on photographic plates in a survey to determine their distribution. In need of the plates for his own research, Baade joined Hubble in this endeavor and together they made discoveries in the constellations of Leo and Coma, among others. These investigations were beginning to unveil the enormous scale of the universe in unprecedented detail for the time. A cluster of 300 nebulae in Leo, for instance, was identified in an area of sky not much larger in diameter than the Moon at perigee. In addition to the counts, distribution and magnitudes of the nebulae, Hubble was working internally on nebulae closer to home, seeking objects from which he and Humason could add layers of detail to the investigation of expansion. A collection of 140 star clusters in Andromeda were found by Humason to have velocities similar to M31; strong evidence that they were constituents of that galaxy.[30]

While Hubble grappled with his galaxy counts, Humason attended to a number of long-term projects that needed to be wrapped up.

Working alongside his friend and fellow spectroscopist Paul Merrill and Mount Wilson computer Cora Burwell, Humason helped complete Merrill's survey of B-type stars. Of the 233 stars they had found during their nearly decade-long survey, all but 11 proved to be B-type. The massive blue hydrogen rich stars (up to 16 times the mass of the Sun) were among the fascinations Merrill had in the field, and the team updated their findings several times over the course of the project.[31]

As of the 1932 report to Carnegie, Humason had finished the classification and analysis of the distribution of 4,066 stars in the Kapteyn Selected Areas, a monumental effort that would contribute to research on stellar evolution for years to come.[32]

Innovation was king at Mount Wilson and Caltech at the time. Advances in technology were occurring nearly at the same rate as discoveries in the field, often aiding and abetting progress in every department of research. Among these improvements were the short-focus camera, the thermocouple measurements by Nicholson and Edison Pettit of the Moon and planets, Michelson's interferometer and the Eastman Kodak advanced emulsion plates. A photoelectric photometer created by Stebbins and Whitford was undergoing improvements including a correcting lens to widen star fields for both direct and spectroscopic research would soon be in regular use at the observatory.

One of George Hale's less visible institutions was the weekly Journal Club meeting that he started years earlier to keep members of the various departments at MWO and Caltech apprised of the broader work underway at the two facilities. Hubble also invited members of both institutions to his home on Woodstock Road for a monthly meeting of cosmological minds. The same cast was often on hand and included Humason, Baade (starting in 1932), Tolman and Zwicky, and occasionally Adams and people from Lick Observatory or visitors like Mayall and Wright.

As his star continued to climb Humason was often summoned to voice his thoughts and opinions on spectroscopy, the redshift phenomenon, optics and the apparent expansion. On the heels of his joint paper with Hubble on the velocity-distance relation he had achieved a new level of respect with the staff at both Mount Wilson and Caltech. His workman-like attitude and humility in relation to his ability as an observer pioneering in deep space were refreshing and appreciated by all who knew him. To Hubble's detractors, Humason was a perfect antidote to his partner in the nebular classification program.

Throughout his career, Humason shied away from offers to write anything speculative, preferring to write only on the subjects he was studying, wrestling as he had done for many years with internal conflicts regarding his educational qualifications. Prompted by friends and colleagues he took his turn speaking with staff at the Monday sessions and at gatherings hosted by Hubble. Conspicuously absent from the meetings at Hubble's house was

Adrian van Maanen, whose rift with Hubble was still ongoing. Van Maanen was more than happy to return the favor at his own parties.

While Humason continued to grapple with self-doubt, Hubble continued to ruffle the feathers of the top brass of Carnegie and Mount Wilson, which by now included John C. Merriam as CIW president. Riding a wave of notoriety and press attention he was getting and shining on the organizations, Hubble had begun to take liberties with his time away shortly after his marriage to Grace ten years earlier. During their latest expedition the couple spent an extra month touring Europe in the spring of 1934 (Fig. 9.2), visiting friends while Edwin spoke at various events and attended a meeting of the International Council of Scientific Unions in Brussels. Adams reluctantly covered for his errant star but was compiling mental notes of the various ways in which he continued to flaunt his notoriety and bully others in the field.

Fig. 9.2 Passport photograph of Grace Burke Hubble and Edwin Powell Hubble from 1934. (Image from Edwin Hubble Papers, Huntington Library, San Marino, California)

In his biography, Christianson noted two instances that illustrate this latter point backed up by correspondence between Adams and Merriam. While in Europe in 1934 Hubble sent Merriam a "scathing letter" berating him for allowing the publication of an article Hubble said he had not authorized for the "Scientific Monthly." In fact, as the Carnegie president later learned through some back-checking, Hubble had not only authorized this publication but had agreed to supply "illustrations" to complement the piece, all the while planning to publish it elsewhere. As Merriam said to Adams privately, the brashness and verbosity of Hubble represented "a very unfortunate attitude of mind."[33]

The second incident came two years later as Hubble was preparing to give the Rhodes Lectures at Oxford in 1936. He had ignored protocol and gone over Adams's head to ask Merriam to extend his salary for the trip he had taken two years earlier. During that three-year period, he had spent "eleven months" away from the observatory, eating up CIW funds meant for his research while touring Europe with Grace and giving the occasional lecture. Merriam wrote to Adams denouncing Hubble's failure to acknowledge the institution that helped pay his way on these trips. In addition to his presumptuousness, the CIW president reminded the director that Hubble had never responded to requests to come to Washington to deliver a lecture for the CIW. Adams replied that Hubble seemed to suffer "an extreme form of individualism and personal ambition" which combined with "obtuseness regarding his relations to the Institution and other scientific men." Nevertheless, both ultimately gave Hubble his way so long as he agreed to deliver a lecture for the CIW.[34]

Hubble clearly thought he had good reason to feel entitled to special treatment given the notoriety his work had brought the observatory. His attitude and his lifestyle were offensive to many of his peers. But during the first 15 years of his career, he worked hard to establish his line of research and its import, going deep in search of answers to longstanding issues. It was a process which required years of tedious work and, unlike his major breakthroughs, delivered comparatively little to the field. Perhaps blinded by ambition, he would carry that work forward into the final years of the decade.

Not long after their return from Europe in the summer of 1934, Edwin learned of the death of his mother, Virginia Lee Hubble. The 70-year-old matriarch had been living with his brother Bill in Louisiana and died at home of a heart attack on the heels of a protracted illness. Having contributed little to her welfare and had almost no contact with his siblings since moving to California, Hubble reluctantly attended his mother's funeral in Springfield, Missouri.

In August 1933 Hubble published the longest and most detailed paper of his entire career in the form of a 70-page analysis of the distribution of 44,000 galaxies from 1,283 Eastman Kodak plates taken northward of a declination of −30 degrees by the 60-inch and 100-inch telescopes.

Unable to resolve stars in the overwhelming majority of the galaxies within reach of the 100-inch at that time, Hubble had to rely on a "mean luminosity scale" using novae and Cepheid variables to get the scale he needed to underscore his conclusions. He understood this scale would likely change as newer and better equipment (most notably the 200-inch) improved the resolution of stars in more distant systems, but the mean luminosity measures were vital to extending the expansion program that he and Humason were working on with a measure of certainty.

With that objective in mind, his paper sought to cross examine his findings in numerous ways, drawing on distribution counts by Shapley for 100,000 galaxies and comparing and contrasting his findings for the distributions in both the northern and southern hemispheres. Compiling his findings in charts, graphs and maps he thoroughly exploited the data to show the relative abundance of galaxy clusters toward the poles of the Milky Way and how this tailed off toward the so-called zone of avoidance, the 10-to-20-degree wide equatorial strip where galaxies were almost totally absent on the plates.[35]

This was, as Mount Wilson and Palomar stalwart Allan Sandage would later express it, "a brute-force approach to cosmology" but the overwhelming amount of data left Hubble and Humason reasonably reassured that the corresponding increase in the velocity-distance relation was generally correct, in scope if not in scale.[36]

The lack of substantive data for the southern sky made talk of moving the 60-inch to a suitable location in South America a momentary topic of discussion at the higher levels of the observatory but the constraints of the Great Depression ruled the idea out. Hubble had promoted newly graduated Nick Mayall to lead such a sky survey if it occurred, but Mayall had gained a firm position at Lick where he was designing a new ultraviolet spectrograph for the 36-inch Crossley reflector (Fig. 9.3), the completion of which would enable him to undertake nebular research on specific bright galaxies within range of that telescope.

Humason, who was in frequent contact with the young astronomer, brought the idea of a collaboration to the attention of Hubble and Adams. The two observatories had a history of sharing data and personnel for various investigations going back to the early days when George Hale and William Campbell ruled the facilities. The idea was to have Mayall use the 36-inch Crossley reflector to gather nebular data for galaxies beyond the northern

Fig. 9.3 Nicholas Ulrich Mayall and the Crossley reflector at the Lick Observatory. (Image courtesy of UCSC Special Collections & Archives)

limits of declination available to the 100-inch telescope (which, as noted earlier, was constrained by the design of its mounting). Mayall, an admirer of the dynamic Mount Wilson duo, was more than willing and Adams and Campbell struck a deal in which the observatories would partner on the remaining program.[37]

The Velocity-Distance Relation For Isolated Extragalactic Nebulae penned by Hubble and Humason was read by Hubble before the National Academy of Sciences (NAS) on April 23rd, 1934, two days before the Hubbles boarded the USS Manhattan heading for England. This paper established the velocity to distance ratio to a far greater limit. Hubble effectively fused the conclusions he had made in his lengthy compendium on galaxies in clusters a few months previously with Humason's findings on 35 new velocities for isolated galaxies and another 50 previously measured. In fact, they had 94 velocities but rejected nine because Humason was approaching the edge of his comfort zone in regard to the resolution of the spectra of the fainter galaxies he was going after and would not allow them to be used. On their chart the 85 galaxies fell on a slope that extended to the new record of 39,500 kps for a cluster in the constellation Boötes.[38]

If van Maanen was dreary over his ill-fated attempts to thwart Hubble's claims about expansion (or indeed the existence of galaxies outside the Milky Way, generally) this latest publication couldn't have helped his sleep patterns. His long dispute with Hubble was soon to come to a head in an episode that will be detailed in the next chapter.

As Hubble sailed across the Atlantic to take more than his share of downtime away from the observatory Humason attempted to settle into his new life as a noted astronomer and spectroscopist. After years of the observatory's engineers repeatedly improving equipment to enable him to keep pushing the giant reflector to its limits the 43-year-old explorer was growing weary.

Nevertheless, he kept going and published a report in September detailing a faint nebula in a cluster in Boötes that both Hubble and Baade had recently found on direct photographs. With Hubble away, Humason asked Baade to measure its magnitude photometrically. The latter estimated magnitude 17.5, noting that there were as many as a hundred still fainter nebulae within 5 minutes of arc of their target. As an example of the lengths Humason was willing to go to firm up his data he used six different wavelength measurements to get an accurate weighted mean velocity for the nebula (normally two were used at the time). The cyanogen band gave the largest redshift and a velocity of 40,751 kps while at the low end the G band gave 38,125 kps. Using these together with measurements of the H and K lines he arrived at a mean velocity of about 39,200 kps.[39]

By now Baade, a close friend of Humason who shared his penchant for practical jokes and storytelling, was filling his famous friend's ear with his ideas about the folly inherent in Hubble's approach to research. While the overall expansion may well be genuine Baade suggested Hubble's bull-in-a-china-shop approach lacked the fine details required to weed out some of the uncertainties in his findings. Humason would become increasingly trapped in this developing animosity between Baade and Hubble. In embarking on the project with Hubble he had thrown his weight behind the expansion problem and the two had forged a healthy mutual respect.[40] In Baade, Humason had someone he could confide in on the depths of the work with Hubble, and with whom he shared common interests and personality traits, but Baade's exacting scientific approach represented a threat to Hubble, and Humason knew it. In coming years Baade's work on the physical nature of the galaxies, and the stars within them, would impact the work of Hubble and Humason both positively and negatively and the relationship would fray.

Ultimately Humason could fall back on the fact that the measurements he had made for the galaxy classification and expansion program were accurate in relative terms and as such would stand for all time. This had been spelled out by Einstein in his press statements after his visit to Pasadena in 1931. That assurance at least was satisfying, and it underscored the very real and fundamental nature of the expansion.

References

1. W. de Sitter, *A proof of the constancy of the velocity of light,* in: KNAW, Proceedings, 15 II, 1912-1913, Amsterdam, 1913, pgs. 1297-1298
2. W. de Sitter, Assoc. R.A.S., *On Einstein's Theory of Gravitation, and its Astronomical Consequences,* (MNRAS 1916), Vol. 76, No. 9
3. W. de Sitter, On the relativity of inertia. Remarks concerning Einstein's latest hypothesis, in: KNAW, Proceedings, 19 II, 1917, Amsterdam, 1917, pp. 1217-1225
4. A. Einstein and W. de Sitter, *On the Relation Between the Expansion and the Mean Density of the Universe,* (PNAS 1932), Vol. 18, No. 3, pgs. 213-214
5. W. de Sitter, *On the magnitudes, diameters and distances of the extragalactic nebulae and their apparent radial velocities,* (BAIN May 1930), Vol. 5, No. 185 pgs. 157-171
6. Allan Sandage, *Centennial History of the Carnegie Institution of Washington: Volume 1 The Mount Wilson Observatory,* (Cambridge University Press 2004), pg. 503-505
7. Edwin Hubble letter to Willem de Sitter, August 21, 1930, mssHUB616
8. *Ibid.*
9. Edwin Hubble letter to Willem de Sitter, September 23, 1931, mssHUB 617

10. Gale E. Christianson, *Edwin Hubble: Mariner of the Nebulae,* (University of Chicago Press 1995), pgs. 259-260

11. *San Mateo Bridge Opened by Coolidge: Autos Crowd it Full,* (San Jose News March 4, 1929), Retrieved Feb. 10, 2021, https://news.google.com/newspapers?id=AD AiAAAAIBAJ&sjid=_aMFAAAAIBAJ&pg=3006,6299044

12. Federal Highway Administration, *Two Bay Area Bridges – The Golden Gate and San Francisco Oakland Bay Bridge,* (www.fhwa.dot.gov 2017), Retrieved Sept. 22, 2020, https://www.fhwa.dot.gov/infrastructure/2bridges.cfm

13. J.A. Swanson, S.H. Williamson, *Estimates of national product and income for the United States economy, 1919-1941,* (Explorations in Economic History 1972), Vol. 10, Issue 1, Autumn 1972, pgs. 53-73

14. Alonzo L. Hamby, *Man of Destiny: FDR and the Making of the American Century,* (Basic Books 2015), pg. 175

15. S.O. Pidhainy, Prof. I.I. Sandul and Prof. A.P. Stepovy, *The Black Deeds of the Kremlin: A White Book,* (The Basilian Press 1953), Vol. 7: Book of Testimonics, pgs. 187-308

16. Timeline of German Economic Status 1910-2003, Commanding Heights, https://www.pbs.org/wgbh/commandingheights/lo/countries/de/de_economic.html, Retrieved June 18, 2019

17. Herbert P. Bix, *Hirohito and the Making of Modern Japan,* (Perennial 2001), pg. 250 Rajmohan Gandhi, *Gandhi: The Man, His People and the Empire,* (Haus Pub. 2015), pgs. 305-326

18. Time, *Crime: Never-to-be-Forgotten,* Time Magazine, Monday, May 23, 1932

19. Smithsonian National Air and Space Museum, *Amelia Earhart Solos the Atlantic,* (si.edu), Retrieved June 5, 2019, https://pioneersofflight.si.edu/content/amelia-earhart-solos-atlantic

20. The Year 1932, (filmsite.org), https://www.filmsite.org/1932-filmhistory.html

21. Richard F. Shepard, *Palace Theater Spruced Up, Ready to Light Up,* (NY Times, March 28, 1991), Section C, pg. 11

22. NY Times October 3, 1932, *Native Art to Lead in New Music Hall*

23. David Lindsay, *George Eastman: The Wizard of Photography,* (pbs.org), American Experience, Retrieved July 2, 2019, https://www.pbs.org/wgbh/americanexperience/features/george-eastman/

24. Donald E. Osterbrock, *Walter Baade: A Life in Astrophysics,* (Princeton University Press 2001), pgs. 24-48

25. W. Baade, *The Resolution of Messier 32, NGC 205, and the Central Region of the Andromeda Nebula,* (ApJ 1944), Vol. 100, pgs. 137-146

26. Edwin Hubble, *A Relation Between Distance and Radial Velocity Among Extra-Galactic Nebulae,* (PNAS 1929), Vol. 15, No. 3, pgs. 168-173

27. Harlow Shapley, Adelaide Ames, *A Survey of the External Galaxies Brighter than the Thirteenth Magnitude,* (AHCO 1932), Vol. 88, pgs. 41-76

28. *Research Astronomer Lost by Drowning,* (Popular Astronomy 1932), Vol. 40, 1932, pgs. 448-449

29. Harlow Shapley, Adelaide Ames, *A Survey of the External Galaxies Brighter than the Thirteenth Magnitude,* (AHCO 1932), Vol. 88, pgs. 41-76

30. Walter S. Adams, Contributions to the Mount Wilson Observatory, (CIWY 1931-1932), No. 31, Pg. 140

31. Paul W. Merrill, Milton L. Humason, and Cora G. Burwell, *Discovery and Observations of Stars of Class Be: Second Paper,* (ApJ 1932), Vol. 76, pgs. 156-183

32. Milton L. Humason, *Spectral Types of Faint Stars in Kapteyn's Selected Areas 1-115,* (ApJ 1932), Vol. 76, pgs. 224-274

33. Gale E. Christianson, *Edwin Hubble: Mariner of the Nebulae,* (University of Chicago Press 1995), pgs. 220-223

34. *Ibid.* pgs. 253-255

35. Edwin Hubble, *The Distribution of Extra-Galactic Nebulae,* (ApJ January 1934), Vol. 79, pgs. 8-76

36. Allan Sandage, *Centennial History of the Carnegie Institution of Washington: Volume 1 The Mount Wilson Observatory,* (Cambridge University Press 2004), pgs. 514-516

37. Interview of Nicholas Mayall by Bert Shapiro on 1977 February 13, Niels Bohr Library & Archives, American Institute of Physics, College Park, MD USA, www.aip.org/history-programs/niels-bohr-library/oral-histories/4766

38. Edwin Hubble and Milton L. Humason, *The Velocity-Distance Relation for Isolated Extra Galactic Nebulae,* (PNAS 1934), Vol. 20, No. 5, pgs. 264-268

39. M.L. Humason, *The Apparent Velocity of a Nebula in the Boötes Cluster No. 1,* (PASP 1934), Vol. 46, pgs. 290-292

40. Allan Sandage, *Centennial History of the Carnegie Institution of Washington: Volume 1 The Mount Wilson Observatory,* (Cambridge University Press 2004), pg. 273

10

Cosmology And The Human Condition (1935-1936)

Hubble and Humason are preparing to publish their third and final paper on the velocity-distance relation at Mount Wilson Observatory in 1936, eager for the completion of the 200-inch telescope at Palomar Mountain. Having nearly pushed the Rayton camera to its limits Humason is provided a new Schmidt camera and spectrograph that allows him to pursue even larger redshifts. An improved photoelectric cell on the 100-inch reflector allows Joel Stebbins and Albert Whitford to draw conclusions about absorption and reddening in space with implications for our understanding of the size of the Milky Way. Repairs to the clock drive and mercury systems of the 100-inch enable it to function smoothly, and John Strong's improved method for aluminum mirroring both decreases costs and improves definition. The completion of a paved highway to the summit of Mount Wilson and news of the organization's success now brings many thousands of visitors to the observatory, justifying a start to work on a new 300-seat auditorium and lecture hall. Adams works to secure Walter Baade's commitment to the observatory for the foreseeable future, despite the German astronomer's insistence that he will one day return to his homeland. The rise of the Nazi Party strains tensions between Baade and Fritz Zwicky, illustrating the fraying of nerves over the rise of nationalism at home and abroad. Fed up with van Maanen's continued denial of his galaxy distance measures, Hubble finally confronts him with the aid of Adams and Seares in an attempt to resolve the issue for good. Hubble and Humason remain steady and on course. Humason finds time to write a popular article on the expanding universe. In his most ambitious year to date, Hubble publishes *The Realm of the Nebulae* and his soaring paper on the distribution of galaxies in search of the curvature of space. To wrap up, the duo publishes their final Mount Wilson contribution to the expansion theory.

© Springer Nature Switzerland AG 2021
R. Voller, *Hubble, Humason and the Big Bang*, Springer Praxis Books,
https://doi.org/10.1007/978-3-030-82181-4_10

In 1935 the average American worker earned around $1,600 a year. A new home cost just under $3,500 dollars and a gallon of gas was 30 cents. A new car cost $600 dollars and a loaf of bread was just eight cents. But 20 percent of Americans were still out of work and the economy was showing little signs of improvement despite the efforts of the Roosevelt administration.[1]

FDR had introduced new programs with immediate effect after taking office, but with public faith in the system flagging many in the workforce had simply abandoned trying to find a job. Adding insult to injury the Black Sunday dust storm blanketed the Plains states on April 14th, capping a month's long wave of such storms in the region that had become known as the Dust Bowl.[2]

The New Deal programs of economist-in-chief Roosevelt included the Federal Deposit Insurance Corporation (FDIC) to protect bank deposits and the Securities and Exchange Commission (SEC) to monitor the stock market in order, between them, to prevent misuse and irresponsibility among bankers and investors.

Far-ranging public works projects were ordered. The Tennessee Valley Authority built dams and hydroelectric power plants to provide jobs and low-cost power to citizens in that state, while the more broadly-based Works Progress Administration (WPA) put 8.5 million people to work on construction of infrastructure, housing, bridges and buildings. The Social Security Act that was passed by Congress and signed into law by Roosevelt on August 14th, created a personal trust for every working American to be used for their retirement.

The economy would continue to see periodic downturns, but the government's efforts and Roosevelt's continuing public addresses began to have a positive effect on the level of public trust in the system.[3]

The first meeting of Alcoholics Anonymous was held not quite two years after the end of Prohibition and just as the first canned beer was going on sale. Parker Brothers released the board game Monopoly under the backdrop of economic uncertainty.

Despite the horrors of the Depression era being endured by people in some parts of the country, most were working, and the country as a whole remained a hot bed for opportunity and innovation.

Boulder Dam on the Colorado River, later renamed Hoover Dam, was opened under the supervision of former president Herbert Hoover, who had been a mining engineer prior to entering into politics.

Amelia Earhart became the first person to fly solo across both the Atlantic and Pacific Oceans and Malcolm Campbell broke the 300-mph mark driving a super-powered car over the Bonneville Salt Flats of Utah.

Americans with radios could dance the night away listening to Benny Goodman's swing orchestra or tune to *Your Hit Parade* to hear the 15 top musical hits of the week.

In the countries of Europe still reeling from the Great War the economic woes increased the public sense of dismay and gave rise to Nationalism with increasingly ominous effects.

After successfully leading public works projects to restore the faith of the Italian people in his government the fascist dictator Benito Mussolini sent his troops to Ethiopia to seize territory in the horn of Africa.

Taking his cues in part from Mussolini, German Chancellor Adolf Hitler introduced the People's Car (Volkswagen) after using his growing popularity to declare himself Führer in 1934 upon the death of President Paul von Hindenburg, then seizing absolute control under the guise of his Nationalsozialistische Deutsche Arbeiterpartei (Nazi) Party. It wasn't long before his Nuremberg Laws stripped Jews of their civil rights. In the spring of 1935 Hitler unveiled a growing and impressive war machine of tanks, artillery and the 2,500 aircraft of his newly established Luftwaffe. He introduced the draft, seeking to create a standing army 600,000 strong. By 1936 the citizens in the Saarland and Rhineland regions had voted to rejoin the German state. None of these developments were permitted under the Treaty of Versailles but in the face of Hitler's intimidation the League of Nations struggled to find a peaceful solution. The populations, infrastructure and economies of England and France had been devastated by the Great War and neither had the military or economic power to intervene. Since the League's diplomacy had no teeth, the dictators of the world knew they could act with impunity.

Emperor Hirohito and his government, facing a shortage of natural resources to fuel the Japanese industrial economic engine, unleashed a massive land grab against the Chinese to seize Manchuria. When this action was scorned by the League in an emergency summit the Japanese representative unapologetically walked out of the proceedings.

In the United States the reaction to the growing unrest in Europe, Asia and Africa was mixed. The ongoing economic doldrums and memories of the loss of American lives in the Great War fueled isolationist sentiments. Roosevelt yielded to Congressional voices and signed the Neutrality Act in August to curtail the sale of arms to warring nations.[4]

The renewed threat of German aggression prompted mixed public opinion. Some argued to act decisively against Hitler. However, many Americans were ambivalent if not actually sentimental to his cause. Anti-Semitism had reached its highest point in the 1920s with the publication of Henry Ford's four-volume *The International Jew: The World's Foremost Problem*. This assembled some ninety articles the aging automobile titan had printed in his periodical The Dearborn Independent, and it sold nearly a million copies.[5]

Hitler revered Ford so much that he was the only American mentioned in the Führer's wildly popular manifesto *Mein Kampf*, which chronicled his long struggle against what he considered the villainy of alien forces, communism and Judaism. He denounced the liberal press as fraudulent and demanded an end to the "victimization of the German people" by "international poisoners."[6]

In an interview with a Detroit News reporter, the Führer called Ford an inspiration and vowed to install Ford's anti-Semitic ideals in Germany. In July 1938, on his 75th birthday, the German Consulate in Cleveland awarded Ford the Grand Cross of the German Eagle, which was the highest foreign civilian award bestowed by the Nazi Party.[7]

In the 1930s a Catholic priest named Charles Coughlin with up to twelve million radio listeners and a million subscribers to his own newspaper, Social Justice, further fanned the flames of anti-Semitism.[8]

By the middle of the decade over half of Americans polled believed Jews to be greedy and untrustworthy. The figure had reached 60 percent by 1938 and anti-Jewish sentiment remained high through the 1940s. (For comparison, the Anti-Defamation League reported that 11-14 percent of Americans held intensely anti-Semitic views in 2020.)[9]

***

Against this global backdrop of racially charged populism the staff and visiting researchers at Mount Wilson lived and worked together almost without incident until 1935. It would be naïve to assume that the scientific community attracts only those whose perspectives are purely inclusive. Nor was it the case that everyone got along all the time, whether on the mountain or at the observatory offices in Pasadena as evidenced by Hubble's periodic fits of temper. But the success of the institution, the financially stabilizing backing of Carnegie and Adams's fair and frugal hand were enough to keep egos in check most of the time. In terms of its technical and experimental output the observatory was soaring.

A development with special interest for Hubble and Humason was the photoelectric cell by fellows-in-residence at the observatory Joel Stebbins (starting in 1930) and Albert Whitford (in 1933).

A photoelectric cell utilizes the photoelectric effect (Chapter 1) to convert light energy into electrical energy. Nowadays it has a wide range of applications from security systems to solar energy to automatic door openers etc. A semi-cylindrical cathode made of cesium oxide or other photosensitive metal is connected to the negative pole of a battery, opposite an anode made of a thin platinum wire connected to the positive terminal. Both are enclosed in a quartz tube. When light at a specific wavelength is absorbed by the cathode it releases electrons that are attracted by the anode. The faint electrical current from this reaction can be measured by an ammeter to discover the intensity of the light arriving at that frequency.

In the hands of Stebbins (Fig. 10.1) and Whitford, this device proved very useful to astronomers in exacting more precise measurements of objects in space. Astronomers had long held that space dust (for lack of a better word) absorbed starlight and therefore dimmed the apparent luminosity of distant objects. The photoelectric cell was not concerned with luminosity but instead focused on the electrical current produced by the light, which gave a more accurate reading of the source's absolute magnitude.

Using the photoelectric cell attached to the 100-inch, Stebbins and Whitford were able to finally trim down Harlow Shapley's enormous Milky Way Galaxy to something closer to its actual size, with a diameter of around 30,000 parsecs or 100,000 light years.[10] The ability of such a cell to eliminate the effects of light absorption in space would prove useful in virtually every

Fig. 10.1 Dressed for success (From left) Adrian van Maanen (Parallactic proper motions); Frank E. Ross (Creator of the Ross correcting lens - visiting from Yerkes Observatory); Milton Humason (Stellar and nebular spectroscopy); Joel Stebbins (Photoelectric cells) in Pasadena ca 1932 (Courtesy, Ann Humason Bernt family)

aspect of the field of astronomy, including the extent and nature of the expansion of the universe.

With reports of exciting discoveries being made by MWO the popularity of the facility was achieving epic proportions. The completion of the Angeles Crest highway, a two-lane paved road that wound its way through the San Gabriel Range right past the observatory's entrance in 1935 made accessing the summit much simpler. Tens of thousands of visitors came each year to view the firmament through the lens of "the 60-inch telescope on Friday evenings." So great was the fanfare surrounding the mountain that Carnegie authorized the design of a 300-seat auditorium and adjacent exhibit room to be constructed on the grounds.[11]

Success breeds success, and George Hale had used the observatory's runaway success to sell the creation of the 200-inch telescope to the trustees of the Rockefeller Foundation, which contributed six million dollars in 1928 toward its completion. The dome for the giant new reflector hailed in the press as the "Big Eye" was under construction on the summit of Mount Palomar, a 45-minute drive south of Pasadena. In addition, an 18-inch aberration-reducing Schmidt telescope was expected to be completed there by 1936 with construction of a larger 48-inch version starting a couple of years after that.

After several early attempts to create a suitable mirror, General Electric had been pulled from the project in 1931 and Hale turned to Corning Glassworks in New York, which had developed a new product called Pyrex. In March 1936 the 200-inch blank was loaded onto a specially fitted railcar for transport to the Caltech optical shop in Pasadena.[12]

The specter of renewed conflict in Europe was beginning to cause tensions to boil over at MWO and at Caltech. The fact that the 200-inch was being financed by the Rockefeller Foundation (a Caltech financier) and would be operated under the auspices of the university added another wrinkle for the old guard on the mountain. The nature of the new relationship between the two institutions forecasted the arrival of a new kind of researcher with a broad knowledge of physics as well as observing skills. The arrival of this kind of researcher was anticipated by Walter Baade.

It had not taken long for Baade to make his mark on the observatory and an indelible impression on its director Walter Adams. A patient, careful observer who was as interested in eliminating any systematic errors in his observing and theoretical techniques as he was in the value of their eventual results, Baade was the antithesis of Edwin Hubble, who leaned heavily on approximations and mean values. In Adams's opinion, Baade, not Hubble, was the true future of the organization.[13]

The issue for Adams was that Baade was a loyal and enthusiastic German citizen who dreamed of returning to his homeland to continue his work on stellar and galactic evolution at the Hamburg Observatory, where he had started his career. It therefore made no sense and would actually be unethical in the German's mind to apply for citizenship in America, as Adams and others at the observatory had recommended.

The only obstacle standing in Baade's way was Adolf Hitler, who had taken control of the German state and whose Nazi Party was forcing its will on the German people. In 1934, when Richard Schorr, the director at the Hamburg Observatory, first began to contemplate retirement, Baade's name was put forward as the only real candidate to take over the helm. Baade initially provided assurance in writing that he fully intended to accept the position, if and when it was offered to him. By 1936 it was becoming apparent that Schorr and the Ministry of Science and Education in Berlin were going to ask Baade to take the director's post. Schorr wrote to Baade suggesting it would be a good time to declare his allegiance to the Nazis in writing. This put Baade in a predicament. He did not sympathize with the Nazi regime, but he loved his country. Like many, including Einstein, he had assumed that Hitler was a flash in the pan and his brand of German Nationalism would soon fade. In a moment of extreme diplomacy Baade resisted on moral grounds but assured Schorr he would work with Hitler's administration if he were to be made director.[14]

Baade fully understood the prospects for a life and career at Mount Wilson. The climate was very good all year around and the seeing on the mountain was far superior to Hamburg. The new observatory at Palomar had one of the world's finest observing instruments in the 18-inch Schmidt and promised the 48-inch Schmidt camera and the 200-inch, although the finishing of the latter was now not expected until 1941. On the other hand, he was confident that he could be just as effective, if not more, as Hubble had been if a small Schmidt camera were to be installed at Hamburg. With its economical spherical mirror and correction plate, this system was starting to attract notice and Baade was becoming one its most outspoken supporters.

Despite the difference of opinion over their methods of research, Baade and Hubble got on well during the 1930s. Baade believed rightly that Hubble's approach, while daring and successful in leading the world into the dark centers of the universe, would require a lot of refinement over the coming decades. But Baade applauded Hubble's overall classification scheme for galaxies and often came to the aid of Hubble and Humason on their observing programs by adding direct photographs within range of the list of targets for his own work. In this way he was acting both as a supportive colleague and as a prying fellow researcher, snooping around for signs of missteps or oversights. Although his interests included every bit as much of the spacetime continuum as the two lords of Mount Wilson, Baade's specific research was centered on galactic dynamics, and their impact on intergalactic dynamics. This preserved and protected the friendship that Humason and Baade had formed, at least for the time being.

Of all the staff at Mount Wilson, Baade was definitely the closest to Milton. Their early lives had been rather different but, in many ways, they were cut of the same cloth and could often be seen clowning around or pranking each other, both on the mountain and off. A photograph of the Humason's and Baade's on a site scouting visit to Caliente, Nevada, shows the couples cloaked in colorful shawls while riding a wagon drawn by donkeys[15] (Fig. 10.2).

It was Humason that Adams tapped to try to persuade Baade to remain when the offers from Hamburg started coming in. As the story goes Humason took Baade on a drive, most likely in the sequoia forests north of Los Angeles, and extolled the value of an observatory with the world's best instruments being used by the likes of Baade, Hubble, Humason and the rest of the prestigious group of outstanding observers at the facility at the time.

The final sticking point was to sort out both the political and financial aspects involved in Baade succeeding to either the directorship at Hamburg or

Fig. 10.2 Fun with friends (From left) Helen and Milt Humason, Muschi and Walter Baade on a site visit in Caliente, Nevada ca 1932 (Courtesy, Ann Humason Bernt family)

at Mount Wilson. At the time, everyone including Baade figured that when Adams retired Hubble would be next in line for the directorship. Adams was set to do just that once the 200-inch was completed around 1941. Baade's salary at Mount Wilson was $3,600 a year and Adams had managed to get him an additional $500 raise starting in 1937, but this was a far cry from the $7,500 he was being offered by Hamburg and he would be in charge of the direction of the observatory, not under the thumb of the camera-friendly Missourian who was thought to be the one to take over in Pasadena. This fact above all was inducing Baade to lean in the direction of his homeland.[16]

A headstrong and fiercely independent thinker, Baade's mind was changed by a number of factors and events. The first was that Hitler's regime was persistent and its policies were making life intolerable for many of his friends, some of whom were Jews and others who, despite having converted to Christianity years before, were still treated as Jews.

One of these fellows was Rudolph Minkowski, who had written to Baade in 1934 asking him for help getting out of Germany while it was still possible. Minkowski was the son of a medical professor known in Germany as the "grandfather of insulin" for his research on diabetes.[17] With an education in physics, Minkowski served as an officer in the German army during the Great War and later collaborated with Baade at Hamburg. None of this mattered. According to the racial laws of the country at the time, Minkowski's Jewish heritage (he had been baptized Christian) condemned him and his family to the fate of all Jewish people living under Nazi control. He was stripped of his rights and was subject to ongoing oppression.

When he received the letter from his friend, Baade immediately went to Adams to secure a position for Minkowski. Adams, desperately clinging to the hope that he could find a way to convince Baade to stay in Pasadena, was happy to help. After some budget-crunching the director scraped together sufficient money from the Rockefeller Foundation to pay the salary and Minkowski was summoned. He would prove his worth five years later with the discovery that there are distinct types of supernovae and would go on to become a member of the fab four stellar and galactic evolutionary research team with Hubble, Baade and Humason[18] (Fig. 10.3).

The second mind-changing event was the 18-inch Schmidt camera at Palomar. A raving fan of the aplanatic system, his native allegiances were mollified by the addition of the new instrument at a research facility, miles from Mount Wilson where he wouldn't feel the grip of Hubble's grasp as tightly. Palomar had the bonus of being farther from the light pollution of Los Angeles, whose city limits were advancing ever closer to Mount Wilson.

Fig. 10.3 The Fab Four of the Mount Wilson Nebular Research Department (From left) Milton Humason, Edwin Hubble, Walter Baade, and Rudolph Minkowski discussing images taken with the 48-inch Schmidt telescope ca 1950 (Image courtesy of Hamburg Observatory)

These earlier factors triggered the final event when, after considering his options, Baade went to Adams with an offer. Baade realized how badly Adams wanted him to remain, and with the worsening situation in Germany and the seemingly monthly improvements to the facilities at his disposal, his mood was changing. He still intended to go home once Hitler was removed but he had no desire to pledge his allegiance to the Nazis, then or ever. With this in mind, he told Adams he would stay at Mount Wilson if he could have a salary raise to $6,000 per year. Motivated by his dedication to the destiny of the organization, the crafty Adams shifted salary money around between Carnegie and Rockefeller to satisfy Baade's request. Adams had his man, for the time being at least.[19]

Baade's time at the observatory had already proven fruitful both for the researcher and the institute. He had befriended Fritz Zwicky not long after his arrival in 1931. Swiss-born Zwicky had come to the US around 1925 and was making a name for himself as a professor of mathematics at Caltech, but he also had interests in astrophysics and astronomy. Baade and Zwicky shared a couple of languages and an interest in novae. They coined the term "supernova" in their 1934 paper *On Super-novae*.[20] In the ensuing *Cosmic Rays from*

Super-novae they suggested these events were the collapse of giant stars, the violence of which produced the mysterious cosmic rays discovered two decades earlier by balloon-borne instruments.[21]

Knut Lundmark had suggested the idea of this special class of "giant novae" as early as 1920 and Hubble referred to them as "exceptional novae" in his 1929 paper.[22] Lundmark had been a visiting researcher at Mount Wilson using the plate collection for his galaxy classification system in 1933 and may have picked up the term from seminars given by Baade and Zwicky.[23]

Baade and Zwicky published three joint articles in 1934 highlighting several discerning factors that distinguished the two types of novae, including the relatively frequent rate of occurrence of novae at 10 to 30 per year in the Milky Way and M31, versus supernovae, which seemed to occur once a century. Baade and Zwicky's conclusion on supernovae was that this type of collapsed star might eventually evolve into a neutron star. Their work predicting the massive gas ejections from these exploding stars would later lead Fred Hoyle to the discovery that they carry with them the heavier elements manufactured by nuclear fusion in their very hot interiors.

It was the success of this work that convinced Baade to push Adams to add the 18-inch Schmidt camera with its wide clear field to the new Caltech observatory on Mount Palomar. Baade had convinced Zwicky, a member of the faculty at Caltech to help with the overtures. The plan worked and the Schmidt camera became the first operational research telescope at Palomar in September of 1936, where it was put to good use by Zwicky in his supernova surveys. It was the success of this instrument that led to the addition of the 48-inch Schmidt camera on the mountain.

Their research program was supposed to involve Zwicky locating supernovae using the 18-inch on Palomar and Baade then tracking them with the large Mount Wilson reflectors. It all worked smoothly until the irascible and defensive Zwicky became suspicious of his German counterpart's nationalistic pride, notoriety and good standing.

Zwicky was very prickly, easily offended and had few loyalties. He often overstepped the bounds of convention and decorum in collegial matters. Donald Osterbrock documents the story of Zwicky's heated response to a copy of an article that Cecilia Payne-Gaposchkin at Harvard sent in which she "mildly criticized" their calculation of the energy released by a supernova's collapse. Baade apparently had not been aware of the letter from Zwicky to Payne-Gaposchkin and on later learning of its aggressive tone he began to distance himself from his Caltech partner.[24]

As a scientist, Baade was as fair-minded as he was thorough and believed strongly that criticism ought to be made only of the work, never of the scientist personally. On the basis of that conviction, he must have felt it necessary to step away from what had been a highly productive work relationship. He would later prove hypocritical when he made disparaging comments against his great friend Humason regarding the allocation of observing time on the 200-inch at Palomar.

The relationship between Baade and Zwicky deteriorated still more as Hitler began his incursion into eastern and western Europe. Although he had decided to remain in America, Baade never bothered to apply for citizenship because he expected he would return to his homeland either at a later point in his career of in retirement. After an argument broke out between the two over a scientific matter at a meeting at Caltech the Swiss astronomer, in full froth, decried Baade as a Nazi. They had to be separated and Humason took his shaken German friend aside to calm his nerves as the others attempted to cool down Zwicky, who had taken flight in a fit of rage. Even though Baade had always made clear his abhorrence of the Nazis the rift with Zwicky lasted for the remainder of their lives.[25]

The looming trouble in Europe was also putting pressure on Hubble's relationship with his good friend Aldous Huxley, whose *An Encyclopedia of Pacifism* was a response to the perceived arms race between Germany, Japan, England and the Soviet Union. As the book was receiving popular acclaim in Los Angeles, Hubble simmered over what he saw as the apathy of the United States in the face of intimidation by a handful of bullies.[26]

Huxley and his wife Maria Nys relocated to Los Angeles in 1937, where he would live until her death in 1955. Although their intellectual capacities and curiosities gave Hubble and Huxley a healthy mutual respect, Huxley's humanist and pacifist tendencies were too strong for his friend's liking. Not long after arriving in Southern California, Huxley began dabbling in Hinduism and established a long friendship with the Indian philosopher Jiddu Krishnamurti, with whom he debated western and eastern philosophy.

As the world began to spiral out of control in the latter part of the 1930s, Hubble would find Huxley's insistence on finding a path to peace less and less palatable. Friendships like that the Hubbles shared with Anita Loos (a close friend to Grace Hubble) and their common love of the British homeland held the two friends together through the difficult years which followed. Loos evidently made Huxley's introduction to Hollywood, where he would earn more than $3,000 per week in the coming decade with screen credits in collaborations for *Pride and Prejudice* (1940), *Madame Curie* (1943), *Jane Eyre* (1944) and other movies.

Huxley's sprawling career and notoriety as a writer and philosopher made him an almost perennial nominee for the Noble Prize for Literature (seven nominations in total) and made him one of the most noteworthy writers of the 20th century.[27]

* * *

With his observing runs growing ever longer and more tedious Humason knew he was fast approaching the limitations of the Rayton spectrograph in his relentless pursuit of redshifts and was therefore eager for something better.

The hype surrounding the new Schmidt camera system convinced Theodore Dunham to try to make one, which he did with advice from Don Hendrix in the optical shop. The result had a focal ratio of f/1 and it performed well with a "quartz correcting plate" in one of the "small nebular spectrographs." The spherical shape of the mirror made it easy to scale up at low cost. Adams asked Hendrix to improve and expand Dunham's camera.[28]

Like his friend Milton Humason, Hendrix lacked an education beyond high school. The native of Fort Worth in Texas relocated with his family to Glendale, California where he finished high school in 1923. When the music company he had worked for since finishing school went bankrupt during the Great Depression, Hendrix answered an ad for an opening at the optical shop at Mount Wilson and was hired despite not having any experience. There he realized his genius for optics and went on to become the observatory's master optician. In an interview Dunham later praised the old master saying, "Hendrix…could do anything that looked impossible."[29]

One year later, the industrious Hendrix had not one but three cameras ready for use in various spectrographs. He had made a more refined "metal mounting and a new correcting plate of ultraviolet transmitting glass" for Dunham's camera, with a 30-inch focal length. It was "designed to work off-axis so that the photographic plate [was] outside the incoming beam."[30]

Dunham remembers how after making his original plate square he was eager to remake it rectangular so as not to obstruct the incoming light. To his dismay, Hendrix brandished a glass cutter and proceeded to score the plate down the middle. As Dunham looked on in horror, Hendrix said, "All right, here we go – bmmpt…" and it snapped in two![31] In addition to the refinements to Dunham's camera, Hendrix had produced a camera with a focal length of 73 inches for the coudé spectrograph. Having successfully finished these two cameras he got down to the real business at hand by creating a "camera of five-inches focal length and a focal ratio of 1.75 that focused a beam of light on a 10 millimeter (about 3/8th of an inch) square plate with a high degree of definition"[32] (Fig. 10.4).

Fig. 10.4 Humason holds a small spectrum plate on a glass slide ca 1955 (Image courtesy of the Observatories of the Carnegie Institution for Science Collection at the Huntington Library, San Marino, California)

The Schmidt spectrographic camera was exactly what Humason needed, an extremely fast camera that had extremely low dispersion. The only terror for a bleary-eyed observer, Humason would later say about the plates, was "if you let loose one of those little things" while developing it in the darkroom "it is liable to flop right over and land on the film side and stick to the bottom of the tray," destroying an entire week's work![33]

Nevertheless, the new spectrograph helped Humason extend his redshift record through the middle of the decade, reaching 42,000 kps for a bright galaxy in a cluster in Ursa Major. By 1936 however, he had reached the end of his tether with the new camera and Hendrix went to work again.

While Hubble and Humason compiled the data for what would become their final Mount Wilson paper on the expansion question, Hubble was in the process of settling his old score with Adrian van Maanen. In the face of mounting evidence lending support to Hubble's claim, van Maanen had dug in his heels and insisted that his measurements of the proper motions of stars in M31 and other nearby galaxies showed these objects to be much nearer than Hubble was claiming.

In 1935 Hubble went to Adams to insist on getting access to van Maanen's photographs so that he and others could independently measure the supposed proper motions.

Wishing to avoid a scene, Adams suggested that Hubble try to settle the matter with van Maanen directly, but Hubble's halfhearted attempt ended in failure. It is hard to imagine Hubble having a serious discussion about the differences in their research when he did not respect the approach in the first place. His counterpart was doing the same experiment time and again and it was Hubble that was expecting a different result. The fact that van Maanen kept achieving the same result without changing his method or considering Hubble's for a moment was driving the Missourian mad. Gnashing his teeth, Hubble next asked Humason to take new direct plates for M33, M51, M81 and M101 for comparison with earlier plates taken by Ritchey which van Maanen had also used in his study. Presumably the hope was that Humason being a friend of van Maanen, his involvement might be a calming influence on the Dutchman.

Under Adams' authorization, Hubble then asked Nicholson and Baade, regarded as the top stellar and nebular photometrists on the mountain at the time, to undertake independent measurements of the objects on the old and new plates, with Adams and Seares monitoring their progress. When the report came back essentially negative on proper motions in these four objects Hubble immediately wrote a report to this effect for publication in the CIWY. However, to his utter disbelief neither Adams nor Seares would put it into print as written. As Adams later pointed out, Hubble's tone was "intemperate" and showed an "attitude of animosity" unbecoming of the moral and ethical code that Hale, and he had cultivated since the observatory's founding three decades earlier.[34]

When Hubble insisted, he wouldn't change a single word Seares told him he was free to publish his report anywhere he wished, just not as part of the observatory's contribution to the CIWY.

As a compromise Adams suggested an arbitrator. Seares would interview both Hubble and van Maanen, then submit a report based upon his own analysis. As resident master of photometry, surely Seares must be able to interpret the data in a fair and equitable manner. If approved by all parties, this report would be published under the names of all four men involved in the recent study. Everyone supported the final analysis except for Hubble who, in the words of his director, "opposed it violently."[35]

Mayall, who was on a visit from the Lick to meet with Hubble and Humason, asked Hubble how he felt about the developing situation. "[Hubble] reflected for a little while, maybe thirty seconds or so, and finally said, 'My attitude is no compromise...' He didn't mention van Maanen by name, but he repeated, 'My view is no compromise.'"[36]

Adams and Carnegie Institution president John Merriam were exchanging letters at this point about Hubble's inability to take the high road in any matter that directly related to his findings. Adams pointed out it was not "the first case in which Hubble has seriously injured himself in the opinion of scientific men by the intemperate and intolerant way in which he has expressed himself."[37]

No one inside the observatory really doubted the soundness of Hubble's findings and most of the scientific community – including Shapley, at least privately – believed the nebulae were extra-galactic. Hubble could have simply let van Maanen's universe fall by the wayside, as Shapley's had a decade earlier. Instead, he bullied and belittled and embarrassed van Maanen whenever the subject of proper motions within spirals arose. The story of the misplaced napkin ring from years earlier was well known and often retold. It clearly showed Hubble's inability to handle the issue in a manner commensurate with the English gentleman that he so stridently sought to exemplify.

With every attempt to exact a peaceful settlement between the two warring astronomers exhausted, Adams forced the issue by overseeing the publication of two articles in a coming edition of the Astrophysical Journal (ApJ), one by Hubble and the other by van Maanen. This put the matter to rest for Hubble, satisfied he had won the war in print. There was general agreement that Hubble was correct on the substantive issue, but everyone agreed that van Maanen had won the battle of civility.

As was the case with many of the men and women around the observatory, van Maanen saw Humason as a colleague and a friend. He frequently sought Humason's opinion or help in providing spectra for his research on the classification and motions of specific stars of extremely low luminosity and other characteristics. The pain of losing out in a race to make a fundamental discovery wasn't exactly new to Humason. The two published a short article in the PASP in 1936 on the spectral type, magnitude and proper motion of the white dwarf star Ross 627 whose apparent magnitude of 14.1 made it one of the faintest of its type at the time. Given their individual expertise, Humason wrote the spectral analysis and left the proper motions to van Maanen.[38]

Hubble's protracted dispute with van Maanen didn't stress his working relationship with Humason. The two men were working like a well-oiled machine, each aware and respectful of the part that the other played as they pioneered together at the frontier of space. As Allan Sandage would later point out, it was "Hubble's practice to include Humason as a coauthor on papers where the redshift data…was of central importance," adding that "it was always understood that the final values in the discussions and the text of this and subsequent joint papers were due to Hubble's analysis alone."[39]

Over the course of the preceding decade, they had combined to introduce the world to the incredible four-dimensional perspective of a universe whose boundary was still beyond the reach of the powerful Mount Wilson reflector. This quest had not just cemented their names in the annals of science, it forged a bond which incorporated attributes similar to those that arise between individuals in extreme circumstances. It was the same type of rapport Hubble might have wished to cultivate under the stress of war years before.

Evidence of Hubble's respect began to show as early as 1932 when he praised Humason in an interview for the New York Times. Leaning back in his chair while examining the image of one of Humason's spectral plates, he opined to his interviewer:

> Do you realize the amount of patience, ingenuity and expert experience that little photograph represents? Besides having to guide the telescope continually, Humason had to control the focus and the comparison spectrum…keep the temperature of the spectrograph exactly right – not for an hour or so, but all night. Sitting in total darkness with his eyes on a slit of dim light little larger than a pinhead, he worked levers and punched buttons night after night for a week without once moving the photographic plate or losing sight of that faintly luminous spot in the sky, and here's the result, a report on conditions in a very remote region of the universe. Talk about the romance of the heavens, why Humason has brought it down to earth.[40]

Hubble, whose observing runs with the 100-inch telescope usually amounted to several nights over which time he would produce as many as a couple of dozen direct photographs, knew as well as anyone what Humason was going through in the pursuit of galaxies requiring as much as 70 hours of exposure to get a single spectrum. To make matters worse, as he sought to stretch the limits of the equipment and of his will and expertise, all too often he came away without a useable spectrum (Fig. 10.5).

Perhaps more telling were the personal reflections and comments in Hubble's hand over the course of the partnership. In a handwritten postscript

Fig. 10.5 Milton Humason and Edwin Hubble during an interview ca 1950 (Image courtesy of the Pasadena Star News)

at the bottom of a typed letter to Mayall at Lick in 1938 he referred to his fellow collaborator as "Milton," an extremely rare note of affection in the world of the observatory where everyone was referred to by his last name.

The note in Humason's personal copy of Hubble's 1936 best seller, *The Realm of the Nebulae*, reads:

> To Milton Humason, whose research contributed so largely to the contents of this book. – Edwin Hubble.[41]

Apart from the occasional dust up, a real camaraderie existed among the men at Mount Wilson and Lick Observatories and Caltech during the 1930s and 1940s. Humason was at the center of life in those years. His frequent visits to the Santa Anita racetrack were often attended by Baade, Minkowski, Adams, Benioff and others from the various campuses.[42]

In the words of Allan Sandage, "Humason was a superlative mule driver, fisherman, imprecationist, drinker, poker player, raconteur, rake and rogue, gentleman and friend," and it is likely that all or most of the people who worked around him shared this sentiment.[43]

One surviving Humason canard involved the mysterious disappearance of the evangelist and founder of the Foursquare Gospel Church Aimee Semple

McPherson in 1926 and her scandalous reappearance weeks later. The self-described faith-healer, having only recently moved to Los Angeles, was feared drowned during a swim near Venice Beach on May 18th. This set off a frenzy of activity as divers tried to recover her body and "her followers" used "dynamite...hoping to raise her" but actually "surfaced only dead fish."[44] After her "miraculous" reappearance in a Mexican village five weeks later, Humason told friends that McPherson had actually been shacking up with him in his cabin in the Sierra Madre hills helping him to find God. The small cabin he and Nicholson rented near the West Fork River was a favorite fishing getaway for family and friends until it was unfortunately swept away in a flood in 1938.[45]

Practical jokes were common in those years. Mayall remembered the time the 100-inch night assistant Tom Nelson set up an elaborate scheme with a local radio jockey. According to Mayall, the radios in the domes were usually on for the graveyard shift and on this particular evening at about 2 a.m., while Humason was observing "on the small Cassegrain platform probably in a semi-comatose condition," Nelson suddenly turned the volume to a level sufficient to "rattle the rivets' in the dome walls. The radio jockey came on and said, "The next number is dedicated to Milton Humason, who is sitting at the 100-inch telescope on Mount Wilson, observing an object no one else can see." Nelson later told Mayall, who was on the 60-inch that night, "Milt nearly fell off his perch."[46] Mayall didn't mention the name of the song.

There are indications that Hubble took at least some part and pleasure in the fun while on the mountain in the early years. But by the middle 1930s, as his fame grew, he preferred to spend more time working at home and fraternizing with the Hollywood and literary elite, cultivating his image as an important man of science. Those close to him insisted his air of superiority wasn't deliberate, he was merely "in tune with this magnificent project" he was in. He was very supportive of Mayall on his dissertation and was frequently filled with angst at the ongoing feud with van Maanen. Humason also noted that Hubble would appear "upset that he could not communicate more personally with the other members of the staff."[47]

After Bill Humason's graduation from Caltech in 1935 Milton and Helen rented a home at 539 East Villa, a couple of blocks from the observatory offices and not far from Caltech where the young postgraduate had taken a job teaching.[48] The younger Humason had spent time on Mount Wilson as an assistant observer during summer breaks while in college to get an upfront and personal view of his father's profession.[49]

During the 10 years 1931 to 1941, Milton Humason was probably most confident and comfortable in his role as the top spectroscopist on the mountain and one of the leaders of his field.

In the autumn of 1936, the 100-inch was laid up for six weeks for repairs to its flotation system, which had developed an increase in friction due to years of use and shaking from earthquakes. The inner liners were removed to allow more mercury to lubricate the system and the bearings were replaced.[50]

In addition to his work with Hubble and others, Humason began writing on topics that expressed his growing expertise in the area of optics and even more speculative subjects. Perhaps spurred by Hubble, Adams and others he started to explore his ability as a writer of topical scientific literature.

Some of these articles were similar in scope to his article on the Rayton camera after its completion in 1931. *The Aluminizing of the 100-inch and 60-inch Reflectors of the Mount Wilson Observatory*, published in the PASP in April of 1935, detailed the work of Dr. John Strong at the Caltech optics laboratory on a new vacuum chamber developed to aluminize the mirrors of large telescopes using the evaporative process. As Adams testified in the CIWY for the year 1934, the resulting "surfaces exceed in reflectivity any other metallic surfaces so far tested."[51]

A native of Kansas who was 14 years Humason's junior, Strong was a bit of a character in his own right. He and the American artist and polymath Russell Porter put together what they called the "100 to 1 Shot Club," so named because the "considerations" of the club in meetings "were restricted to topics that were fantastic by a factor of 100:1 over scientific." Worthy topics included whether water swirled down a bathtub drain in opposite directions at latitudes north and south of the equator. The group met "six or seven times a year" and included Milton Humason and John Anderson from Mount Wilson, George Mitchell of the Mitchell camera used for Hollywood movies, a patent lawyer for Paramount Pictures and others. In Strong's words, "it was a group worthy to go down in history."[52]

The 1935 article described in detail the dimensions and weight of the new machine and highlighted its overwhelming success in achieving more reflective surfaces on a number of mirrors at the Mount Wilson, Lick, Yerkes and Dominion observatories, the latter being in Canada. Strong's intention was to aluminize the primary mirror of the 200-inch telescope, which Humason concluded "should present no serious difficulty."[53]

Next, Humason published an article simply called *Super-novae* wherein he began to flex tentatively his literary wings:

Anyone who becomes familiar with the stars is impressed by the fact that, in general, year after year, they remain essentially fixed in brightness and position. Occasionally, however, an apparently new star suddenly appears in the sky, sometimes reaching a brilliancy comparable to the brightest fixed stars. These new stars or "novae" remain bright for several weeks or months...[54]

Surely, Charles Dickens was resting peacefully in his grave. In Humason's defense, he was writing for the PASP, hardly popular reading. Although the article does lack the kind of flourish more common in the writings of Hale, Hubble or Shapley, who were far more adept at compelling an audience, Humason nevertheless did make an intriguing portrayal of these super star bursts and the plan for researching them. Citing Baade and Zwicky, he detailed the intrinsic characteristics of both novae and supernovae, saying that based on the best records available Tycho Brahe's nova of 1572, "which at maximum was brighter than Venus and could easily be seen in the daytime," was the only known Milky Way nova that might be considered a supernova.[55]

His prose then briefly took comparative flight:

The super-nova of 1885 was...10,000 times brighter than the average common novae and the brighter stars; in fact, it was approximately as bright as the total brightness of the entire system (nebula) in which it appeared...At the distance of Sirius (9 light years) it would appear 100 times brighter than the full Moon. At a distance of 300 light years, it would still appear as a star 100 times brighter than Venus...its intrinsic brightness would be about 100 million times that of the Sun, and that during an interval of one month, at near maximum bright-ness, it would radiate as much light as our Sun does in 10 million years.[56]

In this section of the four-page article, Humason managed to adopt the kind of tone of awe inspired curiosity that he and others studying and reading about these rare and highly dynamic objects surely felt. Who wouldn't be excited at the idea of seeing a star explode!

At the end of a highly productive year, he had published nine articles on topics ranging from modern telescope optics, novae and supernovae, the final joint paper with Hubble on the velocity-distance relation, and two broader perspectives on just what exactly his work with Hubble might be revealing.

Evidence for an Expanding Universe was published in July in The Scientific Monthly and *Is the Universe Expanding* was in the August Leaflet no. 91 of the PASP. They covered the territory of spectrographic redshift velocities that Humason had staked out uniquely for himself, with an expert eye on the overall narrative. In both papers his approach was three-fold: (1) to refute counter evidence against the velocity-distance relationship, (2) to review the same evidence with respect to its impact on the question of expansion and (3) to set out the possibility of reaching a concrete resolution of all of these questions when the 200-inch went into operation at Palomar.

Of the two, *Evidence* was the better and more thorough investigation of the matter. With the notable exceptions of his groundbreaking reports on the velocities of galaxies from 1928-1929, 1931 and 1934, this was probably the best popular scientific article he produced during his long career.

In seven brief but powerful chapters Humason filled in the evidence for expansion like the layers of a cake, starting with the tools at his disposal including the magnificent Mount Wilson reflector and its associated instruments as well as the labs and shops available for corroborating data.

Starting with the measurement of motion, he reviewed Slipher's achievement of getting a spectrum for M31 in 1912 with a specially designed camera and how the direction of the displacement of the strong H and K lines of calcium indicated whether a galaxy appeared to be approaching or receding and at what speed. By the time Slipher wrapped up his study in the early 1920s velocities were newly available for 45 galaxies.

Switching to relativity, he cited de Sitter's 1916 prediction that Einstein's theory gave an exponentially expanding universe. He did not mention the contributions by Friedmann and Lemaître. Next came the 1925 discovery by Hubble's of Cepheids in several galaxies. He explained in simple terms how for Cepheids the duration of their light curve gave their candle power, which could in turn be used in conjunction with their apparent magnitude to calculate their distance. For M31 at the time this was roughly 870,000 light years. He cited the corroboration of these distances using novae and other high luminosity stars in the same galaxies, and the later use of the mean luminosities of the galaxies themselves to determine their distances.

The third chapter reviewed Hubble's determination of the relationship between velocity and distance for Slipher's galaxies using this new methodology. Noting the occurrence of galaxies in isolation, small groups and great clusters made up of hundreds of galaxies such as the nearby Virgo Cluster (which was nevertheless about seven million light years away) Humason outlined how distances were determined, namely by measuring isolated galaxies (where this was possible) and by measuring entire clusters. The very high mean luminosity function of clusters was proving very useful for testing whether the relationship between velocity and distance continued to hold at extreme distances.

The next two sections, *Definite Results Obtained* and *Increasing Instrumental Power*, detailed Humason's work, beginning in 1928, in capturing spectra of deep space galaxies, a difficulty he downplayed with characteristic nonchalance. Citing the development of the Rayton camera in 1929 which, with the spectrograph on the 100-inch, cut exposure times to about an eighth of those previously needed, he provided a short list of the galaxy clusters measured to date, the largest being Ursa Major No.2 at a distance of 240 million light years with a recessional velocity of 40,000 kps.

In the chapter *Recession Velocities*, Humason reviewed the data that he and Hubble had compiled. A total of 189 velocities, 146 of which had been gathered at Mount Wilson since 1928, ranging out to a distance 35 times greater than the farthest known prior to Hubble's initial determination on expansion.

All this evidence had been double checked and cross-examined using isolated galaxies and their attendant novae and bright stars to give the closest approximation possible for the relationship between velocity and distance. He mentioned the tiny photographic plate and the absurdly small spectra at roughly two twenty-fifths of an inch, writing almost whimsically of the improved exposure times of 2 to 60 hours. The two most distant clusters, each 30,000 times fainter than the faintest star visible with the naked eye, had to be photographed by centering the cluster on the spectrograph slit "by setting off the distance of the nebula from the nearest bright star as measured on a direct photograph…" Except for a few nearby galaxies whose spectra indicated velocities of approach, the recessional velocity-distance relationship held with startling congruency.

In the final section, *Interpretations Still Controversial*, Humason made special note of the relationship between brightness and redshift as determining distance as a matter of fact. He then ended this incredibly discursive and compelling argument for the expansion of the universe with a cliffhanger, pointing out, "The interpretation of the red-shifts as velocities of recession is still controversial." His final statement on the matter sums up a bewildering conclusion:

> If the interpretation of velocity shifts as a motion of recession is abandoned, we find in the red shifts a hitherto unrecognized phenomenon whose implications are unknown. The expanding universe of general relativity would still persist in theory, but the expansion would not then be indicated by the observations.[57]

If all this didn't mean expansion, what then? Even Einstein was convinced of the reality of universal expansion by now. Nevertheless, the expansion principle, based on the theory of relativity and effectively proven experimentally by the work of Hubble and Humason, was being resisted by none other than Edwin Hubble himself!

In their paper of 1931 that introduced the first comprehensive evidence for what is now commonly understood to be the Big Bang theory in practical terms, Hubble and Humason included a section called "Effect of Redshifts on Apparent Magnitudes" explaining that the redshifts either indicated velocities and the universe was expanding or they indicated some as yet unknown law of nature.[58]

In an expanding universe both the light energy and the rate of photon arrival would decrease with distance as an object was moving away from an observer. However, if the universe was static the rate of photon arrival would stay fixed and only the radiative energy would decrease with distance. The theory behind this assertion had been introduced a year or two earlier by their colleague Richard Tolman (Fig. 10.6).

The Massachusetts-born Tolman had earned a degree in chemical engineering in 1903 and a Ph.D. in 1910 from M.I.T. Primarily a theorist he also made important contributions to practical chemistry as related to relativistic mass in special relativity and electricity prior to moving to Caltech in 1922 as professor of physical chemistry and mathematical physics, the same year as he

Fig. 10.6 Edwin Hubble and Richard Tolman with a scale model of the 200-inch telescope at the Mount Wilson Observatory offices in 1931 (Image from the Edwin Hubble Papers, Huntington Library, San Marino, California)

gained a fellowship to the American Association for the Advancement of Science (AAAS).[59]

Tolman was primarily occupied with the quantum world earlier in his career but starting in 1928 he began to publish extensively on the effects of general relativity, particularly as it pertained to de Sitter's work. In 1929 he suggested that if the universe was expanding, a galaxy's brightness should decrease with its redshift (z) at the rate $(1 + z)^4$. Conversely if the universe was static then a galaxy's brightness would decrease at the rate $(1 + z)$. Tolman suggested an experiment where the "surface brightness of a set of standard galaxies" was measured "at different redshifts" to find the true nature of the universal dynamics.[60]

Hubble was no fool and he knew that Tolman's grasp of the field equations of relativity had gained him Einstein's respect. In attempting to substantiate his 1929 paper indicating a possible relation between distance and radial velocity for extra-galactic nebulae, Hubble had been barreling into deep space as fast as Humason could nab larger redshifts. He now began to think it might be wise to take a step back and attempt to find out what the redshifts really meant.

The problem was that the test Tolman was suggesting was impossible to perform at the time (and for many years to come) because there was not yet "a set of standard galaxies" available to be tested. Hubble and Humason hoped the 200-inch telescope might make such a test possible in the not-too-distant future but right now they required another method, and Hubble's solution was to embark on the massive galaxy count program that would support a variety of research objectives, including a redshift test:

> Eventually it may be possible to test the matter by counts of nebulae to successive limits of apparent magnitude on the assumption of uniform distribution, an assumption which appears to be well established to about the limits at which the additional increment would be expected to become sensible.[61]

The word "assumption" occurs frequently in the work of Hubble and Humason. It was employed as a safe barrier between what they perceived might be going on and what they did not yet understand. This was the plight of the two cosmic mariners as they pushed into the limits of space without a guidebook (or perhaps several less than concrete guidebooks). Making it up as they went along appeared to be the only way forward and with good reason.

Consider for a moment that in 2010 the Smithsonian Institute reported 300 new mammal species that had been discovered in the past decade. So many mammals right here on Earth! In the coming decades, thousands more may be found.[62] And that's the stuff that is relatively easy to find. As the microbiologist Nick Lane put in his great book *The Vital Question*, down at the level of microbes "the biggest questions in biology remain to be solved!"[63]

The point is we hardly understand the nature of our own planet. We can touch it, taste it, hear it and test it all we want to and still we don't have it all figured out. So, it should be more than a little understandable that Hubble and Humason, 90 years ago, using what are by modern standards highly outdated equipment, were not quite sure yet what they were seeing. Equally incredible is that they managed to establish the expansion as accurately as they did, considering the generalized (especially in Hubble's case) way in which they went about doing it.

By now Hubble had precipitated two fairly high profile scientific smackdowns, having blown up Shapley's concept of the Milky Way being the entire universe and sent Einstein's static universe catapulting into oblivion. Now he wasn't totally sure about the latter notion, and he wanted to see if he could figure it out before he laid claim to the practical discovery of expansion. It was the right decision in theory, but future missteps in calculation down the road would lead him to personally abandon the Big Bang for good.

Hubble's extensive galaxy count program that he began in 1931 and ended with a classic paper on the distribution of galaxies in 1934 introduced a new and improved outline of the so-called zone of avoidance lying in the plane of the Milky Way. It discussed the uniformly increasing numbers of galaxies at ever greater distances, a smooth, gradual decrease in the number of galaxies upon approaching the plane of the Milky Way, and the mean density of matter in the universe of roughly 10^{-30} grams per cubic centimeter.[64] It was yet another incredible contribution to the advancement of cosmic knowledge, but Hubble made no mention in this paper of his attack on the redshift problem.

That came two years later in *Effects of Red Shifts on the Distribution of Nebulae*, a paper which Sandage would later refer to as "grand in scope and reach, and effective at cementing Hubble's enormous fame – [yet] nevertheless, wrong."[65]

The difficulty for Hubble started with his incorrect assumption that the distribution of galaxies would be uniform, meaning that their density in space would be evenly distributed, increasing with a magnitude of 0.6m. But his Euclidean notion of space was disrupted when the plotted distribution (Observed Relation) increased in magnitude at the much shallower rate of 0.52m.

Another puzzle arose when Hubble introduced the redshift factor for the relative light energy given off by an object (called the energy effect) to his

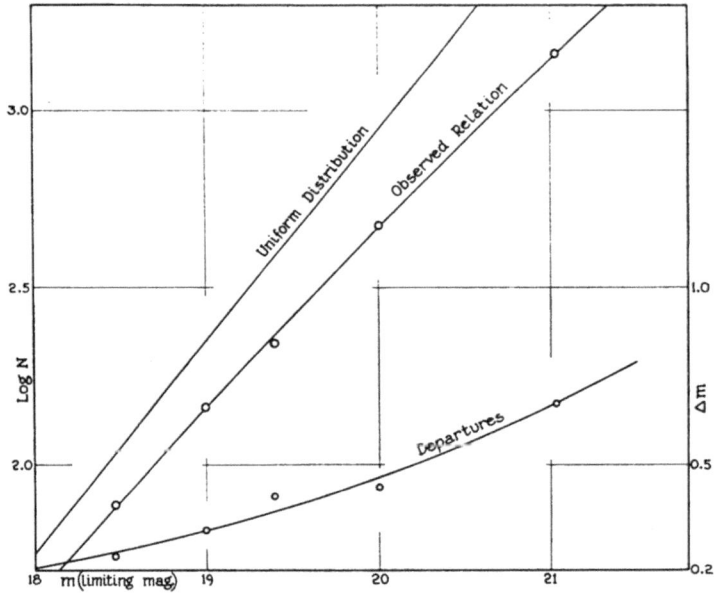

Fig. 10.7 Hubble's galaxy distribution chart from his 1937 paper showing the shallower than expected observed relational magnitude curves that resulted, in part, from his incorrect assumption that galaxies were uniformly distributed.

sample. To his amazement, the data points jumped right onto his assumed Uniform Distribution line. This indicated to Hubble (according to his incorrect assumptions) that redshift might derive from some "new principle of physics," as he called it. When he added the second factor for the rate of photon arrival (the number effect) the plotted points indicated a magnitude even brighter than the Uniform Distribution line (Fig. 10.7).

Dismayed by this new evidence suggesting that he had been wrong about the relation between redshift and velocity Hubble, along with Tolman, calculated the effects of space curvature on the observed data points. This appeared to overcorrect the problem in an even more unsavory way. The amount of curvature was just too great, creating an unrealistically small universe. Satisfied that his extensive number counts, magnitude scales and redshift corrections were correct, Hubble inferred that the redshifts being measured did not indicate true velocity after all.[66]

In fact, Hubble and Tolman's assumptions on both the redshift distance and curved space equations turned out to be incorrect, but Hubble had no way of knowing this in 1936. The corrections would not come until almost five years after his death. And the proper redshift corrections wouldn't be made until the 1960s when Sandage and John Oke recalculated the K corrections (observed magnitudes). These will be discussed later.

Of course, Hubble still held out hope that the 200-inch would enable Humason to push far enough into space to achieve a definitive solution but this plan, too, would prove futile.

It is ironic that the guy who helped lead Hubble to defy his own findings also helped to secure his place in history.

In 1934 Tolman published a book *Relativity, Thermodynamics, and Cosmology* in which he presented the idea that black-body radiation in an expanding universe remains thermal as it cools.[67] This was an indication that cosmic radiation from the Big Bang could still be present (and therefore detectable) in space. This idea would be developed in the 1940s and ultimately prove decisive in the debate on universal dynamics.

The Silliman Lectures that Hubble delivered in 1935 at Yale University were published in 1936 as *The Realm of the Nebulae*. In the book he related the history of development of scientific thought and experimentation which led to the physical discovery of the expansion of the firmament. His prolonged explanations of the scientific terms, inclusion of some of the math involved in the analysis of the subject, and frequent romantic turn of phrase made it inviting reading for laypeople, science enthusiasts and even researchers interested in what he had to say. Its broad popularity was due in part to the excitement being generated by his work with Humason and by the public's need for excitement during the economic doldrums of the 1930s. Here was a book that opened the door to a universe right out of the works of H.G. Wells, filled with millions of galactic systems like our own Milky Way, any one of which might be teeming with life. It was both a commercial success and a personal triumph for Hubble because it made him almost as familiar as Einstein, the ultimate icon of the world of science.

Although Hubble had managed to talk himself out of the reality of the galaxy redshifts, Humason continued his work on the program, now in conjunction with Mayall at Lick with his specially designed quartz spectrograph and the 36-inch Crossley reflector. There was a lot of energy and plenty of anticipation in the future of the project by everyone involved in it.

References

1. What Happened in 1935: Important New and Events, Key Technology and Popular Culture, (thepeoplehistory.com), http://www.thepeoplehistory.com/1935.html
2. Timeline: The Dust Bowl, (pbs.org) https://www.pbs.org/wgbh/americanexperience/features/dust-bowl-surviving-dust-bowl/
3. Alonzo L. Hamby, *Man of Destiny: FDR and the Making of the American Century, Part II: The New Deal,* (Basic Books 2015)
4. *Ibid.* Part III: The World at War
5. Neil Baldwin, *Henry Ford and the Jews: The Mass Production of Hate,* (Public Affairs 2003), pgs. 144-146
6. *Ibid.* pg. 176
7. *Ibid.* pgs. 284-285
8. *Ibid.* pgs. 293-295
9. Anti-Semitic Stereotypes Persist in America, Survey Shows, (ADL 2020), https://www.adl.org/news/press-releases/anti-semitic-stereotypes-persist-in-america-survey-shows
10. Joel Stebbins and A. E. Whitford, *Absorption and Space Reddening in the Galaxy from The Colors of Globular Clusters,* (ApJ 1936), Vol. 84, pgs. 132-157
11. Walter S. Adams, Contributions to the Mount Wilson Observatory, (CIWY 1935-1936), No. 35, pg. 193
12. Caltech, *Palomar Observatory: A History of Palomar Observatory,* (Caltech.edu), Retrieved July 30, 2019, https://sites.astro.caltech.edu/palomar/about/history.html#mirror
13. Donald E. Osterbrock, *Walter Baade: A Life in Astrophysics,* (Princeton University Press 2001), pg. 44
14. *Ibid.* pgs. 71-74
15. Interview with Ann Humason Bernt, 2006, from family archive
16. Donald E. Osterbrock, *Walter Baade: A Life in Astrophysics,* (Princeton University Press 2001), pgs. 75-76
17. *Ibid.* pg. 65
18. *Ibid.* pgs. 65-67
19. *Ibid.* pg. 60
20. W. Baade and F. Zwicky, *On Super-novae,* (PNAS 1934), Vol. 20, pgs. 254-259
21. W. Baade and F. Zwicky, *Cosmic Rays from Super-Novae,* (PNAS 1934), Vol. 20, pgs. 259-263
22. D.E. Osterbrock, *Who Really Coined the Word Supernova? Who First Predicted Neutron Stars?* (AAS 2001), 199th AAS Meeting, id. 15.01; Bulletin of the American Astronomical Society, Vol. 33, pg. 1330
23. Allan Sandage, *Centennial History of the Carnegie Institution of Washington: Volume 1 The Mount Wilson Observatory,* (Cambridge University Press 2004), pg. 204

24. Donald E. Osterbrock, *Walter Baade: A Life in Astrophysics,* (Princeton University Press 2001), pg. 61
25. Interview with Don Nicholson, 2006
26. Gale E. Christianson, *Edwin Hubble: Mariner of the Nebulae,* (University of Chicago Press 1995), pg. 262
27. Sybille Bedford, *Aldous Huxley: A Biography,* (Knopf/Harper & Row 1975)
28. Walter S. Adams, Contributions to the Mount Wilson Observatory, (CIWY 1935-1936), No. 35, pgs. 152-155
29. Interview of Theodore Dunham by David DeVorkin on 1977 April 30, Niels Bohr Library & Archives, American Institute of Physics, College Park, MD USA, www.aip.org/history-programs/niels-bohr-library/oral-histories/4584-1
30. Walter S. Adams, Contributions to the Mount Wilson Observatory, (CIWY 1934-1935), No. 34, pg. 189
31. Interview of Theodore Dunham by David DeVorkin on 1977 April 30, Niels Bohr Library & Archives, American Institute of Physics, College Park, MD USA, www.aip.org/history-programs/niels-bohr-library/oral-histories/4584-1
32. Walter S. Adams, Contributions to the Mount Wilson Observatory, (CIWY 1934-1935), No. 34, pg. 189
33. Interview of Milton Humason by Bert Shapiro in circa 1965, Niels Bohr Library & Archives, American Institute of Physics, College Park, MD USA, www.aip.org/history-programs/niels-bohr-library/oral-histories/4686
34. Gale E. Christianson, *Edwin Hubble: Mariner of the Nebulae,* (University of Chicago Press 1995), pg. 232
35. *Ibid.* pg. 233
36. Interview of Nicholas Mayall by Bert Shapiro on 1977 February 13, Niels Bohr Library & Archives, American Institute of Physics, College Park, MD USA, www.aip.org/history-programs/niels-bohr-library/oral-histories/4766
37. Gale E. Christianson, *Edwin Hubble: Mariner of the Nebulae,* (University of Chicago Press 1995), pg. 233
38. A van Maanen and M.L. Humason, *The Parallax and Spectral Type of Ross 627,* (PASP 1936), Vol. 48, pg. 179
39. Allan Sandage, *Hubble & Humason's Evaluation of the Cosmological Expansion,* (ApJ 1999), Centennial Issue, Vol. 525, pgs. 252-254
40. Edwin Hubble comments on Milton Humason, (NY Times January 16, 1932)
41. Ann Humason Bernt archive
42. Allan Sandage, *Centennial History of the Carnegie Institution of Washington: Volume 1 The Mount Wilson Observatory,* (Cambridge University Press 2004), pgs. 526-527
43. *Ibid.* pg. 496
44. Gilbert King, *The Incredible Disappearing Evangelist,* (smithsonianmag.com June 17, 2013)
45. Interview with Don Nicholson, 2006

46. Draft of a biographical memoir, *Milton L. Humason – Some Personal Recollections,* addressed to Helen Down Humason, Nov. 9, 1972

47. Interview of Allan Sandage by Bert Shapiro on 1977 February 8, Niels Bohr Library & Archives, American Institute of Physics, College Park, MD USA, www.aip.org/history-programs/niels-bohr-library/oral-histories/32867

48. Interview with Ann Humason Bernt, 2006

49. Walter S. Adams, Contributions to the Mount Wilson Observatory, (CIWY 1934-1935), No. 34, pg. 165

50. Walter S. Adams, Contributions to the Mount Wilson Observatory, (CIWY 1935-1936), No. 35, pgs. 193-194

51. Walter S. Adams, Contributions to the Mount Wilson Observatory, (CIWY 1933-1934), No. 33, pg. 154

52. Interview of John Strong by David DeVorkin on 1984 April 20, Niels Bohr Library & Archives, American Institute of Physics, College Park, MD USA, www.aip.org/history-programs/niels-bohr-library/oral-histories/28279

53. M.L. Humason, *The Aluminizing of the 100-inch and 60-inch Reflectors of the Mount Wilson Observatory,* (PASP April 1935), Vol. 47, No. 276, pgs. 81-83

54. M.L. Humason, *Super-Novae,* (PASP May 1936), Leaflet 88, pg. 149

55. *Ibid.* pg. 150

56. *Ibid.* pgs. 150-151

57. M.L. Humason, *Evidence for An Expanding Universe,* (AAAS July 1936), The Scientific Monthly, vol. 43, No. 1 pgs. 80-83

58. Edwin Hubble and Milton L. Humason, *The Velocity-Distance Relation Among Extra Galactic Nebulae,* (ApJ July 1931), Vol. 74, pgs. 69-72

59. John G. Kirkwood and Oliver R. Wulfepstein, *Richard Chace Tolman,* (NAS 1952), pgs. 137-153

60. Allan Sandage, *Hubble & Humason's Evaluation of the Cosmological Expansion,* (ApJ 1999), Centennial Issue, Vol. 525, pg. 520

61. Edwin Hubble and Milton L. Humason, *The Velocity-Distance Relation Among Extra Galactic Nebulae,* (ApJ July 1931), Vol. 74, pg. 30

62. Richard Conniff, *Meet the New Species,* (Smithsonian Magazine August 2010), https://www.smithsonianmag.com/science-nature/meet-the-new-species-748819/

63. Nick Lane, *The Vital Question,* (W.W. Norton & Co. 2015), pg. 51

64. Edwin Hubble, *The Distribution of Extra-Galactic Nebulae,* (ApJ January 1934), Vol. 79, pgs. 8-76

65. Allan Sandage, *Hubble & Humason's Evaluation of the Cosmological Expansion,* (ApJ 1999), Centennial Issue, Vol. 525, pg. 516

66. *Ibid.* pgs. 516-518

67. Richard C. Tolman, *Relativity Thermodynamics and Cosmology,* (Dover Publications 1934)

11

War: Ultimate Disruptor (1937-1945)

As the skies grow more ominous over Europe and Asia, Milton Humason and his wife tour across Europe after he attends a conference for the International Astronomical Union, the highlight being a visit to his sister, Virginia, who is living in the Netherlands. Edwin Hubble enjoys the fruits of fame as a darling of the Hollywood set. The completion of the 200-inch telescope is expected by 1941. But with the perilous slide into war the members of the observatory staff are drawn into service to aid the Allied cause. During this time Humason must come to the aid of his friend Walter Baade, saving him from house arrest and opening the door for his fundamental discovery of the two types of stellar populations. Hubble spends the majority of the war in Maryland, working for the War Department.

On August 1ˢᵗ, 1936, 22-year-old James Cleveland Owens strode into the Olympic Stadium as a member of the amateur athletics team representing the United States of America during the opening ceremony in Berlin, Germany.

By now he was known as Jesse Owens, a nickname given by a grade school-teacher who misunderstood his pronunciation of his nickname "J.C." he went by as a boy. He had been born in Alabama, but his family relocated to Cleveland, Ohio as part of the Great Migration when 1.5 million people of color moved from the South to the industrialized North.

The year before his arrival in Berlin the sophomore, known to his Ohio State University teammates as the "Buckeye Bullet," had broken three world records and tied a fourth in less than an hour at a track and field meet in Ann Arbor, Michigan, a feat still regarded as "the greatest 45 minutes ever in sport." As the story goes, Owens accomplished this while nursing a sore back that he

© Springer Nature Switzerland AG 2021
R. Voller, *Hubble, Humason and the Big Bang*, Springer Praxis Books,
https://doi.org/10.1007/978-3-030-82181-4_11

suffered two weeks earlier falling down a flight of stairs. His injury caused him to make only one attempt in the long jump, but it was enough to establish a world record that would stand for a quarter of a century.[1]

According to the Reich Citizenship Law (one of two Nuremberg Laws the Nazi regime had introduced the year before) the youngest son of a sharecropper and grandson of a slave was regarded as an "enemy of the race-based state" as he entered the packed 100,000 seat stadium, as were athletes of color of all nations. The same laws had stripped German Jews of their citizenship, robbed them of their businesses, lives and livelihoods, defined them to be political enemies or "undesirables" and relegated them to concentration camps in places like Dachau 350 miles to the south in Bavaria.[2]

That week at the 1936 games Owens would humiliate Hitler and the Nazi Party, winning gold in the 100 and 200 meters, the 4 x 100 relay and the long jump, in the process breaking or equaling nine Olympic and three world records. As Will Rogers would later joke, "Mr. Owens…broke practically all the world records…with the possible exception of horseshoe pitching and flagpole sitting."[3]

Along the way the American superstar received a bit of help from one unlikely source. Carl Ludwig "Luz" Long, the German champion in the long jump and Owens's stiffest competition in the event at the games, watched as Owens failed on his first two attempts to qualify in the preliminary round. The two men had become friends in the Olympic village, and now Long went to Owens and suggested that he try jumping from a point several inches behind the foul line. He helped Owens to make two marks one foot apart and Owens made his leap from between them, qualifying for the event. Owens went on to win the gold medal with a final jump of 26 feet 5 inches, which was still more than three inches shy of his own world record.

The first to greet him was Long, the tall blue-eyed, blond-haired runner-up who threw his arm around his new friend in congratulation as they walked along the track in full view of the Führer and the sellout crowd. Owens would never forget his friend's kindness, later saying that Long showed him "a special grace and a special courtesy when I needed help." In awe of his friend's courage, Owens added, "I've experienced many moments in the Sun, but perhaps the most rewarding was to have Luz Long beside me on the winner's platform."[4]

It was a moment of tragic irony that Luz Long died in a British military hospital in 1943 while serving as a private in the German army in Italy, a young man fighting on behalf of his country for a cause which he surely did not believe in. In his last known letter, written to Owens from the field of battle, Long asked his old friend to contact his son Karl (Long might not have known of his second son, Wolfgang, who was born that May and died nine months later) and tell him "what times were like when we were not separated by war" and "how things can be between men on this earth."

True to his friend's wishes, Owens got to know Karl Long and served as best man at his wedding. In a final ode Owens would later write, "You can melt down all the medals and cups I have, and they wouldn't be a plating on the 24-carat friendship I felt for Luz Long at that moment" in 1936.[5]

After his dominant performance at in Germany, Owens returned home to a ticker tape parade in New York City. He was ignored by President Franklin Roosevelt however and faced discrimination that withheld endorsement deals that are so common for amateur and professional athletes today. He was obliged to work menial jobs and run exhibition races against horses for money.

In that same year the Civil War epic novel *Gone with the Wind* was published. Its story followed Scarlett O'Hara, the spoiled young daughter of a white Georgia plantation owner as she battles unrequited love and abject poverty following Sherman's demolition of the South in the final days of the war. It portrayed slaves as docile, happy creatures of the fold and won its author, Margaret Mitchell, the Pulitzer Prize for fiction in 1937.[6]

* * *

Whether Edwin Hubble, a former collegiate athlete in track and field relished in Owens's victory isn't known but what is clear is that the Hubbles were heavily steeped in racist and classist ideologies.

Ensconced in the pristine Caucasian climes of their delightful Pasadena neighborhood, Edwin and Grace were surrounded by mostly like-minded friends who tended to look after each other. While Edwin was on the mountain a neighborhood friend who was a doctor looked after Grace as she miscarried during pregnancy. The Hubbles were frequent dinner guests at local parties and outings. Another neighbor was Homer Crotty, who with his wife were among their favorites. Edwin assisted him to gain membership of the Sunset Club, a group of pipe-smoking, horseshoe-tossing business elites modeled after a group of the same name established in Chicago 50 years prior.[7]

As a product of Deep South conservative Democrats of the 19th century, Hubble styled himself as a sophisticated English gentleman and was careful not to betray his underlying racism to the public. Those characteristics were borne in the pages of Grace Hubble's diary, where some understanding of his Anglophilia may be gleaned from her racist commentary on everyday life. Reflecting on a tour of the Supreme Court on December 11th, 1940, Grace left no question of her anti-Semitic beliefs, recalling a quote from the 17th century English architect, playwright and political activist, John Vanbrugh:

As for these architects and builders of the 'Jew Deal,' Vanbrugh's epitaph will fit:
Lie heavy on them earth, for they laid many a heavy weight on thee.[8]

Following this tour, the Hubbles made their way to a friend's dinner party where Grace noted, "a jet-black enormous chef" was "at the grill, and an Uncle Tom with white wavy hair and the wine list" saw to the food and drink needs of the guests.[9]

As did many in their circle, the Hubbles dismissed those less fortunate citizens suffering the Dust Bowl years in the middle of the country as feckless bumpkins, even as American literature was drawing attention to their plight. John Steinbeck's *Of Mice and Men* made its appearance in February 1937 and Ernest Hemingway published *To Have and Have Not* that October. Both novels dealt with the desperation and anxieties felt by many Americans during the Depression. The Hubbles read these novels of the day with little compassion for the characters portrayed in them.[10]

In stark contrast, the Humasons were of a much more compassionate mindset. Affable and unassuming, Milton and Helen preferred the company of family and close friends.

The loss of Henry Witmer almost 20 years earlier had robbed the two families of their social linchpin. Alice Witmer remained a cherished member of the clan, and by extension the wider Humason family were always welcome guests. The Witmers were grateful to the now retired William Humason for his central role in helping to maintain the various family businesses.

The fact that Milton and Lewis Humason were grown men by the time the Witmer boys arrived in Los Angeles denied them the kinship they might have felt had they been raised in closer proximity. Milton's only other close cousin was Ralph Witmer, who had married but would die shortly after World War II leaving his widow Jacqueline and no children. By far the closest of cousins, Ginny Humason and Joe Witmer, were separated by the Atlantic. Owing to experiencing the harrowing journey home from the Europe in the Great War the two were bonded for life. Ginny was now raising a family with her Dutch husband in the Netherlands.

Joseph was living with his mother as he recovered financially from a small hog farming business which went literally and figuratively belly up in the early years of the Depression. Still in his thirties he was courting Mary Jane Parkison, a UCLA graduate and advertising secretary who was 16 years his junior. They married shortly before Joseph shipped out for active Air Force duty during WWII. Joseph P. Witmer died tragically in 1947 of the same congenital heart condition that had taken his father and uncle so early in life, leaving behind his wife and two children.

As close as the two families had been after the Humasons' arrival in 1902, the death of Letha Lewis in 1950 closed this chapter and placed their fortunes in the hands of the eldest sons of Joseph and Josephine Witmer.[11]

The one political foe that Milton Humason and Edwin Hubble shared was FDR, whose big government proposals and welfare initiatives rankled the conservative in both the rock-ribbed republican and his southern democratic colleague.

Humason's skepticism of Roosevelt was limited to the political realm and never spilled into his social environs, where his calm demeanor and sense of humor were more inviting to all those he encountered. Despite a recession that threatened to erase his economic gains of the past few years he often argued in optimistic terms that a panhandler could still make an honest living working the streams of the nearby foothills for gold nearly a century after the Gold Rush.[12]

In an attempt to reverse the horrors of the Dust Bowl years, Roosevelt signed the Rural Electrification Act to provide electricity to the remotest parts of the country, many of which were suffering protracted drought conditions.

The United Auto Workers Union was formed and in the next two years Roosevelt would back the passage of the Agricultural Advancement Act mandating price support for farm crops in times of low production, and the Fair Labor Standards Act mandating a maximum 44-hour work week, setting a national minimum wage, guaranteeing employees overtime pay and stamping out child labor practices. In a personal move, Roosevelt also established the National Foundation for Infantile Paralysis (March of Dimes) to combat the ravages of Polio, of which he was a sufferer. His steady hand and confident, strident voice during the dismal years of the Depression earned the hearts of most Americans and won him a second landslide election in 1936 over Republican Alfred "Alf" Landon of Kansas.[13]

The harshness of the era left many people pining for attention-diverting stories. J.R.R. Tolkien's fantasy of Middle-earth, *The Hobbit*, debuted in September and Walt Disney's *Snow White and the Seven Dwarfs* became the world's first feature length animated film. Both were instant successes and became classics.

Although naval power was still the driving force of arms development among the major powers, air superiority was beginning to take equal priority in the realm of war strategists. The manifestation of what would become World War Two was as yet unthinkable in most of the world, even as the swift buildup of Germany's forces compelled the governments of Britain, France and the Soviet Union to seek strategic political and military alliances.

On July 7th, 1937, the imperial forces of Japan invaded China, waging an air and ground offensive that wreaked havoc on China's military and its people. Before the end of the year hundreds of thousands of Chinese citizens would be killed or wounded and their villages burned. The ruthlessness of the Japanese assault was terrifying to its survivors as well as to those who bore witness to the atrocities. Several violent attacks against American naval vessels and government personnel diminished public approval of Japan. The US patrol boat named Panay was torpedoed and strafed on the Yangtze River by Japanese biplanes and an American consul named John Moore Allison was beaten by a soldier. Almost immediately, Britain, America and the Soviet Union sent aid to China but by then the cities of Shanghai, Beijing and Nanking had been overrun.[14]

* * *

The combined economic and political stresses of the world were met with the tragic deaths of George Hale and Francis Pease within two weeks of each other in February of 1938. The loss of the spiritual leader and founder of the Mount Wilson Observatory and its legendary instrument designer cast a pall over the mountain and Caltech which wouldn't be lifted for many years.

In the final years of his life George Ellery Hale gained almost mythical status. Although he was scarcely seen at the facilities his presence loomed large over the institutions he had built during the first two decades of the 20th century. The sheer magnitude of Hale's impact on not just astronomy but the whole of science was truly unprecedented. As a student he created the spectroheliograph. He discovered solar magnetism. He drove the establishment of the Yerkes, Mount Wilson and Palomar Observatories with the 40-inch refractor and 60-inch, 100-inch and 200-inch reflectors (the latter, which was still under construction at the time of his death would be named in his honor). He developed and/or co-founded the NRC, ApJ, Caltech and the IAU. He influenced nearly every aspect of the growing super culture at home and abroad. He was responsible for hiring of some of the brightest names in the science world: Adams (spectroscopic parallax), Pease (100-inch telescope and Michelson interferometer designs), Shapley (galactic structure and solar system eccentricity), Hubble (external galaxies, nebular classification and universal expansion), Baade (stellar populations, supernovae and evolution of the universe), Merrill (long period variable stars), Joy (stellar classification), Seares (stat color indices and magnitude classifications), Humason (deep space redshift record and universal expansion) and many more. The great wealth of scientific advancement during the first half of the 20th century owed much to Hale's incredible vision, perseverance and political guile.

Hale's failure to promote and support the advancement of women and minorities, while commonplace at the time, was one of his most poignant shortcomings. We will never know of the lost opportunities for the advancement of science that resulted from this oppression. Look at the contributions to astronomy by Caroline Herschel (comet discoveries), Margaret Huggins (advances in spectroscopy and classification), Henrietta Leavitt (Cepheid period-luminosity discovery), Annie Jump Cannon (large scale stellar classification) and Cecilia Payne-Gaposchkin (stellar spectra and composition, variable stars).

Greater opportunities for women in the field started after Hale's death, at the direction of Walter Adams. Although progress on this reform was slow, their contributions to science and scientific facilities can be seen from the perspective of Henrietta Swope, daughter of MIT engineering graduate and president of General Electric, Herbert Bayard Swope. She came to Mount Palomar in 1952 at the request of Baade and made significant contributions to the accuracy of galactic distances. She later donated a large part of her inheritance to the development of the Carnegie Observatories' Las Campanas facility in Chile.[15]

After the 1936 release of his book *The Realm of the Nebulae*, Edwin Hubble's celebrity hit full bloom. Attending the Motion Picture Academy Awards as the guest of the Oscar-winning director Frank Capra he was presented at the commencement of ceremonies as "the world's greatest living astronomer."

Grace Hubble's great friend, the writer Anita Loos (*Gentlemen Prefer Blondes*) ushered many of her friends to the top of Mount Wilson to spend an evening looking at the stars through the giant reflectors with Edwin as host. Cole Porter and his wife were among them, along with Loos, her husband, and Hubble's friend Aldous Huxley. The Hubbles became friends with Douglas Fairbanks and enjoyed his house parties, which were attended by the Barrymores, Walter Lippmann, and Igor Stravinsky among others. The highlight of this turn with the Hollywood elite appears to have come when Edwin and Grace met Charlie Chaplin at a party hosted by the Huxleys.[16]

While Hubble rode his wave of stardom like a champion surfer Humason contented himself in the shallows of the media world, wading in only as far as he felt he could safely go. After a prolific year of writing, he wrapped up his research on the spectral classification of 16 old novae. Most of them, he noted, showed "strong continuous spectr[a] extending well into the violet, in which neither absorption nor emission lines" could be seen. About half of them showed faint emission lines from either hydrogen or helium, indicating they were probably O-type or early B-type stars.[17]

Reports on novae by astronomers like Humason, Zwicky and Baade led to the discovery, decades later, that these stars were binaries whose constituents usually included a very dense white dwarf (giving the continuous spectrum) and a subgiant or supergiant star. The accretion of hydrogen or helium (hence the emission lines) from the companion on the surface of the white dwarf eventually undergoes nuclear fusion, producing a burst of light.

Following the marriage of William Dowd Humason to Ruth Petty in 1938 Milton and Helen set sail for Europe where they traveled extensively and visited his sister, Virginia Humason Suermondt and her family in Holland (the Netherlands). The trip was scheduled around the meeting of the IAU in Stockholm, Sweden in August, where Milton was greeted enthusiastically by a curious field of researchers seeking, at last, to make his acquaintance.

Political tensions were running high after Germany's occupation of Austria in March. Hitler's veiled rationale for this "Anschluss" was to reclaim land which rightfully belonged to the German people. Having gotten away with it the German dictator turned his sights to the next target.

As the Humasons arrived back in Pasadena in late August the governments of America, Britain, France and the Soviet Union were attempting to dissuade Hitler from invading the "Sudetenland" of northern, southern and western Czechoslovakia where Sudeten Germans supported German annexation.

In the month of September tensions grew by the day as citizens of all nations held onto the hope that cooler heads would prevail. The Polish army amassed along the northeastern Czech border, preparing for an invasion of its own to thwart Hitler's attempt to overrun the country. This prompted a warning from the Soviet government that a Polish invasion would be a violation of its non-aggression treaty. The Soviet army in the meantime was occupying the southeastern Czech border with Ukraine and Romania, which the government vowed to defend with all its military power.

Accompanying these fearful developments was further Japanese incursion into Southern China. By now, Chinese resistance under Chiang Kai-Shek was stronger and the Japanese were halted in several key battles. In an attempt to slow the Japanese, the Chinese flooded the Yellow River, displacing over ten million citizens and drowning hundreds of thousands more.

Under the guidelines of the Munich Agreement, Germany marched into the Sudetenland on October 1st, 1938. The Polish army countered by occupying what it considered to be its rightful territory to the east. One by one the constituents of various territories began to cede and within six months

Czechoslovakia was dissolved in the complex military chess match over its territory.

For the moment public support for Hitler and the Nazi regime was still relatively strong in America, where the aversion to becoming involved in renewed fighting and bloodshed had the same paralyzing effect as it had in France and Britain.

Meanwhile, strong anti-Semitic sentiment fed public ignorance and abnegation toward the atrocities which were being perpetrated against German Jews. The reports of German citizens supported by German paramilitary elements looting and destroying thousands of Jewish businesses and synagogues throughout Germany, Austria and the Sudetenland and killing dozens of their Jewish neighbors while the Nazi regime looked on dispassionately had no effect on public opinion. The massacre dubbed Kristallnacht or Night of Broken Glass for the shards of broken windows from Jewish businesses, led to the incarceration of 30,000 Jewish men in concentration camps.

The wickedness and barbarity against German Jews were brought firsthand to Milton Humason's doorstep on his return from Europe. A young Jewish college student in Vienna named Alice Grosz who had seen Humason speak at the IAU meeting in Stockholm wrote pleading with him to sponsor her student visa application to the US. Her uncle Paul Gross, a commercial insurance agent, had sponsored a visa for her grandmother and the desperate teenager needed one for herself to escape Nazi occupation. The letter survives in Humason's papers, and it seems the famed Mount Wilson astronomer was equal to the task.[18]

Twenty-year-old Alice Gross (the 'z' was changed to an 's' upon entrance into America) sailed with her grandmother from Hamburg, Germany on the S.S. Hansa on January 5th, 1939, arriving in New York Harbor a week later.[19] She and her grandmother lived with her uncle on Riverside Drive in Manhattan where she finished her degree.[20] On August 22nd, 1942, she married Edward Blackman, an engineer and first generation American of Russian descent. The couple lived with Edward's parents in Forest Hills, New York before moving to Manhattan where their only child, Clyde, was born on August 24th, 1949. The family later relocated to California where Alice lived out her days in the warmth of family and friends.[21]

The Naval Act of 1938 that raised spending for battleships in the wake of German and Japanese aggression was a defensive military and political strategy for Franklin Roosevelt, who was facing his first serious test in the 1940 presidential election campaign against the popular Republican nominee Wendell Willkie. Becoming increasingly politically active in the wake of Germany's aggression, Edwin Hubble was prepared to vote across party lines for Willkie due to the former Democrat turned Republican's belief that the US should get more involved in the conflict. Once more the Hubbles found themselves on rare common ground with the Humasons, who were more than willing to vote for Willkie.[22] Roosevelt's promise to keep America out of the conflict in Europe did not yield the same landslide as his past two wins but it helped him get 85 percent of the electoral college and 55 percent of the popular vote.

Unbeknown to most of the world, the discovery of nuclear fission by the physicist Lise Meitner and her nephew Otto Frisch in Germany, and her investigation of the process with chemist Otto Hahn was immediately recognized as a threat within the science community. Discussion of highly charged particles being split in an incredibly explosive chain reaction led the renowned physicist Niels Bohr and others to suspect that Germany was working on an atomic weapon as early as 1939.

That September, Hitler's army invaded Poland provoking Britain and France to declare war on Germany. At the beginning of the new decade most of Europe and Asia was once again at war.

While Roosevelt's administration grappled with the feared consequences of neutrality, Britain bore the brunt of German military might. In October of 1940, as the Hubbles were attending the Hollywood premiere of Chaplin's Hitler parody *The Great Dictator*, London was girding itself under nightly bombardment by the Luftwaffe in the Blitz. In response, the Royal Air Force made bombing runs against German cities including Berlin, Hamburg and Mannheim. With France under Nazi rule, Britain and the Soviet Union strained to keep Hitler in check. The Soviet Union faced the double challenge of preparing itself to counter Japanese threats in Southeast Asia.

The physical, emotional, economic and political realities of life in Pasadena were taking a toll on the staff at the joint Mount Wilson and Palomar observatories. The death of their beloved founder, along with setbacks in the construction of both the 200-inch reflector and the 48-inch Schmidt camera at Palomar were putting a drag on morale.

Fueled by Hitler's continuing aggression emotions were starting to boil over as German-born Baade and Minkowski were increasingly eyed with suspicion by some on the faculty at Caltech and Mount Wilson.

Tensions in the nebular classification camp were no better, where the development of the program on universal expansion was threatened by the slow construction at Palomar. Now entering his fifth decade, Hubble knew that he could ill afford to wait while the world once again tore itself apart. Nevertheless, he was fully committed to American involvement at the outset. Even if he agreed with Hubble that Roosevelt should be more assertive with Hitler and

Hirohito in principle, Humason would have been alarmed at the prospect of his young married son being shipped off to war.

The escalating crisis would eventually embroil much of the planet, pushing back the completion of the 200-inch telescope to 1948 and robbing Hubble and Humason of the last decade of research available between them. During this time Hubble would lose touch with the developments in the field while Humason's total output from 1938 to 1948 would amount to a 12-month period spanning 1937 and 1938.

Mile-a-minute Milt Humason turned 50 on August 19th, 1941 and was presented with a brand-new bicycle by his beloved Helen and family, on which he would speed his way to and from the office, terrorizing neighbors, drivers and pedestrians alike.

The Hubbles' beautiful Spanish Revival home on 1340 Woodstock Road that bordered the boyhood ranch home of future WWII general George S. Patton jr., was a three-mile journey from the observatory offices. Fifteen years after he and Grace moved into the home that they had built shortly after their marriage in 1926 he still preferred to walk to work if the weather allowed. For his turns on Mount Wilson, Grace would drive him to the foot of the hill to take the shuttle to the summit.

When the world was not drawn to the yawning conflict, questions were being raised in other areas of science with embarrassing implications for Hubble. Radiometric dating of rocks in the 1920s had led geologists to conclude that the age of the Earth was on the order of three billion years. Hubble had previously arrived at a total of roughly two billion years for the entire universe.

Brushing such awkward issues aside Hubble continued to lecture widely on the nature of his research, always with Grace at his side. But the scale and scope of the classification and expansion, along with the possibility that the cause of the redshift in galaxies might be some unknown law of physics weighed heavily upon him. His denial of the expansion was due to several factors which can be expressed briefly both in terms of the sheer vastness of space and the fragility of his ego.

Hubble had been dismayed by the results he and Tolman had arrived at in their research, but he was not eager to chase after something that was not real if the math and/or the science did not support it. As has been argued in this book, the preservation of his legacy as a great man of science was nearly as important to Edwin Hubble as his work. His insightful prying vision and open-minded approach had yielded the key discoveries of 1924 and 1929. But what if the universe did, in fact, still have an tremendous mystery to reveal? The prospect of hitting yet another scientific homerun of epic proportions may have been too tantalizing for him to pass up. However, if his initial success had given Hubble the impression that he was the man best qualified to make such a discovery, by the end of 1940 the combination of the scope of the work and his advancing age would likely have caused him to temper his enthusiasm. With the USA seemingly determined to stay out of the war, he tentatively set out to plan his future research.

Humason occupied his time working mainly with Minkowski, Zwicky and Baade. Minkowski's revelation of two evolutionary types of supernovae would be the point of the spear for supernova research for decades to come. Through observations with Baade toward the end of the 1930s Minkowski was able to differentiate between supernovae that possessed broad emission spectra whose luminosity increased with time toward maximum and then decreased, and those with a continuous blue spectrum shortly after maximum that evolved into faint but gradually strengthening emission spectra.[23]

Meanwhile, J. Robert Oppenheimer and two research students calculated how the core of a star would collapse during a supernova, finding that, depending on its mass, it would either produce a neutron star or suffer a runaway process that would end with a singularity now known as a black hole.[24]

When he wasn't feeding Minkowski spectral analysis on exploding stars Humason was publishing reports on the study of blue stars with Paul Merrill and dwarf stars with Alfred Joy. Central to development of astronomical research, this work allowed Humason to clear his slate until he could start using the 200-inch to renew the work on redshifts.

On Sunday, December 7th, 1941, Japan attacked Pearl Harbor on the island of Oahu in Hawaii, killing nearly three thousand civilians and enlisted men and crippling the US fleet. The very next day, in a fiery speech by the thrice elected president Franklin Roosevelt, the United States declared war on Japan.

Since its first incursion into mainland China, Japan's ability to wage unlimited war by land and by sea had been checked by the United States and its allies through embargoes on oil and shipping, and the strategic loan of weapons to China and other nations under threat from the Japanese onslaught. The US had simultaneously and surreptitiously been building its naval and armed forces as well. The Japanese attack on Pearl Harbor was an attempt to cripple America before it could direct its military might on their empire. The US suspected an attack was imminent but didn't know when or where it would take place. Now Roosevelt had the public support he needed, and he was ready to lead the country into war against the Japanese Imperial forces.

So was Edwin Hubble. It was too late to save the world from the damage already done by Germany and Japan, but he felt it was his duty to help the allied cause in restoring peace. He refused a proposal from Carnegie's president Vannevar Bush to participate in the secret Manhattan Project because he wished to be more directly involved in the war effort. Turned down for field duty owing to his age, in April of 1942 he left for Aberdeen Proving Ground in Maryland.[25]

The threat of a Japanese invasion back home in Los Angeles caused public and political paranoia toward Asian Americans to overflow. On February 19th, 1942, Roosevelt signed Executive Order 9066 declaring military jurisdiction over exclusion zones which included Alaska and California and parts of Oregon, Washington and Arizona. It was an act of racist xenophobia from the hysteria of war predicated on a false notion of the security of the state. As a result, 120,000 Japanese citizens (80,000 of whom were second-generation and third-generation Americans) were obliged to give up their homes and businesses and relocate to internment camps for the duration of the war.

Presidential Proclamations 2525, 2526 and 2527 signed by Roosevelt the day after the Japanese attack at Pearl Harbor authorized the internment of people of Japanese, German and Italian descent. But by that time the size of the German population would have meant the relocation of 12 million first-generation and second-generation German Americans and German nationals, which was far too many to manage effectively. Instead, military officials detained and arrested just a handful of particularly suspicious people, most of them German nationals.

One of those slated for detention was Walter Baade, who had not applied for citizenship due to his very real conviction that he would one day return to Germany. After interviews between Baade and Adams acting on Baade's behalf the German astrophysicist was spared detention and arrest but was subject to a curfew that meant he would be bound to his home between the hours of 6 p.m. and 8 a.m. The decision by the Provost Marshall had Baade in a panic. With Hubble off to Maryland on behalf of the war effort he saw his best chance to grab the lion's share of observing time at the 100-inch fading like a supernova. Attempting to aid his prized astrophysicist, Adams wrote strenuous letters to the local War Department vouching for Baade but received little if any sympathy for his efforts. Finally, in May 1942 Adams sent his secret weapon, Milton Humason, with Baade, who was a charming fellow in his own right, to try to convince the Provost Marshall that Baade was worth pardoning. It worked. After some story time and a few laughs with Humason, the Marshall relented, and Baade was given special permission to leave his house after curfew for the sole purpose of working at the telescopes on the mountain.[26]

Seeking to use his almost unlimited observing time to the fullest effect, Baade set about photographing M31 and its satellites M32 and NGC 205 in an attempt to resolve individual stars in these galaxies. He was greatly assisted in these efforts by the fact that there were nightly partial blackouts on the west coast to hamper would-be enemy bombing squadrons. Despite the very dark sky, the resolution still wasn't sufficient. It was normal to use "blue plates" because they were faster, but Baade decided to switch to "red plates" which, while slower, were less susceptible to sky glow. By the fall of the next year, he was able to resolve stars with startling definition for the time.

After some analysis, Baade made his famous designation of the two distinct populations of stars in a spiral galaxy.[1] Metal rich blue population I stars were concentrated in the outer disk or spiral arms. The red and yellow giants of population II, which were rich in hydrogen and helium and low on metals, generally congregated near the galaxy's center.[27]

Baade's first paper on the subject was published in 1944 and it fundamentally changed the course of the discussion of universal evolution.[28] Baade credited the Dutch astronomer Jan Oort for his prediction of two populations in 1926.

Significantly, there were Cepheids in both populations, distinguishable by their spectra. The population I ones like Delta Cephei in the spiral arms were found to be 1.5 magnitudes more luminous than their population II cousins. This meant Hubble's distance scale had to be recalibrated. Baade announced his conclusions at the IAU General Assembly in Rome in 1952, more than doubling Hubble's distance estimated and increasing Hubble's estimate of the age of the universe to more than four billion years, comfortably greater than the age of the Earth that geologists had inferred from radiometric measurements of rocks. After years of relatively friendly rivalry the German had made a key contribution that helped his absent American counterpart.

Humason was able to work only for short intervals during the war and with decreasing frequency as America's role increased. His time was split between home and his travel on orders from the Armed Forces Ordnance Department. He visited optics laboratories around the country to deliver guidelines and offer instruction on the making of prisms for gun and bomb sights that were designed at the observatory lab, and to oversee bombing formations at Aberdeen to improve the effectiveness of allied bombing campaigns.

The war's effect on the civilian population of the US was felt acutely by the Humason family. On March 8th, 1942, Bill and Ruth Humason celebrated

---

[1] In 1978 Population III stars were introduced. These extremely hot stars devoid of metals are predicted by cosmological theory but as yet undetected. One of the goals of the James Webb Space Telescope will be to find them.

the birth of their only child, Ann, with family and friends. Just six weeks later, on April 24th, Milton's mother, Laura, died suddenly at her home in Pasadena.

To make the situation worse, Ruth Humason contracted tuberculosis in October while working as an army nurse and had to be quarantined in a Los Angeles hospital. With their son in distress and in need of support, Milt and Helen took in their infant grandchild. Ann would remain with her grandparents for three years while her mother fought for her life. After a long battle in which she lost a lung, Ruth finally recovered in 1946 and the young family was able to be reunited. Meanwhile, Milton's father William G. Humason died in a boiler explosion at his home on October 20th, 1943. This disastrous chain of events cast a pall over the extended family that continued with the deaths of Ralph Witmer in 1946 and his sister Letha in 1950.

Adding to his stress, letters from Milton's sister Virginia during the war described the hardships that she and Dutch citizens were subjected to as a result of the occupation by the Germans in May of 1940. Her stories told of the chaos and death being suffered under Nazi control. As the occupation dragged on, food became scarce. Ginny and her husband began hiding their Jewish friends in the walls of their family business to save their lives. It was a brave undertaking that meant certain death to all parties should they be discovered.[29]

In the midst of the chaos, Humason was approached by at least one magazine publisher to write an article on the state of the universe from the perspective of the Hubble-Humason project. In a letter to Hubble in Aberdeen, Humason raised the magazine's solicitation and suggested that Hubble write the article. No letter in reply survives from Hubble but it will suffice to say that Humason never wrote the article.

It wasn't the last time Humason would have the chance to offer his take on the expansion problem and his contribution to it in writing. His cousin Tom Humason, an editor at a major publishing firm in the 1950s, wrote him after the war for the same purpose. Not even the gentle nudge from a family member could get him to write such a piece. Humason insisted, as he would continue to do throughout his life, that any claim to the domain over the nature of expansion was Hubble's alone. As has been argued herein this was not entirely true and it would have been well within ethical bounds for Humason to write a piece both discussing the scope and scale of the work and his role in it while maintaining Hubble as its progenitor. In almost every situation Humason's humility, professional courtesy and latent inferiority complex played a role in his decision not to write solely popular works. He avoided public speaking with even greater vigor than writing, even answering a question on this point on a questionnaire for his honorary Ph.D. award from the University of Lund with an emphatic "No!"

Edwin Hubble spent the duration of the war working as chief of ballistics for the War Department, sending for Grace as soon as a suitable home was found in Maryland. She set about making their home cozy in accordance with the standards to which they had become accustomed.

Despite being separated from all astronomy work for the duration Hubble apparently found ways to extend his press coverage, conducting interviews with reporters at Aberdeen on the occasions that they would come visit. He also continued his habit of reading up on arcane subject matter prior to lunch or dinner as a means of appearing knowledgeable on a wide variety of topics. His total investment in the effort to defeat Japan and Germany, the constant pressure of solving the problems involved in that endeavor, and the development of the atomic bomb which ultimately brought the war to an end bore heavily on his health.

The atomic bomb attacks on the Japanese cities of Hiroshima and Nagasaki shocked the world and a wave of anti-nuclear sentiment reverberated from Japan to California. Many of the men from Caltech and Mount Wilson had taken part in the development of the bomb, among them Hubble's friend Richard Tolman of Caltech who served as science adviser to General Leslie Groves, overseer of the Manhattan Project, and J. Robert Oppenheimer, its lead scientist.[30]

The Hubbles returned to Pasadena in the first week of December 1945 and immersed themselves in re-inhabiting the "haunted house" on Woodstock Road. The war was over, but its ravages lingered in the hearts and minds of its survivors. The scent of victory was sweet, but it would be some years before the bitterness of its aftertaste would begin to wane.

References

1. Larry Schwartz, *Owens pierced a myth,* (espn.com 2000), http://espn.go.com/sportscentury/features/00016393.html
2. Greg Bradsher, *The Nuremberg Laws,* (National Archives Winter 2010), Vol. 42, No. 4, https://www.archives.gov/publications/prologue/2010/winter/nuremberg.html
3. Richard Rothschild, *Greatest 45 minutes ever in sports,* (si.com 2010), Retrieved Sept. 24, 2020, https://www.si.com/more-sports/2010/05/24/owens-recordday
4. Mike Rowbottom, *Great Olympic Friendships: Jesse Owens, Luz Long and a beacon of brotherly love at the Nazi games,* (The Independent 2016)

5. *A Friendship that triumphed over racism: Luz Long, Jesse Owens and a lesson for humanity,* (Scroll.in July 15, 2020), Retrieved Sept. 25, 2020, https://scroll.in/field/967461/a-friendship-that-triumphed-over-racism-luz-long jesse-owens-and-a-lesson-for-humanity

6. 1937 Pulitzer Prizes – Novel: *Gone With the Wind,* Margaret Mitchell (MacMillan)

7. Gale E. Christianson, *Edwin Hubble: Mariner of the Nebulae,* (University of Chicago Press 1995), pg. 237

8. Diary kept by GBH on trips she and EPH took to Arizona, mssHUB 72

9. *Ibid.*

10. Gale E. Christianson, *Edwin Hubble: Mariner of the Nebulae,* (University of Chicago Press 1995), pg. 237

11. Virginia Linden Comer, *In Victorian Los Angeles, The Witmers of Crown Hill,* (Talbot Press 1988), pg. 88

12. Allan Sandage, *Centennial History of the Carnegie Institution of Washington: Volume 1 The Mount Wilson Observatory,* (Cambridge University Press 2004), pg. 15

13. Alonzo L. Hamby, *Man of Destiny: FDR and the Making of the American Century,* (Basic Books 2015), pgs. 222-257

14. The Bay Museum, *The Inter-War Period 1938,* (the-bay-museum.co.uk 2019), Retrieved Feb. 14, 2020, https://the-bay-museum.co.uk/2019/11/15/the-inter-war-period-1938/

15. Alfred E. Clark, *Henrietta H. Swope, 78, Is Dead; Helped to Measure Variable Stars,* (NY Times Nov. 29, 1980), pg. 28

16. Gale E. Christianson, *Edwin Hubble: Mariner of the Nebulae,* (University of Chicago Press 1995), pgs. 258-263

17. M.L. Humason, *The Present Spectral Characteristics of Sixteen Old Novae,* (ApJ 1938), Vol. 88, pgs. 228-247

18. Letter from Alice Grosz to Milton Humason

19. US Department of Labor ship's manifest for the SS Hansa, Jan. 5, 1939

20. Sixteenth Census of the United States of America, City and State of New York, April 6, 1940

21. Interview with Clyde Blackman, 2017

22. Gale E. Christianson, *Edwin Hubble: Mariner of the Nebulae,* (University of Chicago Press 1995), pgs. 273-274

23. Rudolph Minkowski, *Spectra of Supernovae,* (PASP August 1941), Vol. 53, No. 314, pgs. 224-225

24. J.R. Oppenheimer and H. Snyder, *On Continued Gravitational Contraction,* (Physical Review Sept. 1939), Vol. 56, Issue 5, pgs. 455-459

25. Gale E. Christianson, *Edwin Hubble: Mariner of the Nebulae,* (University of Chicago Press 1995), pgs. 284-285

26. Donald E. Osterbrock, *Walter Baade: A Life in Astrophysics,* (Princeton University Press 2001), pgs. 98-101

27. *Ibid.* pgs. 101-102

28. W. Baade, *The Resolution of Messier 32, NGC 205, and the Central Region of the Andromeda Nebula,* (ApJ Sept. 1944), Vol. 100, pgs. 137-146

29. Interview with Ann Humason Bernt, 2006

30. John G. Kirkwood and Oliver R. Wulfepstein, *Richard Chace Tolman (1881-1948),* (NAS 1952), pg. 140

12

Aging Companion Stars (1946-1953)

As the dust settles over Europe and Asia, Edwin Hubble and Milton Humason get together with Nicholas Mayall in Pasadena to plan an observing program in expectation of the completion of the 200-inch telescope. During Hubble's absence Walter Baade has ascended to the higher ranks of the astronomical food chain and is awarded more observing time as a result. Hubble's marred reputation leads to his being snubbed by Walter Adams for the directorship of the Mount Wilson and Palomar Observatories being established by Carnegie and Caltech. Attempting to stay astride the broadening field of astronomy, Hubble resists the arrival of young astrophysicists at the observatories. The effects of the war, the advent of the atomic age, and the repositioning of priorities and administration is too much for the aging legend of astronomy and he suffers a massive heart attack on a fishing retreat with his wife, Grace. While Hubble recovers, Humason presses on in the work with help from the young recruit Allan Sandage and Mayall at the Lick Observatory, but the sudden death of Hubble in September of 1953 rocks the nebular classification camp.

The ghastly toll of war can't be compensated by any proportional gain in industrial, medical or technological advancement. No war in history had taken a greater toll of human life than World War II, nor presented the world with more technological advances including the most destructive weapon ever deployed against a civilian population. By its conclusion the deadliest bunch of dictators ever to arise had managed to precipitate the extermination of between three and four percent of the world's population, at least 85 million soldiers and civilians perished. With them went their potential to contribute to the advancement of civilization. Remarkably, for the second time in the

© Springer Nature Switzerland AG 2021
R. Voller, *Hubble, Humason and the Big Bang*, Springer Praxis Books,
https://doi.org/10.1007/978-3-030-82181-4_12

first half of the 20th century the world had faced a devastating confluence of political strife, militarism, destruction of infrastructure and loss of civilian life from the barbaric bombing of cities in Europe, Africa and Asia (including Pearl Harbor).

Resulting from the competition between warring factions for supremacy of the sea, sky and land were a host of newly developed technologies, some of which would lead the world forward through the end of the century and beyond.

Radar and sonar developed during the war were instrumental in detecting the approach of aircraft, submarines and other naval vessels. Jet aircraft, helicopters and rockets would drive a new airline industry and set the stage for the exploration of space. Analog computers presaged the digital computers that have driven almost all modern technological advances. In the field of astronomy, new technologies would provide ever more precise observing instruments and methods for cosmological research.

Just as the Great War delayed completion of the 100-inch telescope at Mount Wilson, World War II put back the 200-inch at Mount Palomar. At the close of the war the giant dome, standing like a great cathedral of science, was empty awaiting the resumption of the assembly of the telescope on its enormous pier.

Much change had come to the MWO in the four years since Hubble left Pasadena for war work in Maryland. In 1942 he still reigned supreme over the astronomy world, sharing the scientific spotlight with the likes of Albert Einstein. By the beginning of 1946 however, although his status in the public sphere remained high, his command of the advances in the field had been diminished. The postwar era in astronomy would belong first to the likes of Baade, Zwicky and Minkowski and later to Sandage, Oort, Hoyle and others whose grasp of the new astrophysics, inventive spirits and access to improved apparatus would steer the field into the next millennium.

By the war's conclusion, Hubble's physical presence had been diminished. He appeared weakened by his administration of the war effort and lamented the use of atomic weapons against Japan. The archconservative who had railed at Roosevelt's reticence toward getting the country into the conflict was now sounding a note of pacifistic alarm at the new threat of nuclear cataclysm.[1]

On a more personally painful note, he learned prior to his return to California that he was being passed over for the directorship of the combined Mount Wilson and Palomar Observatories. His record of aggression, divisiveness and selfishness had finally caught up to him as the power in Pasadena migrated from the astronomers to the physicists.

In a move led by the outgoing director Walter Adams, Hubble's name was replaced for the directorship by Ira Bowen, a recent recipient of the Draper Award for his discovery of spectral "nebulium." William Huggins had coined the name for what he believed to be light from an unknown element in a nebula he discovered in 1864. Bowen found it to be doubly ionized oxygen in gaseous nebulae. To Adams's mind, Bowen was far better suited to the administrative nuances of running a massive observatory facility. Primarily a physicist with a firm grasp of the major threads of astronomical research, Bowen was in a prime position to administer to the needs of the combined staffs at Caltech, Palomar and Mount Wilson.

On catching wind of the impending slight over the directorship that he and almost everyone at the observatories had assumed would be his, Hubble apparently made some efforts to adjust his attitude and demeanor in the hopes of retaining the title. Notes from Humason to Mayall at Lick Observatory told of Hubble's change in attitude in optimistic terms. But it was too little too late. Adams and former Rockefeller president Max Mason had endorsed Bowen. As far as Mason was concerned Hubble was arrogant and self-serving, undesirable qualities for a leader of an organization.

Concerned that Hubble might make trouble for Bowen after the appointment, Carnegie president Vannevar Bush consulted colleagues of Hubble's such as Richard Tolman, who confirmed his apprehensions. Hubble's agenda was always first and foremost in Hubble's mind and would weigh heavily on the scheduling of time on both mountains, especially the soon-to-be-finished 200-inch. With these comments in mind, Bush worried that his initial idea to offer Hubble the same salary as Bowen would be insufficient to quell any pending storm, and so he urged Bowen to be forceful with the veteran astronomer at the outset. The combined observatories were now the marquee department of research at Carnegie and to have chaos and confusion erupt at the top of the administration would, at the very least, be an embarrassment.

Adopting a softer approach, Bowen acted quickly by forming a new Scientific Program Committee and making Hubble chairman. Bowen did this in writing to Hubble, informing him of his appointment and asking for confirmation of his desire to lead the new committee. As expected, the response was sullen, stern and stubborn-headed but the stricken 57-year-old legend agreed so long as he was allowed the freedom to pursue the work as he saw fit. Having quelled tensions for the moment, Bowen agreed to Hubble's demand, all too aware that he might have to intervene if things took a negative turn.

True to his nature, Hubble almost immediately abandoned his responsibilities as chair of the Program Committee, assembling its members only once in its brief and ineffectual history. Having no interest in sparring with Hubble,

Bowen simply distributed the various responsibilities of the committee among its members.

At this point, although his work had been central to its development, Hubble wasn't yet familiar with the term cosmology, which had come into use during the war and was gaining traction. Nevertheless, he sounded off to Bowen like an aging peacock whose tail feathers were clipped, informing the already well-informed director-elect that he had "given much thought to the big problems in [cosmology] and ha[d] rather definite notions on the method of attack."[2]

Hubble was determined to pick up right where he had left his research at the start of the war. Although Baade had made significant progress on stellar populations, as far as Hubble was concerned nothing had changed and to some degree, he was correct. As Allan Sandage would later put it:

> From 1929 until the discovery of the Alpher-Herman microwave background relic radiation in 1965…the field of "practical cosmology…" was…"simply the search for two numbers."[3]

The numbers Sandage was referring to were redshift velocity and magnitude. Everything else fell under the rules set by those two values. Hubble was still fixated on the central questions which he had been grappling with back in 1941, namely the extent of the velocity-distance relationship and the quest to ascertain whether the redshifts were caused by real velocities or some unknown physical phenomenon, which Hubble himself believed must be the case.

Cosmology, the new and vast field of research, involved the study of the evolutionary structure of the universe and everything in it. It was the perfect extension of George Hale's original mission for the observatory. Hubble and Humason had introduced it to the world and now Baade and the new generation of cosmologists were starting to refine it.

When it appeared to Baade that the mode of operation at the observatories was returning to its prewar routine, which had been dominated by Hubble's program, he immediately went to Bowen and complained bitterly. This signaled the onset of a rift between Baade, Hubble and Humason, the latter now ranking third behind Hubble and Baade in the department of nebular photography, photometry and spectroscopy, which also included Joseph Johnson, Minkowski, Edison Pettit, Albert Wilson, Zwicky and Alice Beach. After the retirement of Alfred Joy, Humason was given the job of scheduling time on the telescopes, and it was not long before Baade was complaining that

Humason was favoring his own necessarily time-intensive observing requirements.[4]

Previously in good standing with his longtime friend after helping him escape detention during the war and inadvertently rewarding Baade with almost full control of the 100-inch, Humason could reasonably have felt comfortable in claiming a bit more observing time on the new telescope. He was again preparing to set a course into the deepest reaches of space and no doubt believed that Baade would understand the time requirement for that research. But Baade took him to task calling him, among other things, "a perfect conceited ass!"[5] Baade's bitterness on the subject would eventually envelop Mayall at Lick and Allan Sandage and Halton Arp, the first recruit from the new astronomy graduate program at Caltech, and it would last well into the Palomar years after Hubble's death.

In contrast to the breakdown in relations with Baade, the Hubble-Humason partnership was probably at its strongest in the waning years of their tenures. Having begun his career officially in the fall of 1919, Hubble's mandatory retirement was set for the end of 1954. Humason's turn was 3 years later. If Hubble stayed on in an honorary status with Humason to provide three more years of data collection, the team might still be able to count on some additional years of output. Although the socio-political divide that shaped their early lives was ingrained, their mutual dependence and incredible success had made a working kinship that straddled the social chasm.

Hubble and Humason began to make preparations for the opening of the 200-inch and discussed the tasks that could be done in the interim. Their partner in nebular classification, Mayall, who had worked on rocketry and eventually made his way to the Mohave Desert and the Manhattan Project, had returned to Lick and was ready, willing and able to get back to work.[6]

In his work with the 200-inch telescope, Humason had his sights set on the nice round number of 500 velocities and he hoped to end the 100-inch program of redshifts in the coming year. From there, with help from Mayall and the 36-inch Crossley reflector, they could work toward wrapping up the entire caboodle before the duo retired.

The Mount Wilson and Palomar Observatories, as they were officially called at the time, were brought together on April 1st, 1948, under the combined leadership of the Carnegie Institution of Washington (CIW) and the California Institute of Technology (Caltech). Administration of the facilities would be overseen by an Observatory Committee with six members: Bowen, Adams and Hubble from the Carnegie side and Max Mason, Richard Tolman and E.C. Watson on the Caltech side, with Bowen also serving as chairman.

Having prolonged his directorship to the end of the war, Adams was using Hale's solar lab in Pasadena as his personal facility in retirement. He brought in Mason to ensure there would be a check on Hubble. They then decided to include Tolman, a thoroughly congenial friend and associate of Hubble in the hopes his presence would help to assuage the latter's delicate ego. Unfortunately, Tolman died in September of that year.[7]

The observatories would also embark on an extensive educational program in keeping with Caltech's mission. Guest investigators would be welcomed with authorization from the Observatory Committee. The merger also negated the need for a physical laboratory at the Mount Wilson offices, so these were abandoned, and the lab staff put in new quarters at Caltech where such facilities were already in place and could be utilized for both education and research (The labs at Santa Barbara Street were later modernized and restored under the auspices of the Carnegie Observatories).

June 3rd, 1948, was inauguration day for the 200-inch with 800 representatives of the Mount Wilson and Palomar Observatories, the Carnegie Institution of Washington, the Rockefeller Foundation, Caltech, the Astronomical Society of the Pacific (ASP) and the American Association for the Advancement of Sciences (AAAS) making their way to Palomar. The crowd shuffled along the walk up to the huge white telescope dome rising up into the sky on the 5,600-foot summit. Once seated on wooden folding chairs inside the dome they were treated to a short ceremony in which the instrument was dedicated the Hale Telescope, provoking a standing ovation for George Ellery Hale who had played a vital role in the advancement of science throughout his life until his death ten years earlier at the age of 69.

Above their heads rested the massive new engineering marvel, all 500 tons of it, secure in its huge steel equatorial mounting. The giant 200-inch Pyrex primary mirror, 20 inches thick and weighing over 14 tons (six tons lighter than before it was ground and polished to a paraboloid) sat awaiting the starlight. The mirror was held in place by 36 point supports that helped it hold its shape during observing runs. The mirror's total surface area covered 31,000 square inches and provided a focal ratio of f/3.3 at the primary focus and f/16 at the Cassegrain focus. So precise was its balance and gearing that it was able to be driven across the night sky by a one-twelfth horsepower motor.

The 1,000-ton dome was almost as impressive as the telescope. The slit door weighed 125 tons and the dome completed a rotation of its 430-foot circumference in 4 minutes with four 7.5 horsepower motors rolling it on a set of 32 four-wheel carriages. The plate steel exterior was separated from the aluminum interior by a 4-foot layer of air (a feature carried over from the 100-inch) to help regulate the inside air temperature. Visitors gawked at the sheer size of the dome, 137 feet in diameter and towering more than 13 stories overhead.[8]

Nearly lost in the commotion surrounding the 200-inch telescope was the completion of the 48-inch Schmidt camera nearby on Palomar. This marvelous new wide field instrument employed a 14 x 14-inch plate and covered 40 square degrees of the sky in a single exposure, hundreds of times that of the giant reflector. It was excellent for sky surveys, research reconnaissance and projects like the supernova searches that were already being done by Baade, Humason and Zwicky using the smaller Schmidt camera at Palomar and the large reflectors at Mount Wilson. Its superb optics allowed Hubble, Baade and others to photograph vast regions of space for direct photography and photometric measurements with extremely high precision for that time.

***

With the United States and the Soviet Union squaring off as the world's only superpowers, conflicts broke out between the new states which emerged from the wreckage of the Second World War.

Edwin and Grace felt a hint of Soviet oppression on a trip to Europe in 1948, their first passage since before the onset of the war. Edwin had been awarded an Honorary Fellow of Queen's College in May and so the Hubbles set sail in July to pay a visit first to Oxford to thank his friends in person and then on to Zurich for the Seventh General Assembly of the International Astronomical Union (IAU). Hubble had been selected as one of the American delegates to the meeting and was chairman of Commission No. 28 on extragalactic nebulae. During a banquet at the start of the summit a delegation of Russians suddenly stormed out when they noticed the absence of their flag on the wall. After an appropriate banner was fumbled into place, they marched back in.[9]

The Indian spiritual and political leader Mohandas Gandhi was assassinated during the partitioning of India with the newly created Pakistani Socialist state. Accused by many for being too accommodating in the negotiations between the two countries, Gandhi was shot three times in the chest by Nathuram Godse in New Delhi on January 30th, 1948. Gandhi, the man for whom the honorific title Mahatma or "great-souled" would be bestowed in the years to come, died at Birla House within minutes.[10]

Following its recognition as a sovereign nation, Israel was immediately dragged into a skirmish with its neighbors in what would be named the Arab Israeli War. At the close of this short-lived battle in 1949 Israel would hold an additional 60 percent of the surrounding territory while Jordan held the West

Bank and Egypt the Gaza Strip. It was one in a long succession of battles between the nations of the so-called Middle East that have plagued relations to the present day.

At the close of World War II, the Korean peninsula was split along the 38th parallel by the United States and the Soviet Union. It soon became a battleground for the brewing Cold War as North Korean forces poured into South Korea in the spring of 1950. The two sides jousted for three years, killing as many as three million people and destroying much of the infrastructure on both sides before a stalemate prompted an armistice agreement that reinstated the border.

Despite the heavy toll in terms of loss of lives and the devastation in Honolulu from the Japanese attack on Pearl Harbor, the United States had incurred no damage to its infrastructure. In contrast, most of Europe smoldered in ash and ruin. To bolster the rebuilding of western Europe and disrupt the spread of Communism, the Marshall Plan provided billions of dollars in aid to Britain, France and Germany and several small countries.

Although some prewar economic sluggishness remained in America, there was a sense of national pride as the country worked to convert its industries from wartime to peacetime activities again. Automobile manufacturing and a housing boom led the way, and an influx of mergers of large corporations brought jobs and prosperity. Through the coming decade a growing number of Americans felt they were a part of the middle class of the country. Meanwhile, a range of minority groups would continue to struggle to gain acceptance and share in the wealth.

* * *

In the field of cosmology three themes were set for research by 1946: (1) the expansion of the universe, which most researchers considered to be real; (2) the possibility that the redshifts were caused by an unknown property of physics, which pretty much only Hubble believed; and (3) the cosmology of everything else, which seemingly everybody other than Hubble, Humason and Mayall was starting to work on.

This was the cosmology of the age, and it was a riveting bit of theater for those involved, including in Hubble's camp at Carnegie and Lick. For those seeking to understand the inner workings of galactic structure, there was general consensus that the work done there would ultimately alter or even correct Hubble's earlier work.

In 1948 a fourth thread of scientific theory arose that added a new wrinkle to the debate on expansion, one that utilized both the micro and the

Fig. 12.1 Fred Hoyle ca 1972. (Image courtesy of Caltech Archives)

macrocosmic elements of the field. Fred Hoyle (Fig. 12.1), who was teaching astrophysics at St. John's College at Cambridge in England, was developing bold new theories on stellar and universal evolution.

By temperament Hoyle was a bit like the English version of Zwicky. Brilliant, acerbic and contrarian, he made several important contributions to cosmology. Like Zwicky, a few of his counterclaims were eventually born out experimentally. But, also like Zwicky, many were later rejected or never adopted seriously. In the light of his failures Hoyle would often scoff, "it is better to be interesting and wrong than boring and right."

In 1946 Hoyle began building on the work by Hans Bethe on stellar nucleosynthesis in the previous decade which presented evidence that fusion at the beginning of the universe created hydrogen and helium in sufficient quantities to promote star formation and some of the common elements. This physical condition had been proposed in theory by Arthur Eddington in 1920 based on previous work in physics by Nobel laureates Francis William Aston and Jean Perrin. On this basis Hoyle was able to show that main sequence stars in states of relative equilibrium with core temperatures in billions of degrees, would produce elements heavier that helium, in particular carbon and oxygen. In the 1950s Hoyle would develop this new theory of stellar nucleosynthesis still further, predicting that the elements between carbon and nickel were produced in the expanding cores of late-stage supernovae.[11]

The universe that Hubble and Humason had introduced in 1931 to an astonished world was taking on some interesting evolutionary characteristics that were opening up new fields for research in practical cosmology.

In 1948 Hoyle introduced a concept which challenged the expansion concept directly. During the war he had worked at Portsmouth for Britain's radar project where he met and befriended Hermann Bondi and Thomas Gold. The trio discussed in detail many aspects of cosmology and their implications. Hoyle later spent some of the money he earned for his war work on a visit to Mount Wilson and Caltech, where he developed some of his concepts further. His clash with the expansion theory developed during this period and he later came to reject Lemaître's theory on the basis that it was, in his opinion, "a false process."

In fact, Hoyle had no issue with the expansion. He rejected the notion that it pointed to a beginning and thus a Creator. Hoyle, an atheist, couldn't stomach this idea and, with Bondi and Gold, presented what became known as the Steady State theory as an alternative to expansion.

The theory, which seemed consistent with Einstein's relativity, put forward the notion that the universe could be eternal (that is, have no beginning or end) and unchanging while also seeming to expand. On this point Hubble may have been interested to hear what Hoyle had to say, because he was pondering an elusive force that would explain the redshifts as being caused by something other than velocities.

Basically, Steady State said that as galaxies moved away from each other new galaxies were formed in the intervening space. This necessarily meant that new matter had to form to provide the elements for the new galaxies. Hoyle presented this as a negative pressure "creation field" or C-field with properties consistent with the conservation of energy in the field as predicted by Einstein.

What Hoyle seemed to be suggesting was that spacetime was like a river and that river had a spring, many, many springs, and that these springs provided the matter to form new galaxies in the voids opened between the receding galaxies.

This idea was not especially popular with most of the staff at the Caltech Observatories, but Hoyle's overall reputation meant it attracted more appreciation elsewhere. In particular, on a BBC radio broadcast on March 28th, 1949, he spoke dismissively of Lemaître's concept as "the big bang" and the moniker stuck.[12]

After being derided for this remark by members of the scientific community, Hoyle fired back during another BBC broadcast:

> The reason why scientists like the "big bang" is because they are overshadowed by the Book of Genesis. It is deep within the psyche of most scientists to believe in the first page of Genesis.[13]

This was all very new science, and cosmology was still branching out into various fields of research while trying to play catch up as Hoyle and others produced new theories of how the universe might work. The team of Hubble and Humason had produced a lot of evidence strongly implying the expansion was real, but there were still many questions. To help the duo navigate the difficult waters ahead a new breed of researcher was needed, one with a comprehensive understanding of both the astronomy and the astrophysics of the day.

In 1944, 18-year-old Allan Sandage (Fig. 12.2) was drafted into the navy and spent the remainder of the war as a radio and radar technician stationed in the US. After the war he resumed his college career using the G.I. bill, entering the physics program at the University of Illinois where his father was a professor. He graduated in 1948 at around the same time Caltech was hunting around for bright young talent to train as astronomers with the state-of-the-art equipment at Palomar. Sandage applied to the school and was accepted in 1949. Although Baade was his study advisor he worked as a research assistant to Hubble.[14]

It was a dream gig for a kid from Ohio who counted Hubble among his childhood heroes along with Galileo, Newton and Einstein. Sandage had a grasp of the latest developments in physics needed to enable Hubble and Humason to push the velocity-distance relationship forward and to clean up any related mess they left behind.

Sandage's first impression of Edwin Hubble was something of a letdown. The 60-year-old legend had been slowed by the ravages of war in recent years (and perhaps his affinity for buttered eggs and bacon). From the first Hubble, who Sandage later recalled as "a most complicated man,"[15] insisted on

Fig. 12.2 Edwin Hubble sits at the prime focus position of the 200-inch Hale telescope Oct. 9, 1950 (Image from Edwin Hubble Papers, Huntington Library, San Marino, California)

formality and was essentially humorless in his personal dealings. To the self-described "hick from the Midwest," the aging superstar seemed almost to be more of a persona than a fully realized human being.[16]

Mayall, just getting back to work himself, was some 15 years younger than the stars of Mount Wilson and 20 years older than his new teammate, Sandage.

With a new generation of state-of-the-art telescope technology and researchers with the skills to exploit its full potential assembled, the Hubble-Humason camp was finally able to pursue their research program in the modern era.

Sandage began his work on the project in May 1949 in the plate library in the basement of the offices on Santa Barbara Street. Rudolph Minkowski had launched a new sky survey on the 48-inch Schmidt camera at Palomar and Hubble wanted to use the depth and clarity of the images from the new camera to revise his galaxy counts, magnitudes and distances to tens of thousands of galaxies in the hope of finally being able to detect the curvature of space and infer something about a mysterious process that he thought was responsible for the redshifts.

For his part, Sandage came to Santa Barbara Street in the unwavering belief that the redshifts were true velocities. Humason, who had published widely on the matter (although carefully calling them "apparent velocities" in deference to Hubble) leaned in that direction too. But Hubble loomed large over the observatories and the rest of the cosmology field, so Sandage quietly got on with it, happy to have realized his dream of working at the greatest observatory with some of the greatest astronomers of the century. His reverie would be short-lived, however. He had barely been working two months when the word came from Minkowski that Hubble had suffered a massive heart attack on a fishing trip in Colorado. The altitude and distance to the nearest hospital, along with Hubble's reluctance to admit to himself and others that he was in serious trouble until hours after his symptoms started, exacerbated the severity of his condition.

The news shot through the astronomy world. Most affected was Humason. Through the decades, the symbiosis that existed between the two men had strengthened their bond with Humason not only providing expertise at the telescope but personal qualities that shielded Hubble from the wrath of Adams and various others who found his intrapersonal dealings distasteful.[17]

After months of convalescence Hubble's health improved sufficiently for him to travel. Accordingly, he left for the east coast and then Europe in the spring of 1950 on an extended trip that he and Grace hoped would restore his strength.

Hubble's prolonged absence gave Sandage a chance to get his feet wet on the operation of the telescopes, learning the ropes of the instruments as well as the political ins and outs. There were occasional dustups between researchers but in general the mood was friendly. As jobs went, it didn't get much better than the romance of the heavens (astronomers rank very high in professional happiness ratings). Still, some people around the facilities weren't above a little mischief from time to time.

Sandage noticed how he and the democrats on the staff were often scheduled on the mountain on the first Tuesday in November in election years. Milton Humason, the rock-ribbed republican was in charge of scheduling the observing time for the stellar department for both mountains until his retirement in 1957. Once Sandage sniffed out the scent of foul play he went to Humason to complain about the apparent slight, then visibly blanched when the ace spectroscopist shrugged and said, "Talk to Nicholson, he's been doing the same thing to the republicans on the solar staff for years."[18]

Although each of the large reflectors on the mountains had individual idiosyncrasies, it was customary for recruits to cut their teeth on the 60-inch. As the original large reflector, it was to some the darling of the Mount Wilson instruments. Sandage and his fellow rookie Halton Arp took their turns working with the telescope, learning the error in the drive clock and how to navigate while ensuring the dome and observing platform were kept in sync. A moment of neglect could result in damage to the instrument or its platforms and months of expensive repairs.[19]

Rudolph Minkowski was famous for driving the telescope into the Newtonian platform. So poor was his friend and fellow gambler at operating the telescope that Humason thought it would best if he didn't use it at all. There were plenty of other instruments at the facilities, most of which were easier to operate, making life easier and safer for both the astronomer and the telescope.

By the time Hubble was finally cleared for work on the mountain in October of 1950, Sandage had already trained for several months with Humason on the 200-inch. Humason and Sandage accompanied Hubble on his first run on the giant reflector in two years on a chilly evening. Another benefit of the war was the invention of heated flight suits for pilots of high-altitude aircraft. The body warming overalls had been adapted by the staff at Mount Wilson and Palomar to shield them from the cold winter winds. The new suits likely made Humason wish he'd been born a generation later.

Hubble donned his suit and navigated the stairs to the lift that would take him to the prime focus at the top of the tube, where he climbed down into the telescope to start his run (Fig. 12.2). One of the unique features of the 200-inch was the viewing area was placed at the front of and inside the telescope tube. This allowed the observer to ride along with the telescope while exposing a plate. This arrangement was possible due to the sheer size of the primary mirror and the diameter of the tube.

Although Hubble probably felt "the great joy of using" the Hale telescope, as Humason later put it in his obituary to his late friend and colleague, he had to know too that his time was running out.[20] The war had swept away at least five important years of research.

Hubble's ongoing health issues kept him from observing during the cold winter months. Instead, the legendary observer delivered lectures at Caltech and in May 1953 the George Darwin Lecture at the Royal Astronomical Society (RAS). Near the end of the lecture *The Observational Evidence for an Expanding Universe* he presented a slide showing the negative of what was then the faintest galaxy taken with the 200-inch, remarking, "This is the last horizon."[21]

The remark apparently met with applause from a rapt audience of admirers but was of course off by orders of magnitude. Estimating the 200-inch would be able to go out to about a billion light years Hubble and Humason presumed it would be able to reach the horizon. Baade and a host of younger researchers were beginning to wonder if this might lie much farther out, perhaps as far as fifteen to twenty billion light years.

As for those who flocked to the Caltech halls to hear Hubble lecture, they heard little of cosmological models and related concepts because he was conflicted on the interpretation of the redshifts and in any case, he lacked a deep understanding of such theories.

Carnegie Observatories' director Ira Bowen believed the redshift program was worthy of attack and allowed Humason to proceed to the extent he was able, but he saw little value in revisiting Hubble's faint galaxy counts. He considered this endeavor too time consuming and not guaranteed to deliver much useful data. Bowen, like many at the observatories, was under the impression redshifts were real velocities and that time, equipment and improved methods of research would bear this idea out; Hubble's continued skepticism was simply unwarranted.

For his role in various areas of research Humason collected an honorary Ph.D. from the University of Lund in Sweden in 1950 and selection to the RAS in 1951. In a letter to Knut Lundmark, Humason thanked Lundmark for his role in nominating him for the award. He was still very active at age 62 and continued in his dual role as researcher with Hubble and later Zwicky and as mentor to Mayall, Sandage, Arp and others. His humble down-to-earth charm and love of storytelling appealed more to Sandage than Hubble's stiff, self-centered ambition, but the young astronomer revered Hubble as an icon of science. Over the years, Sandage would continue to extol the virtues of Hubble's genius and Humason's humanity. When they could find the time, he and Humason went fishing in the nearby Sierra Madres and sometimes Mayall would find his way down to join them.

On one of these occasions in 1952 Sandage joined Humason and Don Hendrix, the genius optician of Carnegie Observatories, on a 10-day trip fishing steelhead trout (a Humason favorite) on the Eel River, 250 miles north of San Francisco. The inveterate Everyman, Humason arranged to have their daily catch cooked for them at a local restaurant where the three men feasted on the piscatorial abundance they had reaped. When Sandage queried his sagacious companions on how they figured they could fish in the afternoon when they had each caught their limit in the morning the pair of autodidacts said, apparently without even so much as a smirk, that the rules only applied to the number of fish you had on you at any given time, not the entire day's catch.[22]

On an early run on the Hale Telescope using the new Schmidt spectrograph Humason was able to get a good spectrum for a cluster at 8h 55m 17s and + 30° 21'.3. The dispersion was 430 Å/mm with an error of one to two percent. He measured a recessional velocity of nearly 61,000 kps for the faintest member. The linear relationship between velocity and distance was still holding.[23]

But the 200-inch soon ran into difficulty exposing redshifts for fainter galaxies. It turned out that the ambient light from the sky was saturating the plates and making the spectra too faint to read accurately. Humason and company would simply have to wait until upgrades could be made to renew the assault on redshifts.

On Hubble's return from England in late 1953 those who knew and worked with him noted that he looked relaxed and in good form again. He had a renewed sense of what he thought he could accomplish in the roughly two years he had left before his retirement and met with Humason in his office to discuss a course of action. As he had remarked to Grace on a visit to the 200-inch during his observing run in early September he believed "in two years" he could "have determined the redshift," a value that needed to be reset in the wake of Baade's revelations. As for the rest of cosmology – the state, size, structure and evolution of the universe – he told his beloved wife and nearly constant companion that he would not live to see it.[24] His words couldn't have been more prophetic.

Returning home from the office on September 28th, 1953, Edwin Hubble died of a stroke with Grace by his side. That morning he had met with Humason to discuss details of their combined plan, then decided to walk home for lunch as he often did. Grace found him on her way home from running errands, walking along California Street with his walking stick in hand looking every bit the man she had known for 30 years. She picked him up and they drove home. As they arrived at the house on Woodstock Road, Hubble suffered a cerebral thrombosis.[25] Grace phoned Humason at the observatory, who went over to the house to help her.[26]

Edwin's instructions for Grace were, "I want to disappear quietly." She was determined to carry them out. With a small group of very close friends that included the Crotty's, Grace had his body cremated, and with that Edwin Hubble's star was joined with the cosmos, his remains buried in a copper box in a secret place by Grace with a small entourage invited to see him off to eternity.[27]

References

1. Gale E. Christianson, *Edwin Hubble: Mariner of the Nebulae,* (University of Chicago Press 1995), pgs. 310-311
2. *Ibid.* pgs. 304-310
3. A Sandage, *Hubble & Humason's Evaluation of the Cosmological Expansion,* (ApJ 1999), Centennial Issue, Vol. 525C, pg. 252
4. Ira Bowen, Contributions of the Mount Wilson and Palomar Observatories, (CIWY 1948 1949), No. 48, pg. 24
5. Gale E. Christianson, *Edwin Hubble: Mariner of the Nebulae,* (University of Chicago Press 1995), pg. 305
6. Interview of Nicholas Mayall by Bert Shapiro on 1977 February 13, Niels Bohr Library & Archives, American Institute of Physics, College Park, MD USA, www.aip.org/history-programs/niels-bohr-library/oral-histories/4766
7. Ira Bowen, Contributions of the Mount Wilson and Palomar Observatories, (CIWY 1947-1948), No. 47, pg. 1
8. R.S. Richardson, *Dedication of the Hale Telescope,* (PASP August 1948), Vol. 60, No. 355, pgs. 215-218
9. Gale E. Christianson, *Edwin Hubble: Mariner of the Nebulae,* (University of Chicago Press 1995), pgs. 321-322
10. Robert Trumbull, *Gandhi is Killed by a Hindu; India Shaken, World Mourns; 15 Die in Rioting in Bombay,* (NY Times Sat. Jan. 31, 1948), Front page
11. Geoffrey Burbidge, *Sir Fred Hoyle. 24 June 1915 – 20 August 2001 Elected FRS 1957,* (Biogr. Mems Fell R. Soc. 2003), Vol. 49, pgs. 213-247, https://doi.org/10.1098/rsbm.2003.0013
12. Mario Livio, *Brilliant Blunders: How the Big Bang Beat Out the Steady State Universe,* (pbs.org June 27, 2013), Nova: The Nature of Reality, Retrieved July 21, 2019, https://www.pbs.org/wgbh/nova/article/brilliant-blunders/

13. Adam Curtis, *A Mile or Two Off Yarmouth,* (bbc.co.uk 24 Feb. 2012) The Medium and the Message, Retrieved July 21, 2019, https://www.bbc.co.uk/blogs/adam-curtis/entries/512cde83-3afb-3048-9ece-dba774b10f89

14. Donald Lynden-Bell and Francois Schweizer, *Allan R. Sandage, 18 June 1926-13 November 2010,* (Cornell University 2011), https://arxiv.org/abs/1111.5646

15. Allan Sandage, *Centennial History of the Carnegie Institution of Washington: Volume 1 The Mount Wilson Observatory,* (Cambridge University Press 2004), pg. 521

16. Interview of Allan Sandage by Bert Shapiro on 1977 February 8, Niels Bohr Library & Archives, American Institute of Physics, College Park, MD USA, www.aip.org/history-programs/niels-bohr-library/oral-histories/32867

17. *Ibid.*

18. Allan Sandage, *Centennial History of the Carnegie Institution of Washington: Volume 1 The Mount Wilson Observatory,* (Cambridge University Press 2004)

19. *Ibid.* pgs. 545-546

20. M.L. Humason, *Obituary Notices: Edwin Hubble,* (MNRAS 1954), Vol. 114, pgs. 291-295

21. Gale E. Christianson, *Edwin Hubble: Mariner of the Nebulae,* (University of Chicago Press 1995), pg. 355

22. Allan Sandage, *Centennial History of the Carnegie Institution of Washington: Volume 1 The Mount Wilson Observatory,* (Cambridge University Press 2004), pg. 477

23. Milton L. Humason, *Apparent Velocities of Extragalactic Nebulae in Four Faint Clusters,* (PASP 1951), Vol. 63, pgs. 232-233

24. Gale E. Christianson, *Edwin Hubble: Mariner of the Nebulae,* (University of Chicago Press 1995), pg. 357

25. *Ibid.* pgs. 358-359

26. Interview of Allan Sandage by Bert Shapiro on 1977 February 8, Niels Bohr Library & Archives, American Institute of Physics, College Park, MD USA, www.aip.org/history-programs/niels-bohr-library/oral-histories/32867

27. Gale E. Christianson, *Edwin Hubble: Mariner of the Nebulae,* (University of Chicago Press 1995), pg. 358

13

Shockwave: The Death Of A Star
(1954-1961)

Against the backdrop of the growing Cold War between the United States of America and the Soviet Union, Milton Humason, Nicholas Mayall and Allan Sandage publish the final paper on the Hubble-Humason expansion program. Technological advances are starting to have an effect on experimentation in the field that will eventually yield a victor in the rivalry between the Big Bang and the Steady State concepts of the universe. Fred Hoyle publishes his work on supernova nucleosynthesis, increasing his stature as a serious scientist. The Mount Wilson and Palomar Observatories issue the *Hubble Atlas of Galaxies* in 1961 as Milton Humason continues his work with Fritz Zwicky well into his retirement.

News of Hubble's death rocked the galaxy redshift camp and sent a shockwave through the ranks at Caltech and the Mount Wilson and Palomar Observatories. In his obituary to his fallen friend Humason remembered Hubble's energy, exuberance and persistence in pursuing his work and of the difficulty "a close associate" would have in evaluating "in a detached manner the work of Hubble and the contributions he made toward advancing the frontiers of knowledge." After a succinct account of the multitude and magnitude of his partner's contributions, Humason concluded on a personal and wistful note:

© Springer Nature Switzerland AG 2021
R. Voller, *Hubble, Humason and the Big Bang*, Springer Praxis Books,
https://doi.org/10.1007/978-3-030-82181-4_13

Friends will remember the hospitable welcome which they always found in Hubble's beautiful home in San Marino and the good talk around an open fire.

Humason talked about Hubble's skill as a dry-fly fisherman, a sport for which Humason was known to be a near expert, and how Hubble had enjoyed his expeditions to the Rocky Mountains and in the Test Valley in the Hampshire district of his beloved adopted country of England.[1]

This and other published memoirs and accounts of his life constituted the full extent of the memorials to the great astronomer's life and career.

In keeping with her husband's wishes, Grace held no memorial or funeral services. She would continue to carefully manage her late husband's legacy until her own death in 1980, destroying letters and possessions to portray his life in the best possible light.

For Hubble's young research assistant, Allan Sandage (Fig. 13.1), the troubling question of whether to continue in the work he was doing with Hubble, Humason and Mayall, or disconnect in favor of his own research initiatives loomed large. Sandage had just received his Ph.D. that year and his advisor Walter Baade, while respectfully acknowledging Hubble's enormous contributions, was nevertheless advocating strongly for the young observer to seek his own course. Therefore, when Grace Hubble asked him to continue her late husband's work on galaxy classification and the expansion question, he was uncertain what to do. To this point, he later remarked in an interview:

If you were the assistant to Dante, and then Dante died, and then you had in your possession the whole of *The Divine Comedy*, what would you do?[2]

It was clear to Sandage that a decision to involve himself in redetermining Hubble's law of the redshift and its deeper meanings would likely become his own life's work. Sandage regarded Grace Hubble as "a very marvelous woman" that he credited with helping him to understand "other aspects of life, of learning and achievement." After some soul searching, he concluded that the moment was too big to pass up.[3]

In volume 53 of the CIWY, Walter Baade appears on the Observatory Committee in the position vacated by his late colleague. The responsibility of seeing that the field that Hubble had so ably opened during his thirty plus years in astronomy was thereby extended into the future.

Hubble died on the eve of the Mount Wilson Observatory's 50th anniversary, which was slated for the spring of 1954. His death coincided with the sunset of its reign as the finest astronomical research facility in the world but

Fig. 13.1 Allan Sandage in 1973 (Image courtesy of the Caltech Archives)

the association with Palomar would extend its legacy for a further three decades.

In his brief ode to "one of the most distinguished" of the Mount Wilson astronomers, Ira Bowen refers to Hubble's astonishing first decade of work when he "revolutionized the concepts of astronomers as to the size and content of the universe," pushing its boundaries "to the extreme range of the largest telescopes, a billion or more light-years..." Hubble's revelations and his dogged pursuit to understand the nature of galaxies laid the foundation for the creation of the nebular department at Mount Wilson and, along with Humason, he established "the concept of the expanding universe" experimentally.

This, in turn, had increased the necessity to build a larger and more powerful telescope, precipitating the arrival of the 200-inch at Palomar which was now reigning supreme over the field.[4]

The completion of the nebular classification and expansion program from MWO was now in the hands of Humason and his associates, who were approaching publication of a final report encompassing redshift and luminosity measurements of 600 galaxies.

Five years after the inception of the joint observatories and the completion of the 200-inch and 48-inch telescopes at Palomar, Hubble's demise left Humason to finish the work at hand to the best of his ability in the time he had left as an investigator. Looming over the team was the fact that Hubble had died believing the expansion wasn't real. Whatever their personal feelings on the matter, the work had to be continued to prove it one way or another.

Had he lived, Edwin Hubble would probably not have finished compiling his notes and photos of galaxies in time to publish the results of his overall program on classification and redshifts before his retirement the following year. In that event he would surely have stayed on at the observatories as an honorary researcher (time granted for lost years during the war) to finish the work to the extent that he was able, before setting sail for his own horizon.

The fact that he died before he could finish had compelled Grace to ask Allan Sandage to finish it. Sandage felt he needed Humason's commitment. As a young astrophysicist, Sandage was endowed of considerable ability, but he lacked the confidence, experience and depth of knowledge he felt was needed to finish the work himself. In the ensuing years he would prove his worth as a cosmologist, leading a wave of new advances in theoretical and research techniques that would propel him to the heights of the field and compel his boss, Ira Bowen, to promote him to the Observatory Committee after Walter Baade's retirement.

At this point, however, Sandage wanted the old expert spectroscopist – whom he greatly respected and admired – by his side to assist in that endeavor. This suited Humason, who wasn't yet ready to relinquish his post and wished to be present when the project that he and his fallen friend had worked at for so many years finally crossed the finish line.

Almost immediately after the end of the Second World War the Orwellian creep of cold war between the United States and its only real rival on the world stage, the Soviet Union, started to mold the political landscape of the second half of the century. On the heels of the explosion of the first Soviet-built atomic weapon in August 1949 the tit for tat between the two began to devolve into proxy wars on the Korean Peninsula, Vietnam and later Cuba.

Advances in computer technology and rocketry made during the war were significantly increasing the capability and efficiency of deployment systems. The Soviet Union shocked the world by launching Sputnik as mankind's first satellite into near Earth orbit in October 1957, setting off a "space race" which would drive technological refinement in computing, rocketry and robotics.

The steady thrum of cataclysmic destruction stoked by government propaganda in the US and the USSR fueled a fear of the spread of Communism in the US. This sparked a rise in nationalism that gave voice to Wisconsin Senator Joseph McCarthy and led to the anti-Communism laws. Paranoia engulfed the country as thousands of people lost their jobs and hundreds more were sent to prison for "crimes" ranging from attending meetings, providing support for or attendance of communist leaders or spokespeople, homosexuality and other so-called offenses.

Out of this came a revolt by younger Americans who sought an end to hostilities and a resistance to the façade of postwar idealism led by the music and fashion industries. Rock and roll, rockabilly, gospel and Motown music took root during the latter part of the 1950s, along with rebellious hairstyles such as the pompadour and a trend to ever shorter women's skirts that culminated with the miniskirts of the 1960s.

This new era brought calls for radical social change as well. The Supreme Court upheld the desegregation of public schools with a 9 to 0 vote in Brown vs. Board of Education in 1954. The rise of the Reverend Doctor Martin Luther King jr. began with the Montgomery Bus Boycott in December of 1955 and would continue until his assassination in April 1968. The Civil Rights Movement brought many peaceful (and some not so peaceful) advocates, organizers and leaders to the fore who went on to fight for incremental positive change into the new millennium.

As the Dow Jones Industrial Average reached its first all-time highs since its previous peak before the Stock Market crashed in 1929, the American economy shifted into another gear. The failing Hudson Motor Car Company merged with Nash-Kelvinator (Appliance) Corporation to create an even bigger failure (ultimately) of American Motors Corporation. Television sparked the creation of TV dinners and the fast pace of modern life led to the creation of Burger King and other fast-food franchises in a race to compete with industry leader McDonald's.

At the Mount Wilson and Palomar Observatories where the customary mode of dress was "suit and tie," young researchers in the 1950s started to wear blue jeans and flannel shirts, much to the chagrin of Hubble and the grumpy old men on the mountains. Instead of the big band sounds of the 1920s and 1930s that they were familiar with, the newcomers' radios were tuned to hits like *That's All Right* by Elvis Presley and *Rock Around the Clock* by Bill Haley and His Comets.

The wave of advances in technology hit the observatories just as the careers of some of the biggest astronomical names of the first half century were coming to a close. Humason could easily see from his vantage point that these refinements would soon engulf the work that he and Hubble had pursued over the past three decades.

Motion picture cameras were being used on the big reflectors for planetary photography. Three-color photoelectric photometry would soon make possible deeper penetration of the universe and the new field of radio astronomy was detecting evidence of super luminescent quasi-stellar objects (shortened to quasar) and later black holes. At Caltech, Fred Hoyle published his seminal work on supernova nucleosynthesis in 1954 showing that elements heavier than carbon were produced by the fusion of carbon in the hearts of some late-stage giant stars.

The only credible threat to the Big Bang, the Steady State theory, was still technically on the table as the Hubble classification team continued to compile its comprehensive data. Humason's inability to resolve spectra of galaxies beyond 60,000 kps at the 200-inch due to ambient light flooding his plates led him and Sandage to wrap up the project with a final paper on the velocity-distance relationship as things stood. This demonstrated the massive impact the work of Hubble and Humason had on the field of practical cosmology.

The report *Redshifts and Magnitudes of Extragalactic Nebulae*, referred to later as the HMS (Humason, Mayall, Sandage) Catalogue, consumed most of the April 1956 issue of the ApJ. The 73-page tour de force combined all available redshift and magnitude data for galaxies into a single uniform system for use in cosmological research. *Redshifts* illustrated the unusual symbiosis between the Mount Wilson and Palomar Observatories and the Lick Observatory, two competing organizations working together to present a spectacular guide to the cosmos for use by the entire field.[5]

Such a long partnership was rare then, as today, although government-funded facilities and projects tend to produce far more group-oriented work today. The success of the HMS was due in equal parts to Hubble's vision as an observer and the attention his discoveries garnered, Humason's insight in forecasting the specific usefulness of the 36-inch Crossley reflector in supporting their data, Mayall's enthusiasm and willingness to join the research and the goodwill between Hale, Adams and later Bowen at Mount Wilson and Palomar and William Campbell and Donald Shane at Lick. The central progenitor of the culture which created the environment for such cooperative research was George Hale whose leadership had helped to create the ApJ, the

IAU and other research institutions and publications and spearheaded the contributions by American observatories to the worldwide stellar mapping program instigated by Jacobus Kapteyn.

The absence of Hubble's name as the lead author on the HMS is acknowledged by the authors in the introduction stating that, "had he lived, [Hubble] would have participated as the senior author." Sandage would outline the system and its details in Hubble's place, with guidance from Humason and Walter Baade who continued to caution him that more precise magnitude measurements and photometrics would be necessary for the corroboration and conclusiveness of their results.

Significantly, this was the last report in which the term "extragalactic nebulae" was used to indicate galaxies independent of the Milky Way.

Humason took the lead listing his redshift analysis of 620 faint field galaxies and galaxy clusters in the first part, then Mayall did the same for 300 brighter galaxies and those that were too far north to be within reach of the 100-inch. One hundred fourteen of the galaxies on the two lists were shared, allowing Humason and Mayall to check for any discrepancies in the reduction of data (which was found to be negligible). In part three, Sandage set out the spectroscopic data and magnitudes for 576 galaxies, plotting their derived velocity-distance, Hubble classification and galactic latitude in two separate zones. Assuming the mantle left by his fallen leader, Sandage led a detailed discussion of the various corrections applied as technologies and methods improved. This process had started with the revelation by the 200-inch that what Hubble had believed to be the brightest stars in galaxies were far more luminous OB associations in HII regions. This correction put galaxies farther away than Hubble had estimated with a corresponding change to the slope of the redshift-distance relationship. The process continued with Sandage applying photoelectric measurements to the magnitude scale. Each correction tweaked the value of the Hubble constant and thereby the size and age of the universe. At the time of this final paper, the observational limit was approximately 6 billion light years.[6]

Although separated by 30 years of life experience the friendship between Humason and his young partners and protégés grew stronger over the final years of their combined effort. In letters, Mayall and Sandage express their amazement at Humason's humble attitude and his suggestion that Sandage, not he, take top billing on the final report on expansion.

Although he had contributed an equal share (and many times Hubble's share in terms of sheer observing time), Humason simply wanted no part in claiming co-ownership of the discovery that he so richly deserved. Known for his patience and expert advice as a mentor to many newcomers to the observatories through the years Humason also understood the importance of stepping aside to allow the next generation a chance to shine. But Sandage insisted Humason take the lead in authorship of what would be his own final contribution, placing Mayall second and himself third. This was a fitting tribute to an old master whose work had given Sandage and so many others in the field so much to chew on.

Humason's friend and colleague Robert Robertson, who helped him edit his published works from time to time, calculated the deceleration of the apparent expansion. It supported an oscillating universe that would end in a "Big Crunch" before starting over again, an idea proposed Richard Tolman (and indeed formulated and rejected by Einstein). Deceleration would become one of the critical targets of investigation that Sandage would aim for in his ongoing assault on the expansion issue over the next three decades. Today oscillation is no longer considered plausible.

The last paper established graphically the slope of the relationship between velocity and distance on Hubble's original line from the 1929 paper but offered no definitive conclusion in the debate between the Big Bang and the Steady State theories. The persistent curiosity surrounding this question even found its way into the CIWY. In a recurring section called "Cosmology" Bowen reported on Sandage's continued efforts to test demonstratively the primacy of the Big Bang over the Steady State. Sandage and most others at Mount Wilson and Palomar never really bought into the Steady State concept, but the director's inclusion of it showed that it did have a foothold in both public and scientific spheres.

The continuation of the redshift program in the coming years was taken on by William Baum using the newly adapted three-color photoelectric photometer at the 200-inch. This technology, first developed in the 1940s, superseded the old photographic detectors with a much more accurate photoelectric tube that enabled Baum and others to observer stars as faint as the 24th magnitude. Using this instrument, the image of an object passes through a small diaphragm in the focal plane of the telescope. The image is then filtered through a field lens to a photomultiplier which amplifies the weak light signal. From there the output current can be measured to obtain various results. The extreme accuracy of this technique is due to the precision of the techniques used to measure light currents and the high linearity between the amount of incoming radiation and the signal from the electric current. Later, after development in computer technology, photomultiplier tubes were replaced by charge-coupled devices (CCDs).

While Baum blew past a redshift velocity milestone of 100,000 kps, Sandage specified a four-part test to determine the winner in the contest over the preferred cosmological model. He was to use photoelectric images to identify the mean radiation energy (light) curve function for any given object with much higher precision. The displacement of the resulting curve from its normal position gave the apparent velocity. This would eventually be further refined by the enhanced image clarity that resulted from the introduction of Ritchey-Chretien systems and advances in computer technology and adaptive optics.

The test Sandage devised was: (1) the deviation from linearity of the velocity-distance relation (decreased velocity over distance); (2) the galaxy count-magnitude relation (did it match the decrease in velocity); (3) the angular diameter-redshift relation (to corroborate the distance measurements); and (4) the time scale (to corroborate the time measurements). His objective was to determine the rate of deceleration, which would not be present if the universe was in a Steady State. But despite his best efforts the test could not be performed definitively. The divergence over time was too small to be detected by the equipment at his disposal in 1960. Future developments would be needed to settle the score.[7]

By the end of the decade Baade and Minkowski were retired. Baade's work on stellar populations and on the finite distance measures to Cepheids in M31 with Henrietta Swope had helped Hubble and Humason, and later Sandage, derive the most precise term for the Hubble constant thus far. Minkowski discovered the two separate types of supernovae and instigated the "incalculably important" (in Sandage's estimation) National Geographic and Palomar Sky Survey.

At the end of his long and distinguished career Walter Baade retired and returned to his home in the heart of West Germany in October of 1959. He had expected to spend a healthy retirement writing a series of papers based on results of his long research in the field. Like his old friend Humason he was disinclined to write popular science articles, preferring the science. But his body and his lifestyle cut short his post-retirement. Smoking and drinking heavily and keeping the company of much younger scientists and students, his congenital hip problem and bone spurs in his back soon made it impossible to sit or stand without pain and discomfort. After "successful" surgery to repair his spine in January 1960 in Göttingen he was laid up for five months while his body recovered, standard practice at the time. But the months spent lying around caused disruptions to his circulatory system and Baade died of a stroke at age 67 only a matter of days into his rehabilitation from surgery.[8]

Minkowski on the other hand, lived out his retirement in his adopted home of Berkeley, California until his death in 1976 at age 80 having won a Bruce Medal for his contributions in the field in 1961. As one who benefitted directly both personally and professionally from Baade's compassionate and collegial nature, Minkowski owed much of his good fortune in life and in work to his fallen countryman.[9]

Milton Humason retired officially on June 30th, 1957, together with his old friend and earliest collaborator, Seth Nicholson. Both men had spent their entire 35-year careers at the observatory.

For Humason, nearly his entire life after 1902 was spent on or close to Mount Wilson. The previous year had seen the death of his old friend, mentor and great champion Walter Adams, with whom he had shared family trips in the hills around Los Angeles over the years.

In a lengthy biography of Humason in the Carnegie annals, Bowen reviewed Humason's unique relationship with the observatory that he helped bring to world acclaim, describing his "very unusual proficiency in the photography of spectra of very faint objects." After Hubble discovered the existence of galaxies, Bowen wrote, "Humason turned his attention to the spectrum of these objects and soon accumulated spectra of a substantial number" of them. As the objects grew many magnitudes fainter than could be captured, Humason "had to develop elaborate offset procedures that ensure[d] locating invisible images accurately on the slit of the spectrograph and holding them there during the long exposures." Edwin Hubble had found just the right partner at just the right moment for his assault on universal expansion.[10]

It was surely a matter of the utmost curiosity to astrophysicists, both young and old on the mountains that neither Hubble nor Humason would suggest the term Big Bang or even confirm definitively the expansion during their careers. In Humason's case, he was likely deferring to his partner. No one on Earth saw the redshift-distance linearity more clearly than Milton Humason before 1960.

As the decade came to an end Mayall was preparing to take over as the second director of the Kitt Peak National Observatory west of Tucson, Arizona. In that post for 11 years, he oversaw the installation of the 4-meter-diameter Ritchey-Chrétien Telescope (RCT) that was named in his honor after his retirement in 1971.

Perhaps no one outside the Mount Wilson and Palomar Observatories benefitted more greatly from his relationship with the institution than Mayall. His decision to take a break from study at Berkeley to work at Mount Wilson from 1929 to 1931 found him at the center of the expansion discovery and gave him a chance to work with luminaries like Nicholson, Merrill and Hubble. His friendship with Humason benefitted Mayall as well, providing the opportunity to learn the ropes of astrophotography from the deep space pioneer firsthand, and making him Humason's first choice for an extended partnership.

It was an incredible stroke of good fortune and Mayall made the most of it. Flush with renewed energy for the science, he completed his Ph.D. in 1934 and shortly thereafter went to work at Lick as an observer. It was Hubble who suggested to Mayall, whose advisor was Hubble's friend and matchmaker William Wright, that he focus his thesis on galaxy counts and their relative positions to directly aid Hubble in his research on the curvature of space. Neither man was successful in that determination, but the thesis gave Mayall his doctorate and an introduction to the field.

Shortly after joining the staff at Lick, Mayall proposed the creation of a slitless quartz spectrograph for the Crossley 36-inch reflector that he had been designing while finishing his thesis. He figured correctly that the small, fast spectrograph would enable him to study some of the nebulae in the Hubble-Humason program which had a low surface brightness, thereby complementing the work of the Mount Wilson duo. The plan worked and Mayall's name became forever linked to one of the greatest discoveries of all time. His research and ingenuity made him the obvious choice for directorship at Kitt Peak, in which capacity he continued to show his skill and vision as a leader. In addition to his many contributions to the development of Kitt Peak he promoted participation in the establishment of the Cerro Tololo Inter-American Observatory (CTIAO) in Chile.[11]

While Sandage worked with Mayall on the final piece of the Hubble-Humason legacy, the *Hubble Atlas of Galaxies*, Humason spent his observing time at Palomar assisting his friend and colleague Fritz Zwicky by participating in both his supernova sky survey and cluster galaxy research.

Born in Varna, Bulgaria in 1898, the son of a Swiss industrialist and ambassador, and an ethnic Czech mother from the Austro-Hungarian Empire, Zwicky had emigrated to the US in 1925 as an "international fellow from the Rockefeller Foundation" at Caltech.[12] His first wife, Dorothy Gates, was the daughter of the California State Senator Egbert Gates and her money had contributed to the funding for the Palomar Observatory.

Over his long career Zwicky made many contributions to the development of science in a variety of areas, not the least of which was astronomy and astrophysics. In addition to his work with Baade on supernovae (in the process inventing that term), he had theorized the existence of neutron stars and dark matter, proposed the idea that galaxies could be used as gravitational lenses (using a previously discovered aspect of Einstein's equations), and also held more than fifty patents for jet engines, jet propulsion, underwater and jet thrust motors. There were some bizarre ideas such as "tired light" as an alternative to expansion,[13] eruptive "goblins" as a mechanism for solar flares,[14] the firing of a gun through the dome slit to smooth turbulence while observing, and the rearrangement of the solar system by firing pellets into the Sun to create explosive propulsion. Even if his idea to move the entire solar system to Alpha Centauri within 2,500 years was ill-conceived, he was certainly an out-of-the-box thinker. His pet insult for his enemies (which were plentiful at Caltech and the observatories) was to refer to them as "spherical bastards" because whichever way you looked at them, they were still bastards.[15]

As of 1961, Humason had discovered 16 new supernovae and contributed greatly to Zwicky's work on galaxies in clusters. Zwicky managed to convince Humason to delay his retirement until their joint work, *Spectra and Other Characteristics of Interconnected Galaxies and of Galaxies in Groups and Clusters*, was published in 1963. Among its lines of discussion, the paper called attention to the dearth of dense matter in galaxies overall as evidence for the existence of dark matter.[16]

The *Hubble Atlas of Galaxies* appeared in 1961. As Sandage described in the preface, "the major groupings and the suggested similarities which are discussed in the 'Description of the Classification' are principally due to Hubble. I have acted mainly as an editor...of a set of ideas and conclusions that were implicit in the notes…"

Humason contributed greatly with direct photographs for the volume, while leaving the details of the classification to Sandage's interpretation of the notes Hubble left behind. That same year, Humason published a series of photographs of planets he had taken at the coudé focus of the 200-inch in a volume for the University of Chicago Press which was edited by Gerard Kuiper.[17]

With the publications of the final paper on expansion, the *Hubble Atlas of Galaxies*, and the supernova and galaxy research with Zwicky, Milton Humason finally gave up fishing for objects in deep space in favor of "steelhead and salmon" in the cool waters of Northern California. Hugo Benioff and his wife had retired to Mendocino and the Humasons moved nearby to be close to their friends.

References

1. M.L. Humason, *Obituary Notices: Edwin Hubble,* (MNRAS 1954), Vol. 114, pgs. 291-295
2. Dennis Overbye, *Allan Sandage, Astronomer, Dies at 84; Charted Cosmos's Age and Expansion,* (NY Times Nov. 17, 2010), https://www.nytimes.com/2010/11/17/science/space/17sandage.html
3. Interview of Allan Sandage by Bert Shapiro on 1977 February 8, Niels Bohr Library & Archives, American Institute of Physics, College Park, MD USA, www.aip.org/history-programs/niels-bohr-library/oral-histories/32867
4. Ira Bowen, Contributions of the Mount Wilson and Palomar Observatories, (CIWY 1953-1954), No. 53, pgs. 34-35
5. M.L. Humason, N.U. Mayall, and A.R. Sandage, *Redshifts and Magnitudes of Extragalactic Nebulae,* (ApJ April 1956), Vol. 61, No. 3, pgs. 97-162
6. *Ibid.*
7. Ira Bowen, Contributions of the Mount Wilson and Palomar Observatories, (CIWY 1959-1960), No. 59, pgs. 28-29
8. Donald E. Osterbrock, *Walter Baade: A Life in Astrophysics,* (Princeton University Press 2001), pgs. 208-211
9. Donald E. Osterbrock, *Rudolph Minkowski 1895-1976,* (NAS 1983), pgs. 269-298
10. Ira Bowen, Contributions of the Mount Wilson and Palomar Observatories, (CIWY 1956-1957), No. 56, pgs. 70-71
11. Donald E. Osterbrock, *Nicholas Ulrich Mayall 1906-1993),* (NAS 1996), pgs. 187-213
12. Richard Panek, *The Father of Dark Matter Still Gets No Respect,* (Discover 2008), https://www.discovermagazine.com/the-sciences/the-father-of-dark-matter-still-gets-no-respect
13. *Ibid.*
14. F. Zwicky, *Nuclear Goblins and Flare Stars,* (PASP 1958), Vol. 70, pgs. 506-508
15. Interview with Don Nicholson, 2006
16. F. Zwicky and M.L. Humason, *Spectra and Other Characteristics of Interconnected Galaxies and of Galaxies in Groups and in Clusters III,* (ApJ 1964), Vol. 139, pgs. 269-286
17. A Sandage (ed.), *The Hubble Atlas of Galaxies,* (CIW 1961)

14

The Last Horizon (1961-1991)

Milton Humason and Nicholas Mayall exit the scene, leaving Allan Sandage to try to dig down to the bottom of the expansion problem to better define and prove the discovery definitively. As the decade of the 1960s gets underway, the Steady State theory remains a plausible rival to the Big Bang theory. The late Edwin Hubble's reluctance to truly embrace his own discovery in the face of supporting evidence makes life difficult for Sandage as he starts his assault on the reality of expansion. With a little help from two different and unlikely sources, the theory is proven and Sandage spends the remainder of his career corroborating and fine-tuning aspects of the expansion experimentally.

The premature death of his mentor and hero Edwin Hubble left Allan Sandage the task of sorting out the fine details and unanswered questions from the general conclusions arrived at by Hubble and Humason.

Settling the debate between the Big Bang and Steady State solutions to Einstein's field equations was the great question but there was much work still to be done to better understand the effects of the HMS catalogue on stellar, galactic and universal evolution.

The most perplexing of these problems was Hubble's insistence that the redshifts were probably the result of some undiscovered mystery of physics. Hubble's inductive reasoning had led him to assumptions which in turn caused him to believe that the expansion was an illusion, but various developments before and after his death had revealed fatal flaws in his approach.

© Springer Nature Switzerland AG 2021
R. Voller, *Hubble, Humason and the Big Bang*, Springer Praxis Books,
https://doi.org/10.1007/978-3-030-82181-4_14

For a start, Jesse Greenstein's "devastating paper" of 1938 reduced Hubble's suggested mean spectral color temperature coefficient for the Andromeda Galaxy (upon which much of his 1936-1937 conclusions on galaxy counts rested) from 6,000 to 4,200 degrees, similar to the heat signature in black-body radiation. This rendered Hubble's correlation of redshift with magnitude inconclusive. "The effect of such low temperatures," Greenstein explained, "on the present interpretation of counts of extragalactic nebulae is serious." In other words, it was impossible to say whether or not the effect of redshift on the magnitudes of galaxies was due to actual velocity.[1]

The timing and impact of Greenstein's report no doubt seized on Hubble's mind and imagination, and was among the problems that he and the classification team would have to consider going forward. This plan was disrupted initially by the war and later, most finally, by Hubble's health. Accurately measuring the relationship between color (heat) energy and recessional velocities and interpreting the results vis-à-vis the galaxies and expansion would occupy Sandage and his collaborators John Oke and Albert Whitford at the Lick Observatory for the next several decades. This was only one of several issues Sandage would have to resolve in order to prove definitively the velocity-distance linearity's correlation with Einstein-Friedmann-Lemaître cosmology. The magnitude and distance scales were even bigger problems.

By 1958, Sandage had reduced the Hubble constant to 75 km/sec/Mpc,[2] in close agreement with modern measurements for the distance scale, which range between 66 and 82 with uncertainties of around three percent. This represented a change from Hubble's 1936 scale by a factor of seven based on corrections to Leavitt's period-luminosity relation for Cepheids in the local group of galaxies and the reinterpretation of the brightness in the HII regions for nearby galaxies which Sandage had made in the five years since Hubble's death. This correction was made possible by advances in filter photography which enabled Palomar to resolve what could only be described by Hubble as "bright knots" into hydrogen rich clouds like those in the Orion Nebula.

While Sandage was trying to infuse Hubble's intuitive universe with some much-needed precision, a young student in Hamburg, West Germany, named Wolfgang Mattig came to his (and Hubble's) aid.

In two short papers in 1958 and 1959 that Sandage would later claim were a "watershed for practical cosmology itself" and "a fundamental development that changed the field," Mattig revealed the mistakes in Richard Tolman's math and included corrections that were fundamental to the advancement of expansion theory and cosmology in general.

When Hubble and Tolman made their formulation for redshift distance measurements, the Mount Wilson astronomer had assumed that the distance to a galaxy using its velocity (or redshift) could be derived using a simple equation:

$$d = cz / H$$

in which d is the distance, c is the velocity of light, z is the redshift and H is the Hubble constant.

Mattig realized that the equation itself was dramatically oversimplified. Crucially, no one had ever bothered to use Alexander Friedmann's theoretical equation that established a velocity-distance scale for all redshifts. The methods of that time were only accurate for nearer galaxies with small redshifts. Mattig realized this and corrected the situation in two papers totaling five pages which altered the physics of cosmology and helped Sandage and others greatly in determining the effects of redshift on distance.

Mattig's much more complex formulation required that the redshift z was proportional to the distance d as given in:

$$d = 2c / H \left\{ 1 - (1+z)^{-0.5} \right\}^{3}$$

Combining Mattig's equations with Howard Robertson's 1938 correction for the effect of redshift on apparent luminosity (namely that the apparent luminosity is proportional to the absolute luminosity divided by the square of the Doppler redshift) Sandage planned to revise Hubble's final campaign on galaxy counts that began in 1949 during his own initial involvement with the observatories.

Again, Sandage's efforts were frustrated by the lack of technical means for carrying out his investigations. For a start he believed, as did others, that any definitive study of galactic magnitudes must involve the measurement of independent galaxies, not by the aggregation of magnitudes over an entire plate of galaxies.

The means to do this, first by areal photometry and later by improvements in computer technology, adaptive optics and other means, were not available in the 1950s. The end of what Sandage called the "exploratory period" (the use of observed plate limits to derive magnitudes) wouldn't come until the mid-1970s when technology finally made it possible to begin anew his assault on the Big Bang as initially adumbrated by Hubble and Humason.[4]

The absence of proper technical support to fuse his formidable background in theoretical physics to further correct, quantify and unambiguously prove the Big Bang must have been a considerable frustration for Sandage. But as instrumentation and techniques improved in the 1990s, he was able to apply his perception honed by Hubble, his critical analysis honed by Baade and his persistence and skill at astrophotography honed by Humason and finally obtain definitive observational evidence.

Fortunately, a quarter of a century earlier a timely bit of experimental luck had helped ameliorate Sandage's frustration.

In 1964 a seemingly unrelated event resolved the debate over cosmological models in favor of the Big Bang with a high degree of certainty. The radio astronomers Arno Penzias and Robert Wilson were making final adjustments on a new highly sensitive radio antenna at Bell Labs in Holmdel, New Jersey when they encountered a low energy "noise" that they could not explain.

A test to see if the noise was originating from nearby New York City was negative, so they examined the telescope itself. On inspection they discovered that the horn was filled with what they referred to as "white dielectric material." The telescope had become a perch for bats and pigeons and the inside of the horn was coated by their droppings. They quickly surmised that this must be the problem and cleaned the horn, but the noise persisted, leaving them flummoxed.

In March of 1965 Penzias received a telephone call from a friend, B.F. Burke, who he had encountered returning from a conference in Montreal, Canada in December 1964 and told him of his frustration in tracking down the mysterious hiss. Burke now suggested that Penzias ought to contact Princeton where Robert Dicke was developing a new microwave radiometer to search for the cosmic microwave background (CMB). This radiating remnant of electromagnetic energy from the Big Bang was predicted first in 1948 and independently by Dicke around 1960. They were about to start taking measurements when Penzias called and said they had detected a signal that they couldn't explain. As the phone call developed, it became clear that Penzias and Wilson had discovered the CMB accidentally.[5]

Their subsequent article, published in the ApJ in July of 1965, was the first definitive observational proof of the Big Bang. Their discovery won Penzias and Wilson the Nobel Prize for Physics in 1978. It was the award Edwin Hubble had so coveted and probably deserved for his 1929 announcement on the practical discovery of the expansion. It might otherwise have been bestowed upon both Hubble and Humason. But in the observational era this award wasn't given to astronomers and the whole idea moot.

Notwithstanding this lost object of Hubble's vainglory, 37 years after the onset of their partnership and 34 years after Lemaître first suggested the Big Bang in theory (albeit not using that name) the score had been settled. In 2018 the IAU recommended that the law of proportional velocity to distance for galaxies should be referred to as the Hubble-Lemaître law.

Edwin Hubble did not live to see the outcome of the race to understand the meaning of the velocity-distance linearity, but he would surely have been interested in the results.

As for his longtime partner, by the time the discovery was made only a year or two into his retirement Milton Humason had forgotten more about red-shifts than most people would ever know. He was in frequent touch with Sandage in Pasadena and marveled at the rate at which he and his collaborators were extending the expansion using modern equipment and methods. "The 200 inch," he later said, "was designed with the hope that we could get out to the limit or 'the horizon'…Well there is apparently no horizon, at least as far as the 200-inch goes." [6] Fifty-five years on, at the time of the writing of this book, there is no sign of this zero-point for time and space.

The discovery of the CMB had once again captured the world's imagination, setting off a media buzz and renewed interest in the mystery of the Big Bang. This spurred interest in Edwin Hubble who was no longer available for comment. That task fell to Humason, who was called upon for an interview orchestrated by Zwicky in 1965. Humason spoke almost whimsically of Hubble, who in his words "was greatly excited and interested" in the spectra he was bringing down from the mountain and how upon arrival at the office "the first thing [he] could expect" was for Hubble to "knock on [his] door, and he'd come in and want to know what I'd gotten." This kind of childlike exuberance is seldom spoken of by Hubble's biographers but is readily felt by almost everyone who gets to look through a telescope at the wonders of the cosmos.[7]

Milton Humason lived long enough to witness the beginning of the space race between the USA and the USSR, the lunar landing and the advent of space telescopes. The National Aeronautics and Space Administration (NASA) launched the first Orbiting Astronomical Observatory (OAO) in 1966 but a battery failure killed the mission. In 1968 a second OAO was successfully launched, and the telescope carried out ultraviolet measurements of stars and galaxies.

In 1970, delighted by the results of the OAO, NASA initiated plans for a large reflecting space telescope. In 1983 it was named the Hubble Space Telescope. Delayed by recession, computing and other design details and the loss of Space Shuttle Challenger in 1986 it did not find its way into orbit until 1990. It is still operating and has become the most successful optical telescope ever made. Its incredible capacity for deep space exploration and imaging has simultaneously advanced scientific discovery while evoking public astonishment at the beauty and wonder of the cosmos.

Milton Humason died suddenly on June 18th, 1972, at home in Mendocino, California, with his wife Helen by his side. In the months ahead, friends and family would write to his widow to express their respect and admiration for a man of rare and charismatic character.

He had lived an outsized life for the education he received, and he did it with a measure of grace and gratitude, giving his full attention to anyone who sought him out for research, advice or just a little conversation. When he retired, he reverted to the life he had always led in parallel to that of an astronomer, a quiet, peaceful life with family and friends, fishing the rivers of Northern California.

The history of science has seldom encountered a more unlikely, more productive, more inexplicably dynamic, and simultaneously reticent pairing than Edwin Hubble and Milton Humason. Over many years their contributions to cosmology were broad, far-reaching and helped to reveal the incredible vastness and wonder of the cosmos.

Asked if their failure to conclusively prove the Big Bang or see the last horizon was a disappointment, Humason replied, "Nothing in astronomy is a disappointment…you make an observation to try and find out what it's going to do…all of the interesting problems come from the things that didn't turn out the way you thought they might…it's those things you want to follow up and not the ordinary things."

References

1. Allan Sandage, *Beginnings of Observational Cosmology in Hubble's Time: Historical Overview*, Part of the Space Telescope Science Institute Symposium Series, 1998, Mario Livio, S. Michael Fall and Piero Madau (editors)
2. Ira Bowen, Contributions of the Mount Wilson and Palomar Observatories, (CIWY 1957-1958), No. 57

3. Allan Sandage, *Beginnings of Observational Cosmology in Hubble's Time: Historical Overview,* Part of the Space Telescope Science Institute Symposium Series, 1998, Mario Livio, S. Michael Fall and Piero Madau (editors)

4. *Ibid.*

5. David W. Hughes and Richard de Grijs, *The Top Ten Astronomical "Breakthroughs" of the 20th Century,* (CAP October 2007), Vol. 1, No. 1, pg. 15

6. Interview of Milton Humason by Bert Shapiro in circa 1965, Niels Bohr Library & Archives, American Institute of Physics, College Park, MD USA, www.aip.org/history-programs/niels-bohr-library/oral-histories/4686

7. *Ibid.*

Epilogue

The story of the lives and careers of Edwin Hubble and Milton Humason illustrate the complexity of the human experience as we continue to endeavor to understand and define our natural world. The currents of modern life – political, social, ethical and moral – pervade our collective conscience while our personal experience and character shapes our responses to and progression through these currents in our daily lives. We are, as the saying goes, stardust, and as such no two experiences are alike in terms of disposition, perception or interaction. These platitudes, cliché though they may be, lend an overriding empathy, compassion and appreciation for the ability of some to overcome their personal deficiencies to our common benefit. To this end, Hubble and Humason have surely contributed.

The evolution of discovery at every level of human life happens as a slow progression rather than in prolonged interludes interspersed by moments of fundamental breakthrough. Science does not happen in a vacuum. Only by building on the contributions of others are we able to continue our understanding and come up with ideas to further define our world. Nature is too complex for any one of us to make these discoveries during a single lifetime. We are wholly and completely dependent on one another.

As partnerships go, it is hard to surpass the contribution that Hubble and Humason made to science. This contribution was only made possible by the recognition of theoretical work of the age that was itself built on hundreds of years of theoretical and practical development and straightforward trial and error. The impetus to even attempt to ascertain an expansion would not have

© Springer Nature Switzerland AG 2021
R. Voller, *Hubble, Humason and the Big Bang*, Springer Praxis Books,
https://doi.org/10.1007/978-3-030-82181-4

come without the work of Lemaître and Robertson, who based their work on Friedmann, de Sitter and Einstein, who built upon lessons learned from Planck, Faraday and Maxwell, Michelson and Morley, who built upon Newton and so on through the history of scientific development.

These relationships represent the essence of science, which is to express ideas in theory and prove them experimentally. This synergy, so stubbornly ineradicable in its reliability, persists.

The winners of the 2020 Nobel Prize for Physics are an example of the success of this process. The English mathematician and theorist Roger Penrose used Einstein's relativity to prove the existence of black holes. Using the Keck telescope on Mauna Kea in Hawaii, practical cosmologists Reinhard Genzel and Andrea Ghez established the existence of a supermassive black hole at the very center of the Milky Way. The work of the German and American, respectively, took decades of investigation and innovation in the techniques they used to uncover the monster sucking up all matter and light at the heart of our galaxy.

All three were given the award in recognition of their combined work on revealing the phenomenon. Ghez was only the fourth woman to win the award for physics, but it is likely there are many more young women out there with the same potential.

The roles that Hubble and Humason played in the discovery of expansion indicate that each was a misfit pursuing his own path.

In Hubble's case he was in as good a position as anyone in the field to come across the evidence. His focus on nebular research in his graduate education in astronomy and in his early career in the field made him one of a handful of astronomers around the world capable of making the breakthrough.

It was Hubble's well-documented social deficiencies that put his success at risk. A non-conformist by nature, fueled by ambition, an ego that bordered on narcissism and a purpose driven in large part by paternal dogma and imperiousness, he was both self-conscious and self-preserving in his life and work. His ambivalence toward his siblings and his childhood friends, his reinvention of his past, and the hostility he showed toward those he believed to be jeopardizing his legacy are undeniable.

In the latter case it is likely that only his relationship to Humason made his advancement in the field beyond the Cepheid discovery in M31 possible. Humason's reputation within the ranks served as a buffer between various members of the staff at Mount Wilson and Caltech in the 1930s and 1940s. It is difficult to think of anyone else in the spectroscopy department who would have – or even could have – come to Hubble's aid if Humason had refused to pursue the redshifts of faint galaxies. In this regard, Hubble was truly fortunate.

These obvious personal afflictions have dimmed Hubble's legacy in the eyes of many, both within the science field and without. But they are only a part of a larger personality profile that otherwise shows Edwin Hubble to be a highly sympathetic character. He was a purposeful, dedicated and dogged observer, as passionate regarding his field as any of his contemporaries. Freeman Dyson once categorized scientists as either foxes or hedgehogs. Foxes have a broad scope, know many tricks and work on several problems at once while hedgehogs are more dogged and singularly focused. Dyson once told me that in his view Hubble was a hedgehog. If so, he was a foxy one. His ability to draw conclusions that were borne out so thoroughly in reality, based on broad assumptions, would amaze those like Allan Sandage who were left to straighten out his mess.

Hubble showed great loyalty and respect not only to his closest confidants and loved ones but to those with whom he had a mutual working relationship like Humason, Sandage, Baade, Minkowski and Mayall. Although he never saw combat, he was willing to put his life on the line for his country, not once but twice. He was a devoted husband who sought a family with his beloved wife, Grace. These are the traits of a man who had learned some of the values of family and friendship in his early life, even if they were tainted by his aversion to his father's dogma. In view of the psychological "torture," he endured under his father's hand, however well-meaning, it is easy to see how Hubble might've developed a severe case of inferiority to accompany the narcissism and egoism he so often displayed.

In the case of Milton Humason, the inveterate fish-out-of-water, it is easier to see how his childhood and education were misaligned with the career and the incredible breadth of discovery and contributions to science he achieved. It is safe to say we are unlikely ever to see again a career that began in overalls as the janitor of the observatory in 1918 (his first full year there) and an eighth-grade education go on to advance from assistant observer to astronomer to premier spectroscopist, co-discoverer of the Big Bang, experimentally, and co-author of the HMS catalogue, honorary Ph.D., card-carrying member of the Royal Astronomical Society (RAS), and secretary of the Mount Wilson and Palomar Observatories. "He was nominated for the gold medal of the RAS," Sandage once lamented, "and they turned him down. I think that's wrong."[1]

His contributions to Hubble's work and that of many others during his career, together with his incredible life story and everyman character have given Humason a cult like status. He is a regular fixture in the artworld, the muse of poets, songwriters and playwrights. The fact that a deep space probe or spectroscopic observatory hasn't been named in his honor seems a gross oversight in view of the adoration and respect he has garnered from so many around the world. The science world has never concerned itself much with what is referred to nowadays as "optics," perhaps to its detriment.

The Hubble Space Telescope was certainly aptly named and has lived up to the vision of its namesake. But it must be said that without Humason's partnership the discovery for which Hubble is best known could so easily have eluded him. Such was the plight of the spectroscopists in the first half of the 20th century before the first wave of cosmologists hit the scene with more rounded educations.

The fact that Milton Humason had several close calls in making important discoveries speaks to his eagerness, at least in the early years of his career, to make an impression on the field. His general lack of self-confidence likely propelled this ambition to some extent, although it was equally proportioned with the sheer joy he felt at the controls of the great reflectors.

It is doubtful that either Milton Humason or Edwin Hubble could've made the practical discovery of the Big Bang on his own. One lacked the confidence and connection with the broader scientific community and the other the skill and tenacity to see the most difficult aspects of the work through to their conclusion. It defies logic that anyone discussing the practical discovery could refer to the Big Bang without including Humason's name in their discourse. Only Humason's insistence on Hubble taking sole credit lends any credence to such a slight, and even that seems imprudent and unjustified in view of the gravity of Humason's contribution to the work.

Given the enormous difficulty and sheer physical stamina that some of the work required in the early days of their program it is hard to imagine any one person taking on the project on their own. It is noteworthy that of the pair it was probably Humason who possessed the right skill, character and intestinal fortitude to pull it off, although he would have required guidance and more education in photometry. No doubt he could've learned these skills, so lacking in precision in Hubble's day, if not for his abecedarian status on entering the field. It simply wasn't in the cards.

In considering the Mount Wilson advance, the quality of the equipment and the culture of innovation and progress in the field cannot be ignored. The duo of Hubble and Humason benefitted greatly by being at the site of the most powerful telescope in existence, supported by some of the best technicians in the world and a visionary administration backed by one of the world's richest men. There can be no doubt that the 100-inch telescope was a game-changer.

If the stars had aligned differently, Vesto Slipher might have been capable of exploring the deep reaches of space on his own. A gifted observer with a friendly and collegial disposition, his engineer's mindset, probing vision and skillset led him to the discovery of the nebular redshifts. Were he at Mount Wilson, with its technologically more advanced equipment and the guidance of Hale and Adams, he may well have proven to be up to the task. As it stands, his career speaks for itself. In addition to being a great observer he was a great ambassador for the field.

Other possible candidates include Harlow Shapley, Edward Fath, Knut Lundmark and Walter Baade. All had much of Slipher's combination of excellent skillsets in cosmology with the kind of vision needed to follow the clues to the discovery.

Baade was probably least likely to carry the job through given his physical limitations.

Shapley would've required a lot of convincing by Hale and others to abandon his belief that the Milky Way was the whole universe, although he was likely to have done so if he had remained at Mount Wilson into the 1920s and beyond.

Fath could not get along with Adams, which precipitated his departure in 1914 for rosier circumstances (for both men). But Sandage regarded Fath as the Hubble-to-be if he had stayed.

Lundmark would surely have found the expansion, having been immersed in nebular classification and other attributes himself. Hubble's reputation among friends in the field at Yerkes, Harvard and elsewhere sealed Lundmark's fate when he was shunned by Adams after Hubble's initial discovery was published.

Hubble's ability to seize the essence of a concept in writing was one thing that separated him from his competitors. In an article for the RAS of Canada to mark the 100th anniversary of Hubble's birth in 1989, Sandage discussed this aspect of his former mentor and collaborator as it pertained to his work. In it he reflected that Hubble's "surety of language characterized much of his later writings – a surety which tended (and was intended) to conquer the field by prose as well as by the technical results."[2]

The titles of Hubble's 1929 paper and the joint paper with Humason in 1931 are perfect examples of his innate ability to express his intended statement and underscore its import in writing.

There can be little doubt that some part of the credit for the discovery of the Big Bang should go to Edwin Hubble and his partner in the endeavor Milton Humason. They did the work specifically organized to confirm or refute the proposed theory, tested it over decades and wrote about it in direct and somewhat certain terms. It is also clear that the discovery does not belong to any one individual or group but to a phalanx of observers and physicists who over many years worked to connect the theory to the observed universe.

Since their discovery followed on the heels of theoretical predictions it seems reasonable to conclude that the discovery ought to be hyphenated to include the predictors. Revisiting the timeline of publications has prompted changes in the nomenclature in recent years. The renaming of the Hubble law of redshifts as the Hubble-Lemaître law is an example of this, although one could surely argue for the addition of Humason's name to that law. Similarly, an especially hot and bright B-type subdwarf was found by Humason and Zwicky in the 1940s using the Schmidt 48-inch at Palomar and these objects have been named Humason-Zwicky stars.[1]

Hubble's later denial of the expansion should not undermine the effect he and Humason worked so long and so hard to confirm. The fact that his broadly based, inductive approach and the mistakes in his mathematical derivations with Tolman caused him to misinterpret his own overwhelming evidence doesn't diminish the feat itself, however bizarre his denial seems in retrospect.

The partnership of Edwin Hubble and Milton Humason – however unlikely, distrustful, interdependent, tenacious, probing, visionary, equivocating, and dubious it may have been – opened the door to the unthinkably vast expanses of the known universe and set the stage for modern cosmology. Their legacy adds to a distinguished list of partnerships that, along with contributions of other individuals, has helped to advance scientific knowledge of the world around us and the technology to propel that knowledge into the future.

References

1. Interview of Allan Sandage by Bert Shapiro on 1977 February 8, Niels Bohr Library & Archives, American Institute of Physics, College Park, MD USA,
    www.aip.org/history-programs/niels-bohr-library/oral-histories/32867
2. Allan Sandage, *Edwin Hubble 1889-1953*, (Journal of the RAS of Canada, Dec. 1989), Vol. 83, No. 6

---

[1] On the other hand, the hyphenated Edgeworth-Kuiper region of small bodies in the outer reaches of the solar system is generally truncated to the Kuiper belt!

Appendixies

Appendix 1: Additional Reading

10 Physicists Who Transformed Our Understanding of Reality, Rhodri Evans and Brian Clegg, Running Press, 2015

A Latin Dictionary. Charlton T. Lewis and Charles Short, 2020

A Short Bright Flash: Augustin Fresnel and the Birth of the Modern Lighthouse, Theresa Levitt, W.W. Norton & Co., 2013

Astronomical Discovery, Herbert Hall Turner, (Edward Arnold 1904)

Autobiography of Robert A. Millikan, Prentice Hall, 1950

Beginnings, from Lipperhey to Huygens and Cassini, The, Albert Van Helden, (Experimental Astronomy 2009)

Big Bang: The Origin of the Universe, Simon Singh, Fourth Estate 2004

Centennial History of the Carnegie Institution of Washington: volume I The Mount Wilson Observatory, Allan Sandage, Cambridge University Press 2004

Chemical Revolution of Antoine-Laurent Lavoisier, The, American Chemical Society International Historic Chemical Landmarks, http://www.acs.org/content/acs/en/education/whatischemistry/landmarks/lavoisier.html

Churchill's Ministry of Ungentlemanly Warfare: The Mavericks Who Plotted Hitler's Defeat, Giles Milton, Picador 2016

Double Stars: The Story of Caroline Herschel, Padma Venkatraman, (Morgan Reynolds Pub. 2007)

Edwin Hubble: Mariner of the Universe, Gale E. Christianson, University of Chicago Press, 1995

© Springer Nature Switzerland AG 2021
R. Voller, *Hubble, Humason and the Big Bang*, Springer Praxis Books,
https://doi.org/10.1007/978-3-030-82181-4

Einstein 1905: The Standard of Greatness, John S. Rigden, Harvard University Press, 2005

Ferris Wheels: An Illustrated History, Anderson, Norman D., (Bowling Green State University Popular Press 1992)

Ferris Wheel, The, Jones, Lois Stodieck, (Grace Dangberg Foundation 1984)

Galileo's Daughter, Sobel, Dava, Walker & Company (US), Fourth Estate (UK), 1999

Galileo in Rome: The Rise and Fall of a Troublesome Genius, Shea, W.R., Artigas, M. Oxford University Press, 2003

George Ellery Hale: Explorer of the Universe, Helen Wright, E.P. Dutton & Co., Inc, 1966

God's Philosophers: How the Medieval World Laid the Foundations of Modern Science, Hannam, James. Icon Books, 2009

Heirs of Archimedes: Science and the Art of War Through the Age of Enlightenment, The. Steele, Brett D. The MIT Press, 2005

Johannes Kepler and the New Astronomy, James R. Voelkel, (Oxford University Press 1999)

Light and Video Microscopy, Wayne, Randy O., Academic Press, 2010

Masters of Theory: Cambridge and the Rise of Mathematical Physics, (University of Chicago Press 2003)

Miss Leavitt's Stars, George Johnson, W.W. Norton & Co., 2005

Perilous Times: Free Speech in Wartime, Geoffrey R. Stone, W.W. Norton & Co., 2004

Physical Chemistry: Multidisciplinary Applications in Society, Kenneth S. Schmitz, (Elsevier Science 2018)

Planetary astronomy from the Renaissance to the rise of astrophysics: Part A: Tycho Brahe to Newton, ed. R. Taton & C. Wilson, (Cambridge University Press 1989)

Prince & Pauper: Ritchey, Hale & Big American Telescopes, Donald E. Osterbrock, University of Arizona Press 1993

Renaissance and Revolution: Humanists, scholars, craftsmen and natural philosophers in early modern Europe, ed. J.V. Field and Frank A.J.L. James, (Cambridge University Press 1993)

The Long Thirst: Prohibition in America: 1920-1933, Thomas M. Coffey, W.W. Norton & Co., 1975

The Muleskinner and the Stars: The Life and Times of Milton L. Humason, Astronomer, Ron Voller, Springer, 2015

Unrolling Time: Christiaan Huygens and the mathematization of nature, Joella Gerstmeyer Yoder, (Cambridge University Press 1988)

Walter Baade: A Life in Astrophysics, *Donald E. Osterbrock,* Princeton University Press 2001

Appendix 2: Hubble And Humason Selected Bibliography (Chronological)

Hubble, E.P., "The Variable Nebula NGC 2261," *Ap.J.* **44**, 190-97 (1916) [Discovery of what is now called Hubble's Variable Nebula].

Hubble, Edwin Powell, *Photographic Investigations of Faint Nebulae* (Univ. of Chicago Press, Chicago, 1920). [Ph.D. dissertation]

Seares, Frederick H. & Edwin P. Hubble, "The Color of the Nebulous Stars," *Ap.J.* **52**, 8-22 (1920).

Humason, M.L., "Discovery of Bright Hydrogen Lines in the Spectrum of the Variable Star W Cephei, PASP, Vol. 34, 1922

Hubble, Edwin, "A General Study of Diffuse Galactic Nebulae," *Ap.J.* **56**, 162-99 (1922).

Hubble, Edwin, "The Source of Luminosity in Galactic Nebulae," *Ap.J.* **56**,400-438 (1922).

Hubble, Edwin, "NGC 6822, a Remote Stellar System," *Ap.J.* **62**, 409 (1925).

Hubble, Edwin P., "Cepheids in Spiral Nebulae," *Pubs. Amer. Astr. Soc.* **5**, 261-64 (1925); reprinted in *Observatory* **48**, 139-42 (1925).

Hubble, Edwin, "A Spiral Nebula as a Stellar System: Messier 33," *Ap.J.* **63**, 236-74 (1926).

Hubble, Edwin, "Extragalactic Nebulae," *Ap.J.* **64**, 321-69 (1926).

Adams, Walter S., Joy, Alfred H., Humason, M.L., "The Absolute Magnitudes and Parallaxes of 410 Stars of Type M, ApJ 1926

Hubble, E.P., "The Classification of Spiral Nebulae," *Observatory*, **50**, 276-81 (1927).

Van Maanen, A., Brown, J.A., Humason, M.L., "A Star of Extremely Low Luminosity, PASP, Vol. 39, 1927

Merrill, Paul W., Humason, Milton L., "Note on Very Cool Stars," PASP, Vol. 39, 1927

Humason, Milton L., Nicholson, Seth B., "H.D. 163181, A Spectrographic Binary," ApJ Vol. 67, 1928

Hubble, E., "A Spiral Nebula as a Stellar System, Messier 31," *Ap.J.* **69**, 103-58 (1929). [reprinted in *Ap.J.* **525**, 150 (2000) with commentary by Arthur D. Code.]

Humason, Milton L., "The Large Radial Velocity of N.G.C. 7619," *Proc. Nat. Acad. Sci.* **15** 167-68 (1929).

Hubble, E., "A Relation between Distance and Radial Velocity among Extra-Galactic Nebulae," *Proc. Nat. Acad. Sci.* **15**, 168-73 (1929) [also here with Commentary by Robert Kirshner, 2003]

Hubble, E., "Distribution of Luminosity in Elliptical Nebulae," *Ap.J.* **71**, 231-76 (1930).

Humason, Milton L., "The Rayton Short Focus Spectrographic Objective," ApJ Vol. 71, 1930

Hubble, Edwin and Milton L. Humason, "The Velocity-Distance Relation among Extra-Galactic Nebulae," *ApJ.* **74**, 43-80 (1931) [reprinted in *ApJ.* **525**, 214 (2000) with commentary by Allan Sandage.]

Humason, Milton L., "The Large Apparent Velocities of Extra-Galactic Nebulae," ASP Leaflet, 37, July 1931

Humason, Milton L., "Apparent Velocity Shifts in the Spectra of Faint Nebulae," ApJ. Vol. 74, 1931

Seares, F.H., Humason, M.L., Joyner, Mary C., "Stars of Abnormal Color in S.A." PASP Vol. 43, 1931

Merrill, Paul W., Humason, Milton L., Burwell, Cora G., "Discovery and Observations of Stars of Class Be: Second Paper," ApJ. Vol. 76, 1932

Humason, Milton L., "Spectral Types of Faint Stars in Kapteyn's Selected Areas 1-115," ApJ. Vol. 76, 1932

Hubble, Edwin and Milton L. Humason, "The Velocity-distance Relation for Isolated Extragalactic Nebulae, "*Proc. Nat. Acad. Sci.* **20**, 264-68 (1934).

Hubble, Edwin., "The Distribution of Extra-Galactic Nebulae," *Ap.J.* **79**, 8-76 (1934).

Hubble, E., *Redshifts in the Spectra of Nebulae,* (The Clarendon Press, Oxford, UK, 1934). [Halley Lecture]

Humason, M.L., "The Apparent Velocity of a Nebula in the Boötes Cluster No. 1," PASP, Vol. 46, 1934

Hubble, E., and Richard C. Tolman, "Two Methods of Investigating the Nature of the Nebular Red-shift ," *Ap.J.* **82**, 302-37 (1935).

Adams, Walter S., Joy, Alfred H. Humason, Milton L., Brayton, Ada Margaret, "The Spectroscopic Absolute Magnitudes and Parallaxes of 4179 Stars," ApJ, Vol. 81, 1935

Adams, W.S., Humason, M.L., "The Spectra of Four White Dwarf Stars," PASP, Vol. 47, 1935

Humason, M.L., "The Aluminizing of the 100-inch and 60-inch Reflectors of the Mount Wilson Observatory," PASP, Vol. 47, 1935

Humason, M.L., "Super-Novae," ASP Leaflet 88, May 1936

Hubble, E., "Effects of Red Shifts on the Distribution of Nebulae," *Ap.J.* 84, 517-54 (1936).

Hubble, Edwin P., *The Realm of the Nebulae* (Yale University Press, New Haven, CT, 1936, 1982).

Humason, M.L., "Is the Universe Expanding," ASP Leaflet 91, August 1936

Humason, M.L., "The Apparent Radial Velocities of 100 Extra-Galactic Nebulae," ApJ Vol. 83, 1936

Hubble, Edwin P., *The Observational Approach to Cosmology* (Clarendon Press, Oxford, 1937). [Rhodes Memorial Lectures delivered at Oxford in 1936.]

Humason, M.L., "Evidence for an Expanding Universe," The Scientific Monthly, Vol. 43, Issue 1, 1936

Humason, M.L., "Spectrographic Observations of the Nebula Surrounding Nova Herculis 1934, PASP, Vol. 55, 1943

Humason, M.L., Zwicky, F., "A Search for Faint Blue Stars," (Humason-Zwicky stars), ApJ. Vol. 105, 1946

Hubble, Edwin, "First Photographs with the 200-inch Hale Telescope," *PASP* **61**, 121-24 (1949).

Humason, Milton L., "Apparent Velocities of Extragalactic Nebulae in Four Faint Clusters," PASP, Vol. 63, 1951

Hubble, E., "The Law of Red Shifts," *MNRAS*, **113**, 658 (1953). [George Darwin Lecture]

Hubble, Edwin & Allan Sandage, "The Brightest Variable Stars in Extragalactic Nebulae. I. M31 and M33," *Ap.J.* **118**, 353-61 (1953).

Humason, Milton L., Wahlquist, Hugo D., "Solar Motion with Respect to the Local Group of Nebulae, ApJ, Vol. 60, 1955

Humason, M.L., N.U. Mayall, & A.R. Sandage, "Redshifts and Magnitudes of Extragalactic Nebulae," *Astronomical Journal* **61**, 97-162 (1956). (HMS Catalogue)

Zwicky, F., and Humason, M.L., "Spectrographic Investigations of Multiple Galaxies and Clusters of Galaxies," PASP, Vol. 71, 1959

Humason, M.L., and Gates, H.S., "The 1959 Palomar Supernova Search," PASP, Vol. 72, June 1959

Zwicky, F. and Humason, M.L., "Spectra and Other Characteristics of Interconnected Galaxies and of Galaxies in Groups and in Clusters II," ApJ, Vol. 133, Dec. 1960

Humason, Milton L., "Photographs of Planets with the 200-inch Telescope," University of Chicago Press, 1961

Humason, M.L., Gomes, Alercio, M., Kearns, C.E., "The 1960 Palomar Supernova Search," PASP, Vol. 73, No. 432, June 1961

Humason, M.L., Kearns, C.E., Gomes, Alercio, M., "The 1961 Palomar Supernova Search," PASP, Vol. 74, June 1962

Zwicky, F. and Humason, M.L., "Spectra and Other Characteristics of Interconnected Galaxies and of Galaxies in Groups and in Clusters III," ApJ, Vol. 139, 1963

About the Author

Ron Voller Ron Voller is a writer and producer based in Brooklyn, NY. A native of Chicago, he earned his undergraduate degree in literature from the University of Denver before moving to New York City in 1999. He is currently working on his graduate degree in interdisciplinary studies at Johns Hopkins University's Krieger School of Arts and Sciences. In addition to his ongoing historical research, he is working on several works of children's fiction as well as screenplays for documentary and feature films and the development of various television series. Other literary work includes magazine articles and a memoir. Actively engaged in a variety of artistic disciplines, he is an avid songwriter and sculptor and volunteers as a board advisor for a charitable organization that focuses on community development abroad. When not working, he enjoys spending his leisure time with his large family and friends.

© Springer Nature Switzerland AG 2021
R. Voller, *Hubble, Humason and the Big Bang*, Springer Praxis Books,
https://doi.org/10.1007/978-3-030-82181-4

Index

© Springer Nature Switzerland AG 2021

R. Voller, *Hubble, Humason and the Big Bang*, Springer Praxis Books, https://doi.org/10.1007/978-3-030-82181-4